Modernism and Public Reform in Late Imperial Russia

Also by Ilya Gerasimov

THE SOUL OF THE MAN AT THE TIME OF TRANSITION: The Case of Alexander Chaianov (in Russian)

EMPIRE SPEAKS OUT: Languages of Rationalization and Self-description in the Russian Empire (*edited with Jan Kusber and Alexander Semyonov*)

A NEW IMPERIAL HISTORY OF THE POST-SOVIET SPACE (*edited with Sergey Glebov, Marina Mogilner, and Alexander Semyonov, in Russian*)

Modernism and Public Reform in Late Imperial Russia

Rural Professionals and Self-Organization, 1905–30

Ilya V. Gerasimov

Executive Editor, Ab Imperio Quarterly

First published 2009 by
PALGRAVE MACMILLAN

Palgrave Macmillan in the UK is an imprint of Macmillan Publishers Limited, registered in England, company number 785998, of Houndmills, Basingstoke, Hampshire RG21 6XS.

Palgrave Macmillan in the US is a division of St Martin's Press LLC, 175 Fifth Avenue, New York, NY 10010.

Palgrave Macmillan is the global academic imprint of the above companies and has companies and representatives throughout the world.

Palgrave® and Macmillan® are registered trademarks in the United States, the United Kingdom, Europe and other countries.

ISBN-13: 978–0–230–22947–1 hardback

This book is printed on paper suitable for recycling and made from fully managed and sustained forest sources. Logging, pulping and manufacturing processes are expected to conform to the environmental regulations of the country of origin.

A catalogue record for this book is available from the British Library.

A catalog record for this book is available from the Library of Congress.

10   9   8   7   6   5   4   3   2   1
18   17   16   15   14   13   12   11   10   09

Printed and bound in Great Britain by
CPI Antony Rowe, Chippenham and Eastbourne

To MM

Contents

List of Tables

List of Archival Collections with Abbreviations

Russia

GARF The State Archive of Russian Federation, Moscow
>F. 63, Moscow Division of Secret Service (*Moskovskoe okhrannoe otdelenie*)
>F. 102, Department of Police
>F. 934, Ministry of Agriculture
>F. 5865, E. D. Kuskova

NART National Archive of the Republic of Tatarstan, Kazan
>F. 119, The Kazan District Zemstvo Board
>F. 199, Gendarme Administration of Kazan Province
>F. 256, The Land Settlement Commission of Kazan Province
>F. 580, The Head of Agronomist Assistance to Privately Owned Farms of Kazan Province
>F. 638, The Kazan City Committee for Granting Conscription Deferments
>F. 658, The Chistopol District Zemstvo Board

RGAE Russian State Archive of Economy, Moscow
>F. 328, E. N. Sakharova-Vavilova
>F. 731, A. V. Chaianov

RGIA Russian State Historical Archive, St. Petersburg
>F. 403, "Russkoe zerno"
>F. 776, The Main Administration for Press Affairs

TsGIA Central State Historical Archive of St. Petersburg
>F. 449, Petrograd Agricultural Courses
>F. 450, The Stebut Higher Women's Agricultural Courses
>F. 451, The Evening Agricultural Courses of the Society of People Universities in St. Petersburg

Ukraine

DAKO State Archive of Kiev Region, Kiev
>F. R-989, The Kiev District Union of Cooperatives "Consumer"

TsDAVOVU Central State Archive of Supreme Organs of Government of Ukraine, Kiev
>F. 27, The People's Commissariat of Agriculture of the Ukrainian SSR

TsDIAU Central State Historical Archive of Ukraine, Kiev
    F. 1680, Kharkov Inspector of Press Affairs

USA

BAR Bakhmeteff Archive of Russian and East European History and Culture,
    Butler Library, Columbia University, New York, NY
Hoover Institution Archive, Stanford, CA
Museum of Russian Culture Archival Collection, San Francisco, CA

Introduction

This book is the story of a generation of Russians that sought to improve their personal lives but managed to effectively change the ways of the entire country during the decade following the abortive Revolution of 1905. This happened largely beyond the administrative apparatus of the state and outside the organized, if collapsing, revolutionary movement. The force of social activism of self-coordinating individuals had transformed power relations in the Russian countryside, bringing forward a new social class of rural professionals (agronomists, physicians, educators, instructors, and managers of peasant cooperatives). This numerous group (20,000–30,000 strong by 1914) successfully competed with gentry-dominated zemstvo boards (organs of rural self-government) and government agencies for influence in the villages. More important, it had successfully bridged the proverbial gap between the educated elite and the "people" by establishing an intensive dialogue and partnership with peasants. A significant faction of petty agriculturists that we habitually unite under the common name of "peasants" (at least a one or two million of them) had been involved in the process of rationalization and professionalization. Under the influence of daily interaction with agricultural specialists, they began thinking of themselves as economic producers rather than people leading a particular way of rural life ("peasants"). This transformation may partly account for the unprecedented social activity in the Russian countryside after 1917, which eventually decided the outcome of the civil war.

The generation of Russians involved in these processes was comprised of people of different age. The common trauma of the Russo-Japanese war and the Revolution of 1905 disillusioned a segment of educated Russians both in the capability of the state and the fruitfulness of radical politics. While a more traditionalist stratum of Russian intelligentsia dubbed the period of political demobilization and disillusionment *bezvremenie* (literally, "the absence of time") and began disintegrating,[1] new strategies of social activism began to emerge. Naturally, the most prominent and visible group of a new generation of Russian post-1905 intelligentsia was represented by younger people, who entered adulthood in the wake of the revolution's defeat and were mostly preoccupied with finding a niche for themselves under the new circumstances. One of them documented this transformative experience in a diary entry of October 1911:

> In 1904 [right after graduation from the gymnasium], I was at the History and Philosophy Department [of the Higher Women's Courses]. My recently clarified attitude to the world, to the comprehension of life, and pantheism naturally attracted me to philosophy. But 1905 already directly raised the question of *action*, and of course it was equally naturally that since the immediate impression was formed by the people and popular turmoil, [the people were] brought to the fore. Thus... [the goal] was getting closer to the people. And where to be getting closer if not in the village, where there are so many people, where they stagnate, come to ruin, and where at the same time is *nature*. Obviously, the institution of higher learning could be only the Agricultural Academy. There was no other way to the people.
>
> ...But I did not take into account two things...The most powerful...blow was delivered from the immediate feelings. It is impossible to go forward for the sake of love of those who are unknown, while renouncing love of the one who is familiar.... And here is the origin of the turning point.
>
> ...I cannot start my life and activity anew without defining the *basis* of this transformation. And now I can acknowledge that this basis has remained the same: it is nature and love, only my attitude to them has changed sufficiently. Everything sacrificial, renouncing my own individuality that was in *that* intellectual...reception of life, has gone.... It is hardly accidental that instead of the completed perception of life as a life already [dedicated] not to myself,...new opportunities and a new development of forces without any visible limit opened up....
>
> ...The key to success is, of course, far-sighted organization. First of all, population should be moved to SELF-activity [i.e., initiative, lit. *SAMO-deiatel'nosti*].... Credit [associations], cooperatives, all these undertakings should be sustained by local resources, it is just necessary to wake up initiative, interest, and understanding.
>
> ...The task of the moment is to create these types of organizations.... It is necessary to work to achieve all these. What a weight was lifted from my mind. But there has been no other way [for me] to [dedicate myself to] science. Now everything is clear, thank the Lord.[2]

Ekaterina Sakharova, who outlined her new life strategy in the diary, remarkably interlaced her own spiritual search with an outline for a program of social transformations. Born in 1886, she had prepared herself for a typical intelligentsia life, studying great ideologies and self-sacrifice for the sake of the people. The revolution changed this to the effect that she entered the Moscow Agricultural Institute and was attracted to science. After graduation in 1910 she faced a necessity to balance her new interest in professional knowledge with the old populist ethos. This dramatic intellectual process is illustrated in the above excerpt. Contributing to the picture of interconnectedness of Sakharova's personal life and social activism, we can add that the figure of "the one who is familiar" who stood in the way of her self-sacrificing love "of those who are unknown" was most likely her classmate at the institute, Alexander Chaianov. This well-known figure in the history of Late Imperial and Early Soviet Russia was an original economist and a cooperative theoretician and activist. He was two years younger than Sakharova, and the Agricultural Institute was his first choice of institution of higher learning. However, he also entertained ambitious social goals: in 1909 he admitted to Sakharova, that his true calling was not agronomy but "sociology"[3] (then still understood as a scientifically informed practical sphere). His star as a young leader of agricultural specialists rose in February 1911, just a couple of weeks after his final breakup with Sakharova. In an often-quoted speech delivered at the Moscow regional congress of rural professionals, he outlined a program for modernizing the countryside. Echoing Sakharova, he suggested that all of them should strive "[b]y means of influencing the minds and wills of economic people [*khoziaistvennykh ludei*] to awaken initiative in their milieu, and…to direct this initiative in a most rational manner. In a word, to change old ideas into new in the heads of the local population."[4] At the time of the congress, Sakharova was already in love with another former classmate, the future famous Russian geneticist, Nikolai Vavilov (they would marry in 1912).[5] Both Vavilov and Chaianov opted to pursue academic careers and became graduate students at their home institute: Vavilov concentrated on science, while Chaianov divided his time between economic theory and public activism. At the same time, yet another of their classmates, Ivan Barkhatov, was making a swift career as an agronomist-practitioner in the government service. By July 1912, he was head of the government agronomist network (an important element of Stolypin's agrarian reforms) in Kazan province. Just two years later, almost the entire government agronomist network was handed over to the zemstvo, quite in line with the program advocated by the humble graduate student Chaianov (which resulted in a drastic diminishing of Barkhatov's authority).[6]

Even this brief snapshot reveals that an interest in "love and nature" (Sakharova), science (Vavilov and Chaianov), or a fast career (Barkhatov) brought like-minded people together, and invested their individual pursuits of self-fulfillment with far-reaching social consequences. This book is based on the life stories of some 1,200 individuals that together compose a complex picture of group solidarity and competition. At some point, however, we will have to transcend individual cases to reach the level of group actions, institutional transformations, and statistical aggregations. It is the task of the subsequent pages of the book to show how and why these many individuals formed a society that interacted with the government and changed the socioeconomic landscape in the country.

The book consists of 11 chapters arranged in three parts. Part I, "Structures of Mobilization", outlines the social and cultural structural settings that made possible a particular configuration of a new social movement. Chapter 1 focuses on the mental mapping of the Russian reformist discourse on the countryside. It traces Russian society's rising interest in agriculture through a variety of indicators, including the detailed statistics of agriculture-related publications in Russian periodicals. Next, it describes the post-1905 Russian culture of modernization in terms of the Progressivist approach that dominated American and European reformist circles of that time, as an "apolitical politics" of piecemeal social engineering. Finally, the chapter deconstructs the main authentic social categories of the interrevolutionary Russian rural society: the "peasants" and the "three elements" that were in a position of authority vis-à-vis the peasants. Chapter 2 is dedicated to the system of professional agricultural education that prepared cadres of agricultural specialists. Starting with an overview of different institutional forms and general patterns of professionalization, it focuses on the prosopographic survey of one class of 1910 in one agricultural institute. Incidentally, this is the class of Ekaterina Sakharova and Alexander Chaianov, and its demographic profile suggests the existence of generational bonds in a new reformist movement. Chapter 3 reconstructs the process of elaborating a universal master-plan by countryside modernizers. In the spirit of Progressivist universalism, the project of Russian "public agronomy" emerged as an attempt to customize the Italian model of countryside reformism to Russian local conditions. Besides problems with finding an institutional form for the large-scale social intervention of agricultural specialists-educators, the community of rural professionals had to settle structural internal conflicts in order to take off as a single public modernization movement.

Part II, "Dynamics of Modernization," attempts to present the activity of agricultural specialists as a dynamic experience of complex social interactions and evolving relationships with major social actors. Chapter 4 traces the changing role of the state in the staging of public coordinated initiative: from severe police control to benevolent, if cautious, sponsorship. The "change of heart" of the semi-autocratic regime is explained by the ascendance of a new generation of bureaucrats sharing the general premises of the Progressivist culture of modernization and by the government's structural dependence on the human resources of rural professionals—both intellectually and in terms of practical implementation of its policies. For a while, the government modernizers attempted to overrun public initiatives by cloning the forms and methods of the public modernization movement, but by 1914 the state had acknowledged the leading role of the zemstvo-based network of rural professionals. Chapter 5 shows how a combination of quite mundane factors, such as the dynamics of the job market, contributed to the rise and sustainability of the reformist movement in the Russian countryside. Ideological zeal played only an auxiliary role in attracting hundreds of young people into the ranks of agricultural specialists, but once they joined the profession, they interiorized a whole complex of values and life scenarios. The culmination of the public modernization movement and the ultimate proof of its success was the establishment of a dialogue and then partnership relations between agricultural specialists and peasants. Chapter 6 tells the story of this process from both sides. It identifies the group of peasants that became pioneers in this mutual rapprochement, and tracks down a gradual shift in the knowledge–power balance in the countryside, toward a greater self-sufficiency of peasant socioeconomic initiative. By 1914, the initial project of "public agronomy" had been successfully implemented, securing a high social status for rural professionals and recognition of the leading role of the educated progressive public. The success of the technical aspects of the reformist program did not produce an immediate global transformation of the peasant economy, to the dismay of many social activists. Chapter 7 interprets the announced "agronomist crisis" at the peak of success of the countryside modernizers as a crisis of their double identification with the legacy of radical intelligentsia and the new culture of professionalism. The exhaustion of the initial enthusiasm as a driving force of the self-modernization movement was compensated by the stake on professional routine and technological enhancement of the services of agricultural specialists.

Part III, "Patterns of 'Nationalization,'" adds another dimension to the story: the diversity of local experiences in the heterogeneous imperial context and the rise of competing scenarios of nationhood. The successful integration of many opposition-minded intellectuals and locally thinking peasants into a larger society made them aware of their belonging to a countrywide community. In Chapter 8 this community is characterized as the "motherland," as a nation described in terms of patriotic feelings. Exposure to "foreign" experience became the main factor of one's identification with the nation as country: during a trip abroad, or under the impact of war, or just contemplating the significance of foreign trade for the local economic situation. Chapter 9 engages with a parallel development, when the mobilized community of solidarity became reinforced by the commonality of language and ethnocultural bonds. In Ukrainian lands, finding a "common language" with peasants literally meant the Ukrainization of the "public agronomy" movement, while in the Volga-Urals region, agricultural specialists consciously developed the educational system in the Tatar language in order to produce local cadres of modernizers. Thus, the initially anational (pan-imperial) project of piecemeal reformism became an important anti-imperial factor. In Chapter 10, the actions of countryside modernizers in 1917 are presented as an attempt to legitimize the community of conscious subjects of modernizing efforts as the revolutionary nation. The revolution revived archetypal intelligentsia life scenarios that seemed to be buried deeply beneath the new ethos of professional service to the people after 1905. By giving priority to the most demanding version of the "imagined community," many agricultural specialists found themselves alienated from their "constituency" that either could not live up to the high expectations of the idealistic intelligentsia or was too radicalized for the moderate reformers.

The "agronomist crisis" of 1913–14 had already demonstrated that the prospects of the public modernization movement depended on its institutionalization on the national level—whatever the definition of "nation." The process of societal self-organization produced a program of reforms and created the momentum for a broad social movement, but it could not provide for more durable sustainability than that of a mass-scale campaign. The failure of the revolutionary nation in 1917 and the severe damage to different visions of "nation" as a result of the civil war delivered the final blow to the movement of Progressivist-minded public modernizers. Chapter 11 documents their attempts to survive professionally in the turmoil of the postrevolutionary years, and their gradual social disintegration throughout the 1920s. The Stalinist repressions of the early 1930s physically exterminated the remnants of a social group that had already been nonexistent as a distinctive social force. The Postscript to the book draws readers' attention to the long-term consequences of the story of the "public agronomy" movement that are still relevant today. The myth of the weak traditions of independent public initiative and civil society in Russia, and the fixation of the Russian educated elite on the "supreme interests of the state" all originate in the post-1917 crisis of the public modernization campaign. The erstwhile reformers found themselves "nationless" amid the revolutionary chaos and disillusioned with the "people" as a social or ethnocultural category. Many of them believed that the strong state was the ultimate embodiment of the national compound in the heterogeneous and multicultural postimperial society. This explains their support of the early Soviet regime and their moral capitulation vis-à-vis aggressive and interventionist versions of social engineering, such as collectivization. This also explains why the story of the large-scale public reformist movement of the interrevolutionary decade has been forgotten.

This book again makes us hear the voices of thousands of Russians who attempted to improve their society by themselves a century ago, and it seems that they have much to tell us even today.

* * *

This book is the result of more than ten years of research and writing. The manuscript has gone through several major revisions, and it thus seems unrealistic to attempt to mention everyone who has influenced my work in one way or another. Still, my special gratitude goes to those most helpful readers of the remote ancestor of this text: to Ziva Galili, who directed my Ph.D. dissertation at Rutgers University; to Kim Matsuzato of Sapporo Slavic Research Center, who tried to explain to me what a good Russian historian should write; and to Seymour Becker, who taught me how a good historian of Russia should approach his object of study. Later, David Macey helped me to see the hidden political agenda behind the generational enterprise of countryside reformers, which led to the first major revision of my study, after additional archival and library research. I wrote the second version of the book using all of the resources that the Davis Center for Russian and Eurasian Studies at Harvard can offer a "regional fellow." Alexander Semyonov generously commented on the manuscript. Eventually, my new research interests and experience caused me to see my study in a new light, which resulted in the present text written within the collective research project "Languages of Self-Description and Representation in Russian Empire," supported by the Volkswagen Foundation. After all these years I finally feel that the text is ready to become a book. You are reading it thanks to Michael Strang of Palgrave, who was not intimidated by the perhaps not so fashionable topic of my study, and to Therese Malhame, who thoughtfully yet decidedly turned "Russian into English" in my text, where necessary.

My research experience is hardly different from that of my colleagues: the efficiency and friendliness of the Hoover Institution and Bakhmeteff archives in the United States, not infrequent harassment and virtual sabotage at the Russian federal archives in Moscow and St. Petersburg and, to a lesser degree, at the central Ukrainian archives in Kiev. I am truly grateful to the staffs of the regional archives, particularly the National Archive of the Republic of Tatarstan (Kazan, Russia), and the Municipal and Regional archives in Kiev. It is not accidental that more and more scholars come to appreciate their rich collections and supportive staff. I also want to take this opportunity to thank Irina Lukka, the guardian angel of all those coming to the safehaven of all "Slavists"—the Slavonic Library at Helsinki University.

This book was written in a dialogue with and under the influence of Marina Mogilner, one of the most interesting historians of Russia today, and her approval was my main source of "quality assurance."

Part I
Structures of Mobilization

1
Becoming "Progressive": Structural Settings and Mental Mapping of Reformism

Agrojournalism and a new turn toward the countryside

How unusual was the decision of the 20-year-old Ekaterina Sakharova to drop her literature and philosophy studies and dedicate herself to agriculture? She was certainly not alone in this evolution. Ivan Emel'ianov was born in 1880 in Siberia to the family of a poor priest. At the age of 20, after graduation from the Tobol'sk Seminary in 1900, he rejected the career of clergyman and enrolled in the History Department of the Iuriev (Tartu) University—a step rather typical for a *popovich*-turned-*intelligent* of the post-emancipation era.[1] In 1903, however, he changed his mind for the second time and became a student in the Agronomy Department of Kiev Polytechnic, from which he graduated in 1907 with the Diploma of Agronomist of the First Degree. Thereafter, he played an important role in the Russian community of agricultural specialists-modernizers during the next two decades. While we cannot count all those who dropped humanities for agronomy, we can still statistically assess the scale and dynamics of the society's turn to agriculture.

The intensity of public engagement with the needs of rural Russia can be seen in the spread of agricultural societies as a form bridging organized public debate and institutionalized socioeconomic activism. Voluntary associations of people interested in various aspects of land cultivation, the agricultural societies, became the first institutions in Europe to form public discourse on agriculture and seek measures that could improve its ways in the second third of the eighteenth century.[2] While agricultural societies, comprised mostly of wealthy landowners, played a positive role in the spread of modern technologies of land cultivation, their primary significance was in shaping public opinion concerning the countryside.[3] In Russia the first institution of this type was the Imperial Free Economic Society established in 1765.[4] But it was not until 1820 when the first agricultural society in the strict sense was founded in Moscow. By 1861, about 30 agricultural societies had surfaced on the Russian public horizon. It is quite understandable that under the social conditions of pre-reform Russia their practical influence was minimal. After the emancipation, the nature of Russian agricultural societies gradually changed, and after the adoption of the Normal Regulations of 1897 they became cooperative enterprises that constantly increased in number. Thousands of new societies emerged after the Revolution of 1905 and the introduction of Temporary Rules for Associations and Unions facilitated the registration procedure.[5] Their proliferation reflected the growth of public (and state) concern with agriculture, expressed in subsidies of millions of rubles a year rather than any real effectiveness of the agricultural societies, which spent an average 20 percent of their budgets for staff expenses.[6]

Until the late nineteenth century public engagement in the "agrarian question" was extremely limited. The only legal vehicle for such interest, the agricultural societies were mostly the domain of local gentry philanthropy. Illegal forms, such as the famous intelligentsia movement of "going to the people" of the mid-seventies, were even more limited in scope. Numerous attempts by the intelligentsia to found peasant cooperative organizations in the nineteenth century failed without clear evidence of what to blame for the failure: the intelligentsia's preoccupation with socialist ideology or the peasants' weak economic motivation.[7] These dynamics also suggest a possible explanation for the postemancipation intelligentsia's disillusionment with the peasantry in the 1880s, as described by Cathy Frierson:[8] the initial fascination with peasants was generally limited to discursive projections, and only occasionally led to actual engagement (hence the insignificant number of agricultural societies). That is why a generation later, a new and much more practically oriented upsurge of interest in peasants and agriculture became possible.

A more precise device to measure Russian society's general interest in agriculture can be found in the statistics of special periodicals and even individual articles, dedicated to agriculture in any year from the reign of Catherine II to the present. Their dynamics suggest the scale of public interest in agrarian topics, the social composition of readers, and even the political forces standing behind the spurt of "agrojournalism" at the beginning of the twentieth century, for very few of the agricultural periodicals were profitable, indicating that ideological concerns rather than revenues stimulated publishers. It was the mass printed word that contributed enormously to changing the social climate, elaborating the new tropes to be used in subsequent literary debates, and preparing the stage for a new type of social activism.

As one may expect, the first periodicals dealing with agriculture were the *Transactions* (Trudy) of the Free Economic Society, the first issue of which was published on December 7, 1765. It was 56 years before a second periodical appeared, this time dedicated exclusively to agriculture. This was the *Journal of Farming* (Zemledel'cheskii zhurnal), founded in 1821 by the

Moscow Agricultural Society. In 1830 the *Leaflets of the Agricultural Society of Southern Russia* appeared. By the end of the first century of existence of the Russian agricultural press, some 20 periodicals across the Russian Empire were dedicated to various aspects of land tilling: farming, stock-breeding, and forestry.[9] So far, the dynamics of the periodicals repeated those of agricultural societies. Characteristically enough, until the 1890s the majority of these periodicals targeted a very narrow circle of readers interested in the theoretical aspects of agriculture. Few titles were published by the government, while most of the others were published by imperial societies specializing in separate branches of rural economy (sheep-breeding, forestry, etc.)[10] and during the last third of the nineteenth century, by zemstvos.

It was probably the impact of the 1891 famine and the united relief efforts by the intelligentsia that changed the face of agrojournalism (as it changed the pattern of public activity of the entire "educated society").[11] In the 1890s, a number of new, mainly weekly, periodicals appeared that targeted a new type of reader—still highly educated and well-to-do, but now having a practical interest in agriculture (hence the spread of weekly editions in contrast to the monthly and even yearly publications of previous epochs).[12] At this stage, local zemstvos and provincial agricultural societies were the leading investors in the agricultural periodical press, demonstrating the decentralization of the emerging public discourse on the agrarian question. The real boom in agrojournalism took place in the interval between the 1905 Revolution in Russia and the World War I, as Table 1.1 suggests.

The actual figures may vary depending on the selection criteria adopted by different statisticians, but other sources confirm the basic trend. By 1917 almost half of all agricultural periodicals were less than 5 years old, and 75 percent of all publications were founded after 1905.[14]

If we equate the number of specialized periodicals to the popularity of their topic, we can reconstruct a popularity chart of the Russian press. In 1911 agrojournalism accounted for 5.5 percent of the periodicals published in the Russian Empire, which gave it an honorable third place among 28 other topics.[15] By 1912 its share had grown to 6.8 percent at the expense of its immediate rivals.

Still, "politics, publicist writing, and literature" traditionally occupied first place in the minds of Russians, accounting for almost a third of the entire market. "Official publications" and "theological, religious-moral, and church issues" shared second place, each represented by approximately 8 percent of

Table 1.1 The number of agricultural periodicals in Russia, 1907–1914.[13]

Year:	1907	1908	1909	1910	1911	1912	1913	1914
Number of titles:	96	101	120	129	150	177	186	352

all titles. Agricultural periodicals held a firm third place, creating a niche of their own. This was a significant success: advertising and humor magazines could not master even 1.5 percent of the market, occupying seventeenth and twentieth places, respectively. And "philosophy and psychology" were at the very bottom of the list, along with "publications for the troops" and "insurance and fire fighting."[16] Library statistics of readers' requests corroborate this evidence.[17] Agriculture-related social interactions were becoming a matter of broad public concern in the early twentieth-century Russia. Agrojournalism spread to a significant segment of the broader public sphere, as different interest groups supported the discussion of agriculture-related topics in public.

While Russian was the predominant language of that segment of the public sphere, at least a dozen other languages were represented in the world of agrojournalism. In 1912, about three-quarters (73.14 percent) of all of the empire's periodicals were published in Russian, and we find exactly the same proportion of Russian-language publications (73.12 percent) among the agricultural periodical press. This means that the Russians were responsible for the growth of both the general and the special agricultural press in equal proportion. On the contrary, some nationalities (Jews, and to some extent Poles) showed much less interest in agriculture than in other topics, while others (Estonians, Lithuanians, and Ukrainians) were much more enthusiastic about agrojournalism than the average.[18] By 1917 the share of non-Russian periodicals had decreased dramatically (about threefold),[19] largely due to the German occupation of Poland and the closing of the German-language press. However, prewar trends suggested a gradual increase in the number of non-Russian publications.

An important dimension of the "agrarian" side of the Russian public sphere was its spatial organization. The dominant tendency to regionalization of the agricultural press was evident to contemporaries as early as 1913.[20] This represented a dramatic departure from the legacy of the classical intelligentsia grouped around the "thick journals," published either in Petersburg or in Moscow. The most popular agricultural periodicals of the 1910s were published elsewhere: in Kharkov (*Agronomicheskii zhurnal, Iuzhno-russkaia sel'skohoziaistvennaia gazeta*), Samara (*Zhurnal obshchestvennoi agronomii, Samarskii zemledelets*), and Perm (*Permskaia zemskaia nedelia*). In general, by 1917 Moscow and Petrograd together controlled just over a third of the market (35 percent), while the leading role in publishing agricultural periodicals belonged to the "provincial capitals," or central cities of provinces (46 percent). Major district towns (*uezdnye goroda*) and important villages could boast only one-fifth of the entire industry.[21] This meant an ongoing lack of resources and social infrastructure that might make chances in the provincial backwaters more equal to those of the more cultured centers. For provincial activists, this imbalance between the center and the periphery was a painful issue.[22] On the other hand, these figures testify to the hitherto unthinkable decentralization of professional (in this case, agricultural) periodicals.[23]

The next feature of the Russian prerevolutionary agrarian side of the public sphere was its pluralism in terms of the social actors involved in its formation. Looking through the prism of agricultural periodicals we can estimate the "weight" of those actors by the number of periodical titles published by each of them. By 1916, the structure of ownership was very diverse: government agencies, zemstvos, and private publishers each accounted for around 15 percent of the agricultural periodicals (i.e., 45 percent altogether). The remaining 55 percent belonged to a loose conglomerate of various associations: cooperatives, agricultural and professional societies, and so on.[24] This means that no political or institutional lobby was able to control the periodicals market simply by virtue of ownership.

That had not always been the case in previous decades. Looking through a somewhat distorting prism of statistics that fixes the publisher and time of foundation of existing periodicals (but excluding those that had disappeared by 1916), we observe the following dynamics: the last decade of the nineteenth century and the years immediately preceding the First Russian Revolution were marked by a steady withdrawal of the state from the business of establishing new agricultural periodicals. Out of the all government periodical publications existing in 1916, only 7.7 percent had been founded during the decade of 1897–1906. Precisely during this period, the zemstvo agrojournalism market took off, establishing 17 percent of all of the zemstvo titles that survived to 1916. During the years of active implementation of the Stolypin agrarian reform, private publishers and agricultural societies (not government agencies) were most active in the dissemination of new ideas and knowledge. Almost a third of all private periodicals active in 1916 had been founded between 1907 and 1911.

It is only natural that the majority of the surviving periodicals were quite young in 1916. Still, zemstvos had proportionately more newly established titles than any other group: 57.5 percent of all the zemstvo agricultural periodicals had been founded after 1911 (in contrast to some 42.2 percent of private editions). Keeping in mind that these statistics discarded all titles that had vanished by September 1916, they still accurately detect periods when the efforts of a certain group were persistent enough to make the then established periodicals survive in higher proportions (compared with any other period or any other group).[25]

It is difficult to estimate the total number of copies of newspapers, magazines, and journals pouring daily into the stream of agrojournalism. Such information usually was not published, and police records preserved in archives are the main source of such data. Another obstacle to even approximate calculations involves frequent shifts in the size of runs. For instance, print runs in 1910 included 2,400 copies of issues 8–10 of the magazine *Nuzhdy derevni* (Village Needs), 2,500 copies of issues 6 and 7, and only

2,200 copies of issue 23.[26] Ranging from 800 to 3,000 copies per issue (10,000 copies of *Derevenskaia gazeta* and 8,000 of *Khutorianin* were rare exceptions), the average run for an agricultural periodical was somewhere between 2,000 and 2,500 copies per issue.

We have information on 302 out of 310 periodicals published in 1916. One-third (101 titles) were monthly publications, usually targeting specialists and well-educated landowners. Another third was split more or less equally between weekly and fortnightly editions (55 and 67 titles, respectively) of a rather popular nature. A few dailies and a number of journals with one to ten issues a year comprised the last third. Predominantly magazines and journals, these 302 periodicals together published about 6,872 issues. Multiplying this figure by the minimum estimated run of 2,000 copies per issue yields the considerable quantity of 13,744,000 copies per year. During the war, in a situation of paper shortage, hundreds of thousands of people still may have read those 14 million copies of periodicals (not counting hundreds of popular brochures and dozens of special monographs that appeared every year). The intensity of public discourse on agriculture in Late Imperial Russia was obviously very high, as at least 90 percent of the published copies were read.[27]

Finishing our brief overview of Russian agrojournalism, which shaped the worldview of agricultural specialists and activists, and at the same time was itself shaped by their collective efforts, we shall note that it did not exist in a vacuum. If professional periodicals were read predominantly by specialists, there was an intermediate sphere bridging the world of professionals and that of a broad public. A considerable number of articles discussing agriculture-related themes were published in general periodicals, affecting even those readers who were not professionally engaged in agriculture. For over 35 years, until his death in 1925, Alexander Pedashenko composed lists of all pieces published on agriculture, regardless of the source of publication.[28] He also made lists of periodicals that published articles on the topic during a given year. Some of those periodicals appeared only occasionally on his list, for their interest in the topic was only temporary. Still, their presence is very important as an indication of public involvement in the discourse on agriculture. The grandiose taxonomical project of Pedashenko documented the trends in the involvement of Russian periodicals in the discussion of agricultural issues, and the fluctuation in the numbers of publications (books and, predominantly, articles) on those issues.

According to the data collected by A. D. Pedashenko, the number of articles and essays on different aspects of the agrarian question published in various periodicals had doubled between 1905 and 1912 and showed a peak of public interest in agriculture in 1913, when more than 23,500 agriculture-related articles were published, that is, a new piece on agriculture appeared every 22 minutes.[29] The seemingly permanent growth of public interest in agriculture was brought to a halt by the outbreak of the war, but by 1916, a steady decline in the number of publications turned into a virtual collapse. A new and rather short era of mass interest in agriculture was coming, and it would differ very much from the prerevolutionary decade.

Pre-1917 agrojournalism did not have a purely academic character and did not exist in a political vacuum. The Russian press was affected by both governmental control and the influence of the radical opposition. However, my study of the periodicals (both their contents and occasional incidents with the authorities) shows that "direct" politics played a minimal role in the functioning of the agropress. New titles were usually registered without problems, within the timeframe determined by law.[30] Censorship interfered rarely, and in the majority of instances that I have encountered, publications were acquitted of all charges by higher authorities.[31] Criticism of government policies, especially the fierce attacks on Stolypin land reforms, were common in agricultural periodicals but rarely got them into trouble. Equally rare were explicitly revolutionary publications, and in a single case of which I am aware, the authorities exercised rather remarkable restraint.[32]

Thus we can tentatively reconstruct the intellectual climate that influenced the decisions of Ekaterina Sakharova, Ivan Emel'ianov, and Alexander Chaianov at the beginning of the twentieth century. They entered adulthood during a new stage of Russia's engagement with the "agrarian question": this time, interest was much more practical (almost "technological"), widespread throughout the country (i.e., not limited to the elite in the capitals), and intensive (as evidenced by the stable increase in the number of special periodicals and individual articles on the topic). Characteristically, even populist-minded Sakharova chose to study agriculture professionally in order to "get closer to the people," rather than to pursue the traditional and much easier strategy of "going to the people" by becoming a village teacher. Thus, the very understanding of the intelligentsia's mission vis-à-vis the people had changed: the former plan of modernization through political emancipation gave way to a new formula of emancipation through modernization. This was a crucial paradigm shift in the dominant "culture of modernization," to borrow the concept of Esther Kingston-Mann.[33]

A new culture of modernization and the "apolitical politics" of *obshchestvennost'*

Modernity is one of the most contested concepts in the social sciences, and to avoid the trap of choosing one normative model of "true" modernity over rival models we turn to its original meaning at the beginning of the twentieth century. "Modernity," as many other keywords of the twentieth century, was introduced in Russia by Peter Struve, the former founding father of Russian Marxism, turned dean of Russian liberals.[34] In January

1907, he literally inaugurated the interrevolutionary decade of a new type of modernization in Russia:

> The Russian Revolution, as I have noted elsewhere, represents a phenomenon very peculiar by virtue of the two conditions that were combined in it: (1) one that can be called "contemporaneity," and (2) one that can be called "elementarity." I use the word "contemporaneity" for the lack in the Russian language of an expression equivalent to the west-European "modern." The Russian Revolution is very "modern."[35]

The new word correlated with a new understanding of modernization shaped by the then prevailing context of Progressivism, both an ideology and a mindset. For the purpose of our study it is important to note that historians characterize turn-of-the-twentieth-century Progressivism as a transatlantic phenomenon, when "university debates and chancery discussions in Paris, Washington, London, and Berlin formed a world of common referents."[36] As Daniel Rodgers aptly points out, "Atlantic-era social politics had its origins not in its nation-state containers, not in a hypothesized 'Europe' nor an equally imagined 'America,' but in the world between them."[37] The second key feature of Progressivism was its proverbial technocratic approach to solving the global problems of society. It was related to, and partially resulted from, the Progressives' indifference, even hostility, toward politics. The general mood was that "[p]olitics as a governing device had become outdated, falling prey to the mass appeals and backroom deals frequently thought to characterize it. ... Antitheoretical theory begat apolitical politics."[38] Frustrated with the limitations of the nineteenth-century type of democracy of restricted enfranchisement, Progressivism advanced a more efficient scheme of "network mobilization" as a system of multiple campaigns for individual causes.[39] This "apolitical politics" implied a de facto different concept of citizenship, based not on guaranteed formal belonging to the enfranchised political community, but on optional and active participation in a public self-mobilization campaign. This version of citizenship was institutionalized in the form of grassroots associations and clubs.

Russian Empire, though hardly a part of the "North Atlantic world," was actively engaged in this process of Progressive intellectual exchange and dialogue. Every major theme debated by the international reformist community found its prompt response in Russian progressive educated society (*obshchestvennost'*). Sometimes it is possible to measure the intensity of this rapport.[40] We find the same motif of "apolitical politics" in Russian publications, literally praising "socialism without politics."[41] The international Progressivist "culture of modernization" became an integral part of the Russian public sphere and eventually became a dominant force in Russian politics in summer 1915, when the majority of the fourth-state Duma deputies managed to unite in a coalition called the "Progressive Bloc." Both the origins of this bloc and the fact that nobody questioned its name testify to the wide spread of Progressivist political discourse and imagery in Late Imperial Russia.[42]

The ethos of "apolitical politics" of the intellectuals from democratic countries resonated particularly well with the Russian intelligentsia living in the situation of political demobilization in the semi-parliamentary, semi-autocratic country. They could also rely on the domestic tradition of piecemeal social reformism, namely, the "small deeds" approach toward improvement of the people's (primarily peasants') conditions. Not directly challenging the existing regime, back in the 1880s the "small deeds" theory was mocked by the opposition leaders for opportunism, a lack of grand strategy and important goals, and was denied any political significance (except for the negative role of distracting the scarce human resources of the educated elite from radical opposition to the authorities).[43] "Small deeds" became the major synonym for an apolitical venue of social activism.[44] It took more than a decade for the "small deeds" modernization discourse to enhance its position among the intelligentsia, who until the defeat of the 1905 Revolution were very reluctant to give up their political radicalism in favor of economic reformism.[45] Political demobilization after 1905 and the spread of Progressivist ideology and ethos among all strata of Russian educated society, particularly stimulated by participation in the World Exhibition of 1900 in Paris and the takeoff of the city reform movement, gave new meaning to the seemingly compromised "small deeds" ideology. Progressivism in its Russian reading succeeded in uniting the hitherto conflicting social projects of radical populists, with their ideal of servicing "the people," social democrats, focused on economic efficiency, and liberals, mostly concerned with individual success (a development mirroring processes in the West[46]). The same Peter Struve insisted in November 1908:

> Greater [economic] *productivity* is always based on a higher standard of personal *applicability*.... While the eternal idealistic moment of liberalism consists in the idea of freedom and individuality of the person, the eternal realistic moment of a liberal worldview embodies an idea of personal applicability.... The Russian intelligentsia need a crucial reconstruction of their entire economic worldview.... They must understand that the productive process is not a "predatory" one but a creation of the very basis of the culture.... The development of national productive forces should be understood and accepted as a national ideal and a national service.[47]

This was a concise formulation of the political agenda of Russian "apolitical politics:" the building of a new national compound through the development of human capital, to be measured by its economic performance. The vision of Progressivism in Russia as a coordinated movement for society's self-modernization has been obstructed for a while by old-time historiographic clichés that conceptualized the social dynamics of Late Imperial Russia in terms of fixed sociopolitical and economic identities. In the 1970s and much of the 1980s, studies of Late Imperial Russia were greatly influenced by the structuralist formula of "dual polarization" (between the tsarist government and educated society, and between the latter and the working class).[48] The "polarization" scheme implied that imperial society was in a state of complex disintegration, with the Revolution of 1917 becoming a logical and all but inevitable outcome of that process. The revisionist approach of the 1990s instead concentrated on social structures and intermediate groups that "bridged" the alleged polarization of Russian pre-revolutionary society.[49] However, this new research agenda still took for granted the conventional divisions of Russian imperial society along the fixed lines of class, social estate, or party membership. It ignored the fundamental interconnectedness of Russian reformist forces and understood "bridging" as filling the gap between static social groups. Very few scholars would explore the new social structures and practices that began emerging after 1905 to accommodate the "apolitical" politics of social change. Thus, people like Ekaterina Vavilova or Alexander Chaianov could be analyzed in the context of studies of the radical intelligentsia or of economic theory, but there was no way to show their intellectual and practical interdependence with such distinct figures as Petr Stolypin or Peter Struve.

The underappreciation of late imperial society's potential for self-modernization rested, in part, on the influential concept of its "sedimentary" dislocated structure, and on the fatal inadequacy of its frustrated intellectual elite to properly emulate the "true" modernity of the West. The first concept belonged to Alfred Rieber, who replaced Haimson's clear-cut image of Russia as a society composed of (predominantly) capitalist classes with a subtle vision depicting a social fabric woven from a mosaic of social identities, archaic and modern, coexisting simultaneously and overlapping with each other.[50] Rieber's snapshot revealed a "sedimentary society," but could not provide insight into the direction of its *evolution* before 1917. However, this static picture was perceived by many as a diagnosis of social processes under way in early twentieth-century Russia. This is the origin of the popular thesis of the fatal "fragmentation" of Russian prerevolutionary society, which allegedly explains its collapse in 1917[51]—a thesis that became standard in the work of some social historians in the 1990s.[52] Laura Engelstein most powerfully advanced the skeptical vision of the Russian intelligentsia's modernizing efforts. Among the first to introduce the themes and concepts developed by Michel Foucault to the field of Russian history, she opened a new venue for studies of Russia as a partially modernized society of professionals and intellectuals that used modern "techniques" and produced modern "discourses." She also laid grounds for subsequent interpretations of the Soviet regime as "an alliance between the old tutelary state and the new disciplinary mechanisms."[53] Despite her critical and balanced application of Foucault, she built into her model of emerging pockets of modernity in Imperial Russia a rather simplistic juxtaposition of advanced "Europe," where the disciplinary power of professionals (i.e., institutes of civil society) was guaranteed and regulated by law, and "Russia," where "both the reign of law and the ascendance of bourgeois discipline remained largely hypothetical."[54] Engelstein based her persuasive argument about the vulnerability of modern institutes in Russia on metahistorical speculations, using "Europe" as a self-explanatory trope, relying on a then nascent literature on Russian professions, and avoiding any reference to relevant European historiography. As a result, subsequent studies of modern ideologies and social practices in Late Imperial Russia were overshadowed by the sense of their inherited inadequacy and failure to emulate some normative "European" scenario.[55]

Thus, an attempt in the 1990s to revise a "polarization/disintegration" paradigm—which was structuralist in a Marxist or rather Braudelian sense, and was moved by the "trauma of 1917" and the necessity to explain its historical inevitability—was itself bound by the equally structuralist criteria of "normality" (class social structure, institutionalized civil society, party politics, etc.) and a *Sonderweg* vision of Russian development, largely caused by a lack of interest in a truly comparative perspective. Little wonder that, when multiple empirical lacunae were filled in by a new wave of research on professions, local social networks, and the interaction of different social groups at various levels, a conflict emerged between the general methodological scheme of the "postpolarization" tradition and the newly studied, rich body of sources suggesting a different vision of late imperial society—one that is much more dynamic and probably more self-conscious.

An attempt to override this deadlock can be seen in studies of the Russian public sphere and civil society that gained momentum sometime around 2000. Influenced by the seminal work of Jürgen Habermas,[56] the structuralist approach to the problem of social dynamics in Imperial Russia found what seemed to be an ideal vehicle for its description and analysis. The "public sphere" was seen now as a quintessence of "modernity," and "civil society" replaced the industrialization and parliamentary politics of older structuralist theories of modernization as a universal indicator of a country's development. Russian prerevolutionary society did not demonstrate a developed modern class structure or stable parliamentary regime—the two old criteria of modernization quite indifferent to the evidences of burgeoning social activism in the fatally "fragmented" and "porous" Russian social order. However, it now seemed possible to integrate that sedimentary ("imperial") society into a single model of the public sphere, and to interpret the instances of rapid professionalization and political activism as the signs of a nascent civil society. Given the social composition of the empire, the debate about the perspectives of civil society embracing the emerging public sphere in Late Imperial Russia was often referred to as the problem of turning "peasants into citizens."

Most noticeably, this became the theme of Scott Seregny's articles.[57] Following David Moon,[58] Seregny applied Eugen Weber's famous formula "peasants into Frenchmen"[59] to the Russian case in its political reading—"peasants into Russian citizens"—studying the growing involvement of Russian peasants with a nascent rural civil society in the 1910s.[60] He thus joined the ranks of American and European historians who over the past two decades have reassessed the role of local self-government administrations (*zemstvos*) and rural professionals in the service of zemstvos and government in mobilizing and integrating the peasantry into a broader society, and emphasized the amazing success of those attempts in dismantling the traditional peasant isolationism and passivity.[61] Still bound by the powerful legacy of the partial revision of the "dual polarization" historiographic canon, Seregny stopped short of acknowledging the building of a universal civil society in pre-1917 Russia as a success.[62]

The growing literature on Russian civil society became an arena of fierce ideological confrontation: the "optimists" pointed to empirical data suggesting a dramatic growth of public activism and awareness during the last decade of the old regime in Russia,[63] while "pessimists," not questioning the facts, appealed to methodological considerations, first of all the lack of institutional guarantees for a civil society in Imperial Russia and the insurmountable gap between Russian realities and the state of "true" civil society in the "West."[64] Both parties share a common essentialist approach toward civil society as an actual "thing," a formal institution that can be measured against some normative ideal. Joshua Sanborn was one of the first historians of Russia to abandon the archaic structuralist approach toward defining "the norm." He studied social practices in their "eventuality" dynamics rather than distilling "ideal types" from an unavoidably limited pool of processed data.[65] Joseph Bradley has demonstrated the illusiveness of belief in the existence of some ideal "Western" civil society,[66] while Harold Mah has deconstructed methodological essentialization of the public sphere:

> The transformation of social groups into persons who fuse into unity is, of course, a phantasy, and one that is always at odds with an empirical reality of conflicting social identities and interests....

> It seems to me that Habermas and historians are misled if they treat the idea of the public sphere as if it were or could ever be a real institution. Analysis of the public sphere should begin, I would suggest, with a recognition that its location is strictly in the political imaginary....Construing the public sphere as a powerful political fiction would lead historians not to measure institutions and intentions against the criteria of an ideal public sphere....Rather, the historical problem would be to figure out

why and how certain groups are able to render their social particularity invisible and therefore make viable claims to universality.[67]

The dynamics of agrojournalism in Late Imperial Russia suggest that a very considerable number of educated Russians behaved *as if* they were part of an invisible sphere of public debates of the most important issues of the day, which acquired a certain political influence, or even authority. While we can conceptualize this phenomenon in terms of the modern analytical category of the "public sphere," there was a developed category of self-description at that time: *"obshchestvennost'."* The concept of *obshchestvennost'* was firmly built in the language of self-description of educated Russians of the early twentieth century, only the related notion of *intelligentsia* could be compared to it in terms of its universal acceptance in all quarters of imperial society. Reconstructing an intertextual context of its application is equivalent to reproducing texts by Duma deputies, revolutionary leaflets, minutes of professional congresses, and resolutions of public associations.[68] This preponderance of the notion left many scholars quite indifferent to its content, at best equating *obshchestvennost'* with public associations and formal institutions of civil society.[69] Yet there are grounds to believe that the trope of *obshchestvennost'* embodied a certain social and political agenda. Genealogically, it accommodated the semi-rational sociopolitical imagery of social self-organization, previously characteristic of anarchists-Bakuninists and populists of the 1870s with their idea of the ideal society composed of autonomous public associations. Russian leading legal experts in the fields of administrative and civil law during the post-1905 decade acknowledged the fundamental nature of the rivalry between the state and *obshchestvennost'*, which "in fact limits the sovereignty of the state" and steals from it "part of its influence and loyalty."[70] In this respect, Russian lawyers followed their European peers, including luminaries such as Georg Jellinek, who regarded the state and self-organized society as two parallel and even alternative institutions.[71]

A partial explanation for the outstanding status of *obshchestvennost'* in Russian politics and culture may be found in its universal pan-imperial character. While the administration, legal system, and economy of the empire just nominally covered its entire space, being in fact but a hodgepodge of "special regulations" and semi-isolated economic systems, *obshchestvennost'* was one and the same in Tiflis and Harbin, at the district zemstvo board and in the capital, using the Russian language as the universal medium of communication and regarding the boundaries of the Russian Empire as its natural boundaries. This universalism of *obshchestvennost'* made it the most modern social institution in the Russian Empire, and thus authoritative even for those who objected to the leftist political connotations of the broadly defined "Progressivism" of *obshchestvennost'*. The dominant and persistent pan-imperial discourse of *obshchestvennost'* reconfigured and reconstructed the empire as a homogeneous space of equal citizenship in the Russian-language "republic of the letters" (which, however, did not include some borderland regions of the empire). Unlike the imperial schooling system, there were no *numerous clausus* to limit one's access to this emerging national compound;[72] unlike the imperial army, there did not exist any prejudice against certain groups regarded as "unfit" or undesirable for the common civic service,[73] and certainly no privileges for the "well-born." The universality of the *obshchestvennost'* sphere for a while supported the illusion of a similar universality of the Russian Empire itself. Substituting the formal unity of the multifaceted empire, secured by the figure of the autocrat who himself held almost 50 regional titles,[74] by the single community of civic-minded educated public implied that any initiative supported by *obshchestvennost'* had an empire-wide application and meaning. The project of radical populism embraced by the emerging *obshchestvennost'* in the last decades of the nineteenth century was a case in point: how else to explain the "ethnic blindness" of Jewish activists who would agitate for socialism among the Ukrainian peasants, posing as tsarist officials?[75] The dominant mental map of *obshchestvennost'* was some unqualified "Russia," where the universal ideals of enlightenment and modernity were to be put into practice. This unconscious or at least underreflective imperialism of *obshchestvennost'* greatly facilitated its rise as a pan-Russian phenomenon, but made it ill-prepared for the challenge presented by brewing and (for a while) much less articulated and visible alternative nationalist projects.

Russian peasants and the "third element"

When Russian *obshchestvennost'* embraced Progressivism atop its largely eroded populism and once again turned to the countryside, this time with the practical task of organizational and technological improvements, they envisioned their counterparts in the village as an equally homogeneous "Russian peasantry." This can be explained by the powerful populist inertia only partially affected by Marxist schemes. The abolition of serfdom in 1861 launched the process of continuous social engineering and discursive construction of "peasants." From then on, any ideological projection on the peasantry could be followed potentially by an actual change in the peasantry's socioeconomic or legal conditions. Historians studying debates among the Reform's architects—top bureaucrats, noble members of the provincial Editing Commissions (drafting legislation proposals for consideration in St. Petersburg), and general educated public in 1857–1861—may disagree in their interpretations of the nature of the reform and intentions of the actors involved. However, nobody questions its significance in the process of differentiating a legally and economically amalgamated coomplex of the gentry–serf estate into separate entities: the nobility and private land propriety vs. the peasantry and communal landholding of individual households.[76] It was not before the emancipation of serfs in 1861 that the peasant emerged as a universal category, not defined solely by his legal bond to the owner and economic dependency on the landlord. In fact, as Mikhail Dolbilov, who studied the Reform of 1861 as a discursive event par excellence, noted, it was the legislation that eventually constructed a holistic vision of the "peasantry."[77]

Even nowadays the most careful and informed scholars believe in the fundamentality of this social category. In the book that has set a new benchmark for studies of rural Russia, David Moon discusses the "Russian peasantry" as a social entity with clear characteristics and boundaries, and even calculates their quantity as separate from other "Slavic peoples (Ukrainians, Belorussians and Poles), and Finnic, Turkic, and Baltic peoples."[78] Quite aware of the challenges he faces, Moon selected a number of criteria of "Russianness," yet he did not explain why Orthodox "Ukrainian peasants" differed from "Russians" more than "Russian" Old Believers; why individualistic peasants of the Russian North or Siberia were closer to commune-minded peasants of Central Russia and not to Ukrainian farmers; why life "inside the borders of what had been the realm of the Muscovite tsars"[79] did not make "Russians" of the numerous Turkic and Finnic peoples of the Volga region. The general problem is, of course, an attempt to ascribe a modern national identity to a social group that is premodern by definition. But the idea that all varieties of nonprivileged agricultural populations, from Poland to Sakhalin, from Arkhangelsk to Turkestan, could be lumped together under a single category of "peasants" seems to be equally problematic. Even if limited to "Russian" peasantry (which could be possible in some situations empirically, but never for a single methodological reason), the analysis of this category would encounter such different cultural, social, economic, and technological patterns that any generalizations would have been limited to a few meaningful regularities.

We may add that the emancipation reform also overshadowed the theme of non-Russian agricultural populations (i.e., the majority of non-Russian ethnoconfessional groups) in the public debates for decades to come and even in subsequent historiography. Only Orthodox Slav peasants had been enserfed, hence postemancipation discourse fixed predominantly on these groups of petty agriculturists. That is why from the very beginning the new holistic notion of "peasantry" meant "Russian, Ukrainian, and Belarusian peasantry" by default. Non-Slav agriculturists were also included in this notion when it was used in broader all-imperial contexts such as famine relief, education, economic productivity, and so on. The particular circumstances of emerging national movements in the empire in the last third of the nineteenth century also allowed for the ethnically unqualified usage of the term "peasants" (which in practice meant ascribing indigenous "Russianness" to it in the dominating discourse of the Russian-language bureaucrats and public): the largely illiterate rural population was excluded from the public sphere of urban nationalists, and government pressure would not permit the active interaction of the latter with countryside residents. In the East European context, many ethnoconfessional groups lacked a complete social structure, while different social strata could be integrated into different sediments of the complex imperial society, thus hampering ethnic mobilization along traditional social hierarchies. For example, nationalist Polish gentry in the western provinces of the empire found it difficult to rely on a local peasantry that was predominantly Belarusian and Ukrainian by language and (less often) Orthodox by faith.[80] On the contrary, Ukrainian nobility long since been integrated into the privileged stratum of imperial society,[81] and in search of the glorious and unique past that would have differentiated them from the Muscovites, Ukrainian nationalists appealed to the legacy of Cossackdom and distinctive Ukrainian culture to be found among the town literati.[82] The nonintegrated part of the nobility of the Turkic peoples in the Caucasus and Middle Volga had much closer relations with the common villagers, but protonationalist mobilization in these societies was centered on their Muslim identities and their possible modernization, and thus had little interest in the most traditional and benighted segment of their brethren.[83] As a result, until the beginning of the twentieth century, the trope of "peasant" had pan-imperial legal or Russian (East Slav) ethnoconfessional connotations.

Thus by the turn of the twentieth century, neither top bureaucrats nor professional economists nor *obshchestvennost'* doubted the "Russianness" of themselves or the peasants, but were at pains to harmonize the ever-differentiating connotations of "people," "agriculturists," "peasant social estate," and "villagers"—once captured in the single epistemological and (perhaps) social entity of the "peasantry." No longer seen as icons of spiritual virtues and champions of land cultivation, peasants were still perceived as a whole, as a homogeneous stratum of "defective" economic producers. Different political factions within Russian educated society shared a common awareness that a "true" peasantry was yet to be cultivated through coordinated social politics—whether by endowing them with all of the arable land in the realm as a result of revolution and egalitarian agrarian reform, by rationalizing and intensifying agriculture, or by replacing communal landownership with private petty landownership. Hence, to fit Russian realities a 100 years ago, Eugen Weber's formula would have been adapted to "from Russians into peasants"—such was the order of priorities in public discourse of the time. Of course, "Russianness" was problematized in the multiethnic Russian Empire, as Russian nationalism failed to separate itself from Russian imperialism and imperial loyalism and to evolve into some form of a modern nationalism, whether political, racial, cultural, or other.[84] Moreover, it seemed that the very "national" principle (often understood in ethnographic terms) contradicted the idea of rationalizing the peasant economy.[85] Adhering to traditions (and often literally inventing new ones) hampered the reorganization of household production and reallocation of labor resources for purely "ideological" reasons.[86] Not unlike the old populist view of the peasantry, the new Progressivist approach of *obshchestvennost'* was universalist and largely insensitive to national specificity. In the words of Vasilii Ferdinandov, who in 1905 was a young instructor of poultry farming in the province of Voronezh and an organizer of local agricultural societies,

> We [the intelligentsia]…have joined in the universal progress; we have used much of others' labor and time to improve ourselves, while he, that prehistoric ploughman, the salt of the Russian earth, he was consciously left by us and our predecessors outside of culture, outside of life.[87]

By "culture" and "life," apparently, Ferdinandov implied the culture of professionals formed as a result of Westernization, and the enlightened life as structured by the standards of transnational modernity. While there is no doubt that agricultural specialists were imposing their standards on peasants as some universal norm,[88] it would be misleading to squeeze this group into the Foucauldian-type model of a corporation of bourgeois professionals producing hegemonic discourses in their striving for public authority. They were not quite "bourgeois," still being largely a part of the intelligentsia and effectively limited in their ideological outlook by the general worldview of *obshchestvennost'*. Contemporaries characterized this hybridity as "new 'bourgeois' forms of intelligentsia service."[89] The analytical model of professionalization as developed by modern social sciences is an important frame of reference that allows us to deconstruct general patterns of this group's formation and evolution. At the same time, to understand the motivation of the people who joined the ranks of agricultural specialists and the choices they had to make, we should turn to the tropes of their language of self-description, first of all, to the concept of the "third element."

The concept of the "three social elements" emerged at the turn of the twentieth century as an attempt to categorize new types of social identities in the modernized sector of rural society. The first element was represented by the appointed officers of the government agencies; the elected deputies of the relatively new zemstvo structures were called the second element; and the title of the third element was ascribed to the hired specialists in zemstvo service. This scheme obviously contained an allusion to the social hierarchy of the French ancien régime, with the interests of the landed gentry supposedly represented by the first two elements. Unlike the French model, however, the merchants and entrepreneurs were not regarded as a unity with any common agenda to represent (a point extensively elaborated by Alfred Rieber).[90] Hence, the third element was seen as a new social body of the faction of the intelligentsia that was compelled to earn a living by means of professional service, rather than to concentrate solely on literary or political activity.

The very term "third element" was introduced in 1900 in a speech delivered by the Samara vice-governor V. G. Kondoidi at the opening of the annual congress of provincial zemstvo deputies. He warned the representatives of the "second element" about the arrival of a new factor in Russian social life, a third element:

> ...it consists of specimen [sic!] with a large stock of scientific theoretical knowledge, who use it to raise and conquer authority for themselves in the local social milieu....It happens that the [zemstvo] representatives hearken to the word of *intelligenty* without sufficient justification, only because of a reference to science or the teachings of newspaper and journal writers, while these [*intelligenty*] are no more than employees of the [zemstvo] board.[91]

This episode proves that before the Revolution of 1905–07, it was the *first element* that dominated the sociopolitical landscape and quite insightfully defined its configuration, both institutionally and symbolically. In fact, Vladimir Kondoidi, a well-read and shrewd administrator,[92] all but anticipated a classical sociological description of professionals as a group that derived its social status from special knowledge provided by science.[93] Probably because of the ingenious accuracy of Kondoidi's definition, professional *intelligenty* ignored the pejorative overtones in the speech of a high-ranking official and enthusiastically accepted a new denomination—"the third element." Until the Revolution of 1905, the third element was totally subdued and controlled by the second element, not to mention the state agencies. Those who had started their professional careers before 1905 later retold numerous stories of the humiliation and assaults they experienced as young specialists in the zemstvo service.[94] They were treated as potential political delinquents who could not be trusted. In December 1903, Sergei Fridolin, a recent graduate of the Moscow Agricultural Institute, came to the St. Petersburg district zemstvo board (*uprava*) to be interviewed for the position of district agronomist. He was affronted from the very beginning by a zemstvo deputy who was also a senator (which was technically illegal):

> I remember, how the first question addressed to me by that senator was: "Aren't you a socialist?" And before I replied, the other Privy Councilor [*tainyi sovetnik*], also a deputy, answered for me laughing: "Well, what agronomist is not a socialist?"[95]

The situation of specialists in state service was no easier, for instead of personal assaults they were confronted by the system of official political screening. In 1906, the 48-year-old agronomist Mikhail Shaternikov was fired from the State Administration of Land Settlement and Agriculture (GUZiZ) for refusing to sign an affidavit of nonmembership in leftist political parties. In Shaternikov's police dossier there is no evidence that he was actually a member of any party at this time. Apparently, he just protested against the interference of political concerns in purely professional matters.[96]

Yet, time was on the side of the third element. The number of professionals in zemstvo service had increased to 65,000–70,000 by 1908,[97] thus making the third element de facto a significant social force with enormous political potential. It did not take them long to turn the tables, fighting for recognition by, and independence from, the first and second elements. Sharing a very general democratic ethos and a program of practical measures basically confined to professional activity and "small deeds," members of the third element characterized themselves as an "active part of the servicemen in public institutions who... influenced the democratization of the zemstvo organs."[98] The second element, the elected zemstvo deputies, played a dubious role in the elevation of the third element. On the one hand, a result of their efforts—a rapid expansion of the zemstvo budget and activities after the First Russian Revolution—contributed to the rise of the third element.[99] On the other hand, the history of professionals in zemstvo service is a history of constant conflicts between the zemstvo authorities and the third element. By the 1910s, the third element went from the defensive to the offensive, demanding the right for initiative and professional expertise of the zemstvo programs.[100] Previously, speakers for the third element questioned the moral authority of the second element and insisted that the existing oligarchic zemstvos should be replaced with democratically elected zemstvos.[101] In the 1910s, the third element already claimed that it was they who represented the "true" essence of zemstvo institutions (even though institutionally they were salaried employees, and not elected deputies), and demanded that the second element yield part of their control over zemstvo politics.[102] The constant pressure for the introduction of county-level (*volost'*) zemstvos could be viewed as an attempt to shift the balance of power in favor of the third element that would have totally controlled the most numerous lowest level of the zemstvo hierarchy.[103]

Judging from the sources, by 1913, specialists in the zemstvo service felt themselves in a position to challenge even the authority of the landed gentry who only a decade earlier had treated agronomists or veterinarians as their personal employees.[104] It was sometimes said that gentry deputies to zemstvo boards no longer represented the true interests of the class of landowners, for they lacked the necessary knowledge and training to identify those interests.[105] Now the second element could not survive without the expertise of the third element, and the latter was ready to apply all available leverage in a bargain for influence.[106]

From what we know about the climate in zemstvos on the eve of the World War I, the third element and the zemstvo patricians had reached a compromise. Here is a description of a quite typical celebration by the Birsk district zemstvo board (Ufa province) of the fiftieth anniversary of the zemstvo institution in the Russian Empire. Even a brief glimpse at the proceedings reveals much about the distribution of authority in this zemstvo. The grand meeting of the zemstvo board began on February 19, 1914, with religious ceremonies: separate services for Orthodox and for Muslims took place about noon. Then followed a brief presentation on the history of zemstvo institutions in Russia in general, and in the Birsk district in particular. After a break came the time for official addresses. The first speaker, who could be expected to be the most important figure in the audience, was not a board chairman or deputy, but a zemstvo physician, A. A. Smorodintsev, who said,

> Allow me to address this jubilee meeting with brief greetings on behalf of the zemstvo employees, the so-called third element, upon their authorization. We are happy that life as personalized by all of you [*zhizn' v vashem litse*] has recognized, apparently, our modest work for the benefit of the population of the Birsk district.[107]

When the representatives of the third element succeeded in securing their rights and status, a bitter rivalry between the elements gave way to partnership. At least in the sphere of economic assistance to the rural population, the role of specialists was recognized by both the state and the zemstvo.[108] By 1914, the third element had become indispensable in sustaining agricultural productivity and successful procurement campaigns. In their turn, rural professionals greeted the rapprochement between the first and the second elements. This situation gave them a freedom of maneuver between those major employers resulting in a higher degree of independence from both of them. Rural professionals even criticized those who

> are living by the principles of the past, pre-liberation [i.e., pre-1905] opposition of "zemstvo and bureaucracy," who cannot see that during the postliberation epoch both of these elements…have been brought together to such an extent that it is difficult to say where one ends and the other begins…. The most democratic organs of self-government will be unable to cope with the present abundance of their functions without help from the state budget. And the most convincing example of this is America, where individual states…use the aid of the federal budget, and precisely in the sphere of agronomic measures.

> That is why we believe that financing and support [of the zemstvo] by the state…suggests the rise of new, more correct relationships between the state and the zemstvo.[109]

As the gap between the major sediments of Russian society was being gradually bridged during the 1910s, a new social force appeared to challenge the seemingly certain harmony. Without much exaggeration, the rapidly growing cooperative movement can be recognized as the "fourth element" in the modernized sector of Russian society.[110] It was a genuinely mass phenomenon that challenged the authority of the first two elements on different grounds than the third element had a decade earlier. While agricultural specialists could not survive without the favorable attitude of the state agencies and the solid budget of the zemstvos as their employers, cooperatives were much more independent of the government and the zemstvos. Cooperative ideologists claimed that only the voluntary economic associations really met the needs and aspirations of the population, while the zemstvo was a compulsory institution built upon a highly restrictive franchise system. Hence, leadership in representing the interests of the rural population must belong to cooperatives.[111]

Thus, the Progressivist-minded cohort of intelligentsia-turning-professionals can be presented in the language of the epoch as occupying a social niche between the "establishment" of the first and the second elements, and the rising "fourth element," while partially overlapping with the "third element." The post-1905 "turn to agriculture" in the public discourse imparted a prominent role to the hitherto neglected third element, while the dominant trend in the schooling system promised further expansion of the cadres of rural modernizers.

2
Bringing Up a New Generation of Intelligentsia

As we have seen in the previous chapter, the interrevolutionary decade had witnessed an upsurge of wide-scale public interest in agriculture as reflected in the great number of publications on the topic. The quite unromantic and "down to earth" profession of agronomist suddenly appeared in the spotlight of public attention. This came as a surprise to the intelligentsia, whose old dream of achieving a better lot for the common folk now found an absolutely legal outlet recognized even by the government. As late as 1904, the leading liberal newspaper *Russkie vedomosti* expressed a doubt that trained agricultural specialists could have been expected to influence the Russian countryside in any significant way due to their small numbers.[1] These numbers began to grow after 1905, as intelligentsia disillusioned in radical politics rushed to join the ranks of rural professionals. In April 1910, when Ekaterina Sakharova and her classmates were finishing studies at the Moscow Agricultural Institute, a columnist eloquently expressed the dominant mood among youth, writing in a magazine for students: "Damn politics! Purishkevich, Dumbadze... Dumbadze, Purishkevich—thus one can go mad and end up in the St. Nicholas hospital. I decided to engage in agriculture."[2]

New social groups do not appear overnight, and the fresh agricultural specialists ventured into uncharted waters. The social instincts and habits of the traditional intelligentsia were of little help to them. It is only natural then that the leading role in the new social movement was played by the graduates and professors of a few Russian agricultural colleges. Long before the advent of agrojournalism in the early 1900s and the public demand for agricultural expertise, the agricultural higher school had formed a well-informed and scientifically validated critical attitude toward the realities of agricultural Russia and elaborated scenarios of their change. The content of this "wisdom in a capsule" also changed over time.

In the 1880s, even liberal professors saw the role of agricultural higher education in "special training of independent managers, organizers, top administrators."[3] Insofar as agricultural specialists were trained to manage large estates (private or state-owned), nobody would expect them to deal with individual peasant households, or to elaborate and carry out any agricultural program on a national scale. This attitude changed, however, soon after the famine of 1891. One of the government officials in charge of agricultural education in the empire, I. I. Meshcherskii, in 1893 formulated a new task for agricultural specialists. In his opinion, they had to become scientifically educated activists, "putting Russian agriculture on the right track and directing it under complex and diverse soil, climatic, economic, and ethnographic (*bytovye*) conditions."[4] At this point, agricultural specialists had been assigned a task of nationwide importance, which brought them into direct contact with the peasants as economic subjects in their own right, rather than an element of complex latifundia economies. The next 20 years witnessed a radical transformation of Russian agricultural education, under the dual influence of a new state policy and a new public attitude. The institutions of higher learning embodied this new social paradigm in a most explicit way.

In the 1890s there were only five state agricultural colleges (institutes) in Russia, with a total of fewer than 1,400 students, and four veterinary institutes with approximately 1,100 students.[5] By the outbreak of the World War I these figures had grown twofold.[6] In fact, students graduating from state institutions of higher learning with agriculture-related training (agronomists, veterinarians, foresters, agricultural engineers) were only the fifth largest group out of 14 professions, after lawyers, physicians, teachers, and factory engineers. True, such important and socially visible professions as military officers or railway engineers lagged far behind agricultural specialists,[7] despite all the government concern in rearmament and development of railway communications. Still, it is instructive to look at the pace of changing numbers of graduates as an indicator of the social demand for a certain profession. Interestingly, agriculture students were the fastest growing category during the decade 1898–1907, but after the First Russian Revolution the dynamics of their increase lagged far behind the other popular professions, particularly physicians (see Table 2.1.).

However, these figures reveal only part of the story predetermined by government policy: the number of students was regulated by the existing size

Table 2.1   Increase in numbers of students graduating from state institutions of higher education (in the five most popular professions).[8]

	1898–1907 (%)	1907–1913 (%)
Agricultural specialists	+93	+38
Factory engineers	+83	+113
Teachers	−11	+108
Physicians	−7	+149
Lawyers	+9	+72

of staff and available funding. It is scarcely accidental then that the cost to the state budget of an average student's education was lowest for law students and highest for military cadets. Engineers cost less than agricultural specialists.[9] Taking into consideration that only a small fraction of the imperial budget was allocated for education, the steady increase in numbers of agricultural students comes as no surprise. As statistics shows, only a portion of those who wanted to become agricultural specialists were admitted to state institutions of higher learning. During the first half of the 1910s, the competition for admission to state agricultural colleges was constantly increasing: in 1911, 66 percent of the applicants were admitted; in 1912—56 percent; and in 1913—only 45 percent, despite the increasing admission quotas.[10]

Testifying to the truly popular nature of the interest in agriculture, at the beginning of the century, a considerable number of private institutions of higher learning burgeoned, compensating for the inflexibility of the state educational system. In the sphere of agricultural higher education, the number of private institutions by 1917 was equal to the number of state institutions, and in terms of student enrollment, the state-owned schools lagged even behind independent "agricultural courses."[11] According to some estimates, during the period of 1898–1917, independent institutions had trained 1,900 agricultural specialists, but also 1,000 lawyers, 4,000 physicians, and 17,000 teachers.[12] While the actual figures may significantly differ from these (we can estimate that 10,000–20,000 Russians studied agricuture professionally at private colleges without formally graduating with diplomas),[13] the overall proportions seem to be accurate. They reflect the compensatory function of private professional schooling, which was oriented toward training the most needed specialists.

Still, the question remains, why the ranks of would-be agricultural specialists did not make an abrupt takeoff after 1907, as was characteristic of physicians, factory engineers, and teachers. The answer may sound paradoxical: the sudden success of the agriculture-related professions prevented the number of students in agricultural institutes of higher learning from doubling between the years 1907 and 1913. Only 10–25 percent of the increasing demand for agricultural practitioners, especially agronomists, could be satisfied by college graduates,[14] which placed the mid-level agricultural school (*uchilishche*) in an exceptional position. Up to a certain point in their careers, its graduates enjoyed the same professional recognition and benefits as the graduates of the institutions of higher learning. At the same time the mid-level professional education was far less demanding in terms of the educational background required for admission, and notably cheaper. Many students even received scholarships from their local zemstvos or from the institutions themselves.[15] Besides, the practically oriented professional training in the mid-level schools was more relevant for the future practitioners— agronomists, husbandry and horticulture specialists, and so on.

Throughout the early 1910s, the public demand for mid-level agricultural education was stable and high: only one-quarter of all applicants were admitted annually to the *srednie uchilishcha* (despite the quantitative growth of the latter).[16] The competition for seats in the elementary agricultural schools was also high, fluctuating from two to three applicants for every available place.[17] Thus the booming job market removed some pressure from the agricultural institutions of higher learning at the expense of mid-level and even elementary schools, which accounted for the steady, rather than dramatic expansion of agricultural higher education. In general, during the period of 1907–17, the most significant changes in the agricultural institutes were associated not so much with the figures of student enrollment as with their social composition.

The composition of the student body characteristic of the 1890s changed abruptly after the Revolution of 1905–07. The share of students of noble origin declined from 55 percent to 25 percent, while the proportion of peasant sons increased almost fourfold (from 8 percent to 31 percent). In 1915 the social composition of students in the four oldest agricultural institutes mirrored structurally that of the 1890s. Only now the bulk of students was constituted by sons and daughters of peasants and town commoners (together close to 60 percent) instead of children of noblemen and high-ranked officials (formerly represented by about 55 percent). Coupled with an increased representation of the clergy estate and merchants, this made the social portrait of the agricultural institutes' student body by the end of the first decade of the twentieth century very different from that of the 1890s.[18] New people had brought a new spirit of professional public service to agricultural higher education.

This generalization is not based merely on the formal social estate affiliation of the students before and after the Revolution of 1905. We have statistics, although incomplete, of the occupations of the Moscow Agricultural Institute's graduates, from the moment of its foundation in 1865 (as Petrovskaia Agricultural Academy, aka Petrovka). Professor A. F. Fortunatov collected data for 629 (73 percent) out of 860 former students of the institute who graduated during the first 28 years of its existence. According to his statistics, fewer than 6 percent of the students whose diplomas read "learned agronomist" became rural professionals. Almost as many students dedicated themselves to "art and literature."[19] The most popular occupations of those who had graduated from Petrovka before 1894 are shown in Table 2.2.

Perhaps, some of those who reported being employed "in the zemstvo service" worked as agronomists. Still, even if we lump together all categories that could possibly include those who were employed as agronomists by various agencies, their share does not exceed that of estate managers.

After repeated occurrences of student unrest in the early 1890s, the Agricultural Academy was shut down and reopened in 1894 as Moscow

Table 2.2 The distribution of graduates from Petrovskaia Agricultural Academy along major professional tracks, 1865–1893.[20]

Managers of gentry estates	17%
In the government forestry service	17%
Gentry landowners	14%
Educators	13%
Joined bureaucracy	12%
In the zemstvo or town councils service	11%
Other five occupations	15%

Agricultural Institute. Professor K. A. Verner collected information about the consequent employment of 117 out of 269 students (43.5 percent) graduated from the institute during the decade prior to 1905. According to his data, 29 percent of graduates became educators, 27 percent ran their own estates, 25 percent became agronomists, and 19 percent became managers of private estates.[21] Since the occupation of more than half of the graduates remained unreported, the actual picture could significantly differ from these numbers. Still, one cannot but notice the closeness of the figures for managers of private estates (17 percent in Fortunatov's, 19 percent in Verner's statistics).

The post-1905 period reveals a very different pattern. Out of 125 men and women who graduated from Moscow Agricultural Institute in 1910, at least 77 later occupied agronomist-related positions: 62 percent as compared with 6 percent prior to 1894.[22] Many students became foresters, some left the sphere of rural affairs for good, and a few returned to their estates. But the figure of a private estate manager trained in an agricultural institute became obsolete in the interrevolutionary Russian social landscape. In 1914, a survey was held among 229 students (15 percent of the whole student body) to determine their future professional plans. The absolute majority of the respondents were interested in professional careers as agricultural specialists: about a quarter of the total wanted to concentrate on research, more than half felt a calling to "public" (zemstvo) service, and only 1 percent were thinking about private service.[23] Clearly, this was a radical departure from the career preferences and actual choices of students during previous decades.

Let us return to the question of how different social agendas were supported by different social compositions of the agricultural students. We now have grounds to claim that before the Revolution of 1905, the Stolypin agrarian reforms, and society's turn to interest in agriculture, students of humble social background could aspire to careers as estate managers, government specialists, or academics. The descendants of wealthy landowners and bureaucrats studied in agricultural institutes to learn how to reorganize their estates, or because they could not enter a more prestigious institution that would allow them to join state service. The new social composition of the alumni and the new social climate in Russia after the Revolution of 1905 had affected the students' career prospects and the associated educational agenda. Students born to the families of peasants and (mainly rural) clergy brought to higher agricultural education their knowledge of peasant life and needs, while the townsmen contributed a sense of professional service. This audience must have been especially sensitive to the neopopulist views expressed by many professors teaching in agricultural institutes. The combination of these three elements would bring into existence a new social movement of intelligentsia-turning-professionals. A close-up view of the first 15 years of the twentieth century, and just one college, the Moscow Agricultural Institute, allows us to reconstruct a picture of sociability networks that brought people of different background together and allowed for a cumulative effect of their individual efforts.

Moscow Agricultural Institute was the largest among the state agricultural higher-level schools in Russia, accounting during the interrevolutionary period for about 40 percent of all students.[24] The institute, located in the Petrovskoe-Razumovskoe suburb of Moscow, was famous for its spirit of public service and the scholarly accomplishments of its faculty and graduates. As a certain V. Kh. put it in an article published in *Zemledel'cheskaia gazeta* in 1915, referring to the preceding decade:

In Razumovskoe not only was agricultural life being created, but also Russian agronomist thought was crystallizing. The Department of Agronomic Social Science [*agronomicheskogo obshchestvovedeniia*], represented by Professor Aleksei Fedorovich Fortunatov and his pupil A. V. Chaianov, was the crucible where not only the farm organization was studied, but the modern principles of agronomist assistance to the population were developed.[25]

At the beginning of the twentieth century, Moscow Agricultural Institute was a rather compact and elitist school. In spring 1900, a total of 204 students were studying and living on the premises of the institute, on a spacious estate, Petrovskoe-Razumovskoe.[26] Today we would call it the Petrovskoe-Razumovskoe campus, for all the necessary infrastructure, including athletic and recreation facilities, was located nearby. The arrangement was unusual for Russian higher education; with the exception of a few elite schools no other university or institute provided such a unique opportunity for students to build bonds of friendship and a sense of fraternity. Only students were permitted to use boats on the pond, while during winter they could take advantage of the skating rink in Petrovskii Park, or borrow the institute's skis. Annual fees were high, totaling 400 rubles, but this sum also covered accommodation in a single room and a very generous meal plan.[27]

Only men were admitted to the institute before 1905. The typical *petrovets* (i.e., student of the Moscow Agricultural Institute) in the spring of 1900 could be expected to be the son of a nobleman (as were 33 percent of students), or a high-ranked bureaucrat (19 percent), or a city middle-class representative (29 percent).[28] Surprisingly for a young man of such a high social profile, he was most likely a graduate of the "real school" (vocational school modeled after German *realschule*), which trained its pupils for modern professional careers (almost 35 percent of all *petrovtsy* studied in real schools). There were equal chances that our typical *petrovets* had graduated from a military school (*kadetskii korpus*, 15 percent), a classical high school (*gimnaziia*, 14 percent), or even a university (14 percent).[29] Thus we can speak rather about two major types of Moscow Agricultural Institute students: a wealthy middle-class graduate of a vocational school and a nobleman with a classical Russian upper-class education.[30] Seventeen students formally belonging to the peasant estate could not seriously affect this image of the typical *petrovtsy*, while the presence of three princes and two barons added a certain scent of aristocracy to the institute. By January 1906, when the number of students studying at Moscow Agricultural Institute had increased twofold, reaching 429, the same two types represented about 80 percent of all of the *petrovtsy*. Only now, the proportion of city middle-class representatives grew to 36 percent of the entire student body, and the share of vocational-school graduates increased to 48 percent.[31]

Five years later, in January 1911, the tendency toward democratization of the students of Moscow Agricultural Institute became even more obvious, particularly when the cost of education was reduced two times after the abolition of on-campus living. The proportion of students belonging to the old regime's elite (nobility and bureaucracy) had already been reduced below 31 percent, and in January 1915 this figure fell even lower (27 percent). Accordingly, the share of middle-class representatives had jumped to 37 percent by January 1911 (though it decreased slightly to 34 percent by 1915). While the two formerly dominant social groups were reaching a state of balance, the picture became complicated by an increase of students of peasant social background (their share had grown threefold between 1900 and 1915). In 1911 peasants accounted for 22.4 percent of the *petrovtsy*, and in 1915 for 27.4 percent, thus becoming the second largest social group, and leaving the elite representatives behind.[32]

As these figures show, sometime around 1910 a very profound change occurred within the social composition of Moscow Agricultural Institute students. The process of diversification of the student contingent quite paradoxically resulted in a merger of the two major types of *petrovets* into one: a young man—or also a woman now, for example, Ekaterina Sakharova— whose future career was predetermined not by his or her social origin, but by the professional ethos and public agenda formed during study at the institute. As we mentioned, at least 62 percent of the students graduating in 1910 chose careers as agronomists, whose major task by this time was teaching and assisting the peasants.

No less significant was the difference between the students graduating in 1910 and those just entering the institute that year as the class of 1914. In a sense, the class of 1910 was a landmark in the history of the Moscow Agricultural Institute. It was the last generation of *petrovtsy* living "on campus," for in its cost-cutting efforts and because of the constantly increasing number of students, the institute closed its dormitories. The class of 1914 would never experience that intensive sense of community characteristic of the previous generations of *petrovtsy*, coupled with the thrill of living in a social melting pot. For students entering the institute after 1910, it was merely a place of professional education. They inherited an already formed social agenda, which was to be carried out in a society already stabilized on tracks of semi-parliamentarism and gradual reformism. In a sense, they were more profession-oriented than their more intelligentsia-type predecessors.

There is a clear indication that even the demographic structure of the 1910 *petrovtsy* was quite different from the 1914 group. With an approximately equal number of students enrolled (1,166 in December 1910 and 1,186 in December 1914), in 1910 the concentration of *petrovtsy* born in a certain short period was higher than it was in 1914. Of all the students registered in 1910, 58.4 percent were born during the five consecutive years between 1884 and 1888, each year yielding over 100 students (115, 133, 143, 159, 131). Only three years were of comparable significance for the *petrovtsy* of 1914, a short interval from 1890 to 1893 with a less impressive annual rate of births (117, 120, 120). Thus the maximum concentration of the coevals in 1914 did not exceed 30.1 percent.

These statistics suggest that studying at the Moscow Agricultural Institute sometime around 1910 could form a generational experience. People born between approximately 1884 and 1888 and maturing during the First Russian Revolution were more conscious than any other age group about becoming agricultural specialists and improving the ways of rural Russia. It is important to distinguish the concepts of an age group and a social generation. Age merely helped those young people to join the movement and reinforced their sense of unity, but it could not predetermine their choice.

A brief note on terminology seems pertinent at this point. Modern sociologists distinguish several approaches toward the phenomenon of generation. One is functionalist, treating age groups as an element that secures the stability and continuity of the social system.[33] The second approach, political economy, views age as a factor of power and resource allocation.[34] Finally, there is a social generation perspective that emphasizes common historical experience as the basis for keeping coevals together.[35] The demographic aspect is of only marginal interest for this research, as people of all ages were responsible for the mobilization of Russian society around the problems of agriculture. Hence, an extensive functionalist literature following the tradition founded by Talcott Parsons will be of limited usefulness here. Equally irrelevant for the purpose of our study is the political economy approach, for age was not the dominant prerequisite for getting access to power or prosperity in the group under discussion. The conception of generation elaborated by Karl Mannheim,[36] which still holds significant explanatory potential,[37] seems to be the most promising methodological framework for this research.

Mannheim attributes the key role in a generation's formation to a common historical experience: a generation can thus be defined as a group of people who were born during approximately the same time, share the same life experience, and differ from other generations by a particular ideological outlook. The sociological tradition of studying social generations established by Karl Mannheim and José Ortega y Gasset underlines the importance of historical context, which makes it particularly popular with historians seeking to analyze collective experiences.[38] Though not sharing some extremes of the Mannheimian tradition,[39] I do subscribe to its fundamental hypothesis that a generation is shaped by common experience and identical reaction to it by members of the generation-to-be. Three major aspects of this approach, as elaborated by subsequent sociologists, should be mentioned here. First, any distinguished generation is always a "young generation" driven by a youth movement. In modern society "the school and the conscription army were institutions which measured time,"[40] and hence they contributed to the formation of self-aware age groups sharing a particular common social experience. Second, it seems more accurate to talk about generation as a community of seniority rather than one of pure age. This is an important distinction that stresses the voluntary character of joining a social generation as opposed to a predetermined belonging to an age group by virtue of one's date of birth.[41] Finally, the concept of generation is necessarily elitist, that is, it does not denote all individuals theoretically qualified to become a generation's members.[42] In his classic study of the Generation of 1914, Robert Wohl provided a powerful and comprehensive formulation of the modern vision of historical generations:

> A historical generation is not defined by its chronological limits or its borders. It is not a zone of dates; nor is it an army of contemporaries making its way across a territory of time. It is more like a magnetic field at the center of which lies an experience or a series of experiences. It is a system of references and identifications that gives priority to some kinds of experiences and devalues others—hence it is relatively independent of age. The chronological center of this experiential field need not be stable; it may shift with time.

What is essential to the formation of a generational consciousness is some common frame of reference that provides a sense of rupture with the past and that will later distinguish the members of the generation from those who follow them in time. This frame of reference is always derived from great historical events like wars, revolutions, plagues, famines, and economic crises, because it is great historical events like these that supply the markers and signposts with which people impose order on their past and link their individual fates with those of the communities in which they live.[43]

Students of the Moscow Agricultural Institute of the class of 1910, who finished high school during the stormy period 1905–07, formed a newer generation of Russian intelligentsia simply because they were very young. To become members of a particular social generation, however, they would have to make a series of conscious decisions, including choosing a profession and a certain social agenda (as did Ivan Emel'ianov or Ekaterina Sakharova). One may notice a correlation between the rise of broad public interest in agriculture, as reflected in the statistics of special publications, and the appearance of a new ideal type of agricultural professional. It was the appreciation of professional expertise that distinguished the post-1905 generation of educated Russians from the traditional Russian intelligentsia.[44] In his seminal essay on Russian intellectual tradition, Michael Confino wrote that the heirs to the classical intelligentsia "were intellectuals, or professional revolutionaries, not another generation of the intelligentsia."[45] Indeed, in contrast to the classical radical intelligentsia, the new post-1905 generation of Russian intelligentsia that was shaped by the collective experience of disillusionment in the revolution was composed of people who saw their task to be professional assistance to the presumably archaic peasants to integrate them into the modern society and economy:

If before the introduction of the constitutional regime, progressive elements of society, even including many bourgeois elements, were thinking in a revolutionary way, now an evolutionary point of view has begun to prevail where hitherto it was imagined that even the most radical solution of the agrarian question would take the lifetime of a generation to achieve.[46]

A unique combination of structural factors and individual experiences resulted in the formation of a phenomenon that we may conceptualize as the new generation of Russian intelligentsia, simultaneously a distinctive social group and a coherent social movement. We may assume that the young professionals including members of the class of 1910 constituted the most dynamic and coherent part of that social generation. They were joined by people belonging to different age groups, who in the years following 1905 realized that professional expertise could become a vehicle for achieving certain social ideals. It was the synthesis of professional ethos (and skills) and a particular social agenda that distinguished these people from the traditional *intelligenty* or narrow-minded specialists.

Different paths led people of various social backgrounds and ages into the ranks of the new generation of Russian intelligentsia. Among the most prominent figures of that social group was Vladimir Vargin, whose twenty-fifth anniversary in public service was celebrated in 1913 by agriculturally concerned Russian society.[47] Born in 1866, his father was a serf who after the emancipation of 1861 became a town commoner (*meshchanin*) and died when Vladimir was very young. Nevertheless, the boy graduated from the Ekaterinburg gymnasium with a silver medal, and in 1884 was accepted to Petrovskaia Agricultural Academy. He was among the 11 percent of the academy's graduates who, according to Professor Fortunatov's statistics, joined zemstvo service as rural professionals. When many other students were writing their theses on West European subjects, Vargin's topic was "Green fertilizer and its significance for the peasant farms of Perm province." He was obviously eager to serve the people, but no social mechanisms were available at that time. Vladimir Vargin taught in the Krasnoufimsk technical school until 1899, when he was offered a position as provincial zemstvo chief agronomist.[48] In the early twentieth century he contributed to the rapid expansion of the Perm agronomist organization, which became one of the best in Russia. Apparently because of intensive professional activity, the health of the 47-year-old Vargin rapidly deteriorated, and in 1913 he stepped down from his position to concentrate on the experimental farms of the Perm zemstvo.[49]

At the time when Vladimir Vargin was becoming acquainted with his new job as provincial agronomist (the top position in the zemstvo hierarchy of agricultural specialists), Sergei Fridolin was taking his first steps toward a professional career. In fall 1899, he became a student at the Moscow Agricultural Institute. His social background differed significantly from Vargin's. As the son of a successful Petersburg physician, Sergei studied in the prestigious Second Classical Gymnasium in Petersburg. Yet, a boy of mediocre abilities, he was at the bottom of his class. He showed little interest and even less ability to continue his education at one of the institutes that were considered prestigious (e.g., the Military College or Medical Academy) among his friends from the Petersburg upper middle-class. However, he felt some attraction to agriculture.[50] This interest had nothing in common with the sense of public service characteristic of Vargin. Fridolin failed to present a decent graduation thesis at the institute, which indicates that an interest in scholarship was not a decisive argument either. His attraction to agriculture, as we can understand from his memoirs, was similar to the attraction to crude working-class women characteristic of middle-class gentlemen in Victorian

Britain.[51] It was the thrill of coming into contact with an unknown and certainly inferior *something* that provides even mediocrity with a sense of superiority, and with the respect he did not receive from the peers. Young Fridolin was excited by the discovery of manure during the summer break in the village where his family rented a summer house. Even writing his memoirs 30 years later, he could not escape using almost erotic vocabulary in describing the pleasure he received from riding a cart full of manure.[52] Only the harsh discipline dominating his middle-class family, his prestigious high school, and the imperial capital in the 1890s can explain the city boy's fascination with the "simple and low" rural life. Fridolin was a quite typical *petrovets* in terms of educational and social background, and his decision to seek an agronomist position was not very original either (according to Verner's statistics). In 1904 he became the agronomist of the St. Petersburg district zemstvo, but it was not until 1910 that Sergei Fridolin, whom his school friends teased as the "manure man," finally found his calling. After a series of failures as district agronomist, in 1911 he became an instructor of husbandry in the Moscow province zemstvo.[53] The popular brochures on stock-breeding he wrote during this period were hailed by his colleagues as excellent learning aids for the peasant audience.[54]

While Fridolin was struggling with the role of the universalistic agronomist, which he hated,[55] Alexander Chaianov became a student at the Moscow Agricultural Institute in the ranks of the class of 1910. Born to a Moscow middle-class family, he, like many of his fellow *petrovtsy*, graduated from a vocational school.[56] Unlike Vargin and Fridolin, Chaianov did not become an agronomist-practitioner. He chose an academic career, later combined with the role of top manager and theorist of the cooperative movement. As early as 1913, a programmatic study, in which the 25-year-old scholar articulated his vision of the agronomist-as-modernizer, became a handbook for many practitioners.[57]

Despite their different previous life stories, these people became members of the same social movement of the time around 1910, and their consequent paths were in many ways parallel. They opposed the October Revolution in 1917 as a politically and economically destructive coup, anticipating the rapid downfall of the Bolsheviks.[58] Yet their agenda of social activism prevailed over political principles, and they provided the new regime with desperately needed professional expertise on agricultural and procurement problems. Their efforts were highly appreciated by the Soviet authorities. Alexander Chaianov held an important position in the People's Commissariat of Agriculture in the early 1920s.[59] In 1924, Vladimir Vargin was awarded the title "Hero of Labor," and in 1927 he became a candidate-member of the Ural Regional Soviet Executive Committee.[60] Sergei Fridolin obviously sought a new boost for his career from the authorities, publishing highly polemical pro-Bolshevik memoirs, which virtually cleared the way for the Cultural Revolution in the community of agricultural specialists by denouncing the political values of the rural professionals with pre-1917 job experience.[61] Their cooperation with the Soviet regime turned out to be suicidal, as the Cultural Revolution in the late 1920s and the follow-up purges triggered by the trial of the "Labor Peasant Party" in 1931 literally terminated the remnants of the new generation of Russian intelligentsia.

While this parallelism in career tracks can be interpreted as a consequence of belonging to the same social generation, there is no evidence that this belonging determined a particular "destiny" for its members from the very beginning. Rather, it provided them with a particular set of discursive strategies for any given situation, and a number of ideological explanatory models. People were free to choose between these available strategies and models, or they could even turn to an absolutely different set of values and objectives. The bonds of generational solidarity did not create any institutionalized structures but rather a virtual reality of emotional rapport, cultural affinity, and common life experience that was crucial in drawing people together in the absence of strong institutional channels of mobilization. As personal contacts became a basis for group solidarity, the coordinated purposeful professional activities of individuals amalgamated into a distinctive political force that left its mark on Late Imperial Russia.

3
Transfer of the Italian Technology of Modernization and Birth of the Russian "Public Agronomy" Project

The Italian remedy for the Russian problem

The rise of agrojournalism and the public concern for agriculture in the first years of the twentieth century was prompted not only by the paradigm shift in the mindset of the Russian *obshchestvennost'* but also by the deteriorating performance of Russian agriculture. Overshadowed by the stormy political events of 1905–07, crop yields all over the empire decreased dramatically, and the land under crops diminished significantly. In 1905, the land under winter crops averaged only 76.5 percent of that area in 1904.[1] In 1906, the rye yield in 50 European provinces was only 71.5 percent of the average yield.[2] With a statistically calculated norm of 15 poods of grain a year per person as a threshold of starvation, the cumulative harvest of 1906 provided an average of 15.7 poods per person, putting the country on the brink of famine.[3] Only in 1907 did the situation gradually begin to improve, with the crop yield lagging behind the average for the five-year-period 1902–06 by only 6 percent.[4] In 1908, the crop yield of almost all cereals was up,[5] and again in 1909, when for the first time since 1904 the area under crops began to expand.[6] But it was too late to avert a humiliating blow to the national myth of "Russia—the granary of Europe." In 1909, many local periodicals reprinted the information published in the central *Economic Gazette*:

Ekonomicheskaia gazeta reports that steamships loaded with grain are sailing from Argentina…to Russia…to deliver American wheat to the mills on the Volga. This is not accidental: for a number of years already, due to constant crop failures, our export of grain has decreased…in 1902–5, various cereal products, chiefly rye and wheat, were imported into Russia with a value exceeding 2 million rubles; in 1906, almost 6.5 million rubles; in 1907, 12 million [rubles]; in 1908, by December 1, just three major cereals—wheat, rye, and oats—more than 13 million rubles.[7]

The defeat of the Revolution of 1905–07 put on hold the populist dreams of the universal "black repartition" of all arable land in the empire among the peasants. This also drew attention to the studies of the peasant economy and techniques of land cultivation by growing numbers of scholars who argue that the causes of agrarian crisis in Russia were much more complex than the assumed shortage of land in the peasants' possession.[8] After the revolution, the partial democratization of the regime opened new possibilities for legal activity focused on the modernization of the Russian countryside. Yet, the Russian public did not have any ready concept of legal social engineering to replace the old, but still influential, ideal of the "black repartition."

Despite the existing methodological and conceptual differences, Russian rural scholars had bridged many of their old, most fundamental contradictions. After 1905, no serious economist would have repeated the old populist arguments, rejecting the universal character of capitalist development. At the same time, the orthodox Marxist analysis of agriculture lost its former popularity. In 1900, Sergei Bulgakov published his extensive doctoral dissertation *Capitalism and Agriculture*. Bulgakov had begun this research as a leading Russian Marxist theoretician, whose orthodoxy was recognized by none other than Vladimir Lenin himself in 1899.[9] Bulgakov intended to elaborate a complex Marxist theory of the agricultural sector of the economy, based on worldwide empirical data. It is fascinating to trace the evolution of Bulgakov's views throughout the two-volume treatise. At first, he struggled to confine the empirical evidence to the Marxist methodological framework, and to reconcile the postulates of the third volume of *Capital* with the theoretical premises of the first volume. By the end, Bulgakov admitted that by no means was it possible to describe agriculture as an integrated part of the capitalist economy in classical Marxist terms.[10]

In the early twentieth century, Marxists and populists could fiercely dispute the nature and perspectives of social differentiation among the peasantry, or the applicability of certain theoretical categories to the rural economy. However, there were some concepts upon which the most serious scholars agreed. The "toiling peasantry" (*trudovoe krest'ianstvo*) was one of the most important notions broadly discussed in early twentieth-century Russia. Initially it was used by the populists to denote the humanistic nature of the peasantry, as opposed to the predatory essence of capitalism based on exploitation. Sergei Bulgakov reformulated the definition of this concept: "on the peasant farm, land is regarded as a possibility for a producer to apply labor for the sake of survival."[11] Hence the "toiling peasantry" became the term to indicate a type of production unit that seeks primarily to employ all of the available labor resources, instead of maximizing profit from invested capital.[12] The same approach toward the concept of the toiling peasantry was expressed by another prominent Russian economist, Alexander Chuprov, who was more closely affiliated with the populist tradition: "The petty landowner looks at the land as a means of applying his labor and feed-ing his family."[13] Both populists and Marxists based their analysis on the data provided by zemstvo statisticians, and were apparently influenced by the implicit explanatory schemes built in the zemstvo statistical surveys.[14] When, after the defeat of the First Russian Revolution, any projects of land redistribution appeared to be unrealistic, a new public discourse on the agrar-ian question was shaped by discussions of the perspectives of the toiling peasantry.

To post-1905 discussions of the agrarian question, Russian Marxists contributed their understanding of the economic rationality of peasant behavior, while populists added their vision of the peasantry as a particu-lar way of life. This synthetic approach recognized that the main goal of the peasants' activity was not agricultural production "feeding the entire world," but a quite unproductive accommodation of the available labor resources in the archaic realm of peasant self-sufficient culture/economy.[15] After 1900, more and more scholars came to the conclusion that the main cause of peas-ant troubles was not the land shortage per se, nor even the paternalistic land commune, but the obsolete three-field system of crop rotation. According to economic models of the time, the optimal size of a three-field farm would be about 400 hectares,[16] with a ratio of plowed field to fallow to pasture as 1:1:1, and rich fertilization with manure.[17] The typical peasant farm was very far from this ideal. Even the "black repartition" of all arable land in the empire among the peasants would not make the extensive three-field technology on an average peasant plot profitable. In striving for maximization of the sown area, late nineteenth-century peasants virtually transformed the three-field system into a two-field one by plowing up the pasture. As a result, peasant stock-breeding was put on the brink of liquidation. The consequent deficit of manure left the peasant fields without desperately needed fertilizer, which resulted in a further decline of crop yields. To compensate for low produc-tivity, peasants tried to increase the area of plowed land at the expense of what was left from the pasture, or to acquire more land elsewhere. Thus the vicious circle of the archaic peasant economy started a new cycle.

Rural professionals believed that the way out of that vicious circle was the "road—little known to … the farmers, but well studied by people of science and rational practice."[18] In the spirit of the Progressivist approach of the "technological" amelioration of social problems, the solution was found in the intensification of production, diversification of cultivated crops, and introduction of a multifield crop rotation scheme. Indeed, the peasantry of Central Russia cultivated rye as a "monoculture." Crop yield statistics at the beginning of the twentieth century directly indicated the mono-cultural composition of peasants' crops as the main reason for the crisis of peasant economy in 1905–07. When in 1905 the average yield of rye in Russia was almost 20 percent below normal (and in many regions even 70–80 percent), the average yield of potatoes was exceptionally good.[19] In

1906, the rye crop failed all over the empire, this time averaging 30 percent below the normal yield, while root crops thrived.[20] On the contrary, when in 1908 peasants enjoyed a good rye harvest, the potato yield was worse than during previous years.[21] Russian economists, many employed as faculty at agricultural colleges, came out with a prescription for modernizing the peasant economy. They believed that by changing only one element of the old vicious circle—the three-field, rye-centered crop rotation scheme—the whole peasant economy would be turned upside down. The multifield system of growing labor-consuming diversified cultures, including root crops, corn, or flax, would secure peasants from total harvest failures, provide their cattle with much needed fodder, and their fields with natural fertilizer. This intensification of production would have eased the peasants' hunger for land and directly involved them in the national market system. "And no matter how long peasants stubbornly continue to think that the three-field system has existed almost from the creation of the world, they will [eventually] have to introduce a new order and move to grass-cultivation and the multifield [system of] production."[22] This was a really good plan, promising to solve Russia's most burning political and economic problems by means of a purely technical procedure, which looked quite realistic and effective to its supporters. There was only one "but" in this plan: how to compel millions of peasants to adopt it. The "Archimedes' lever" had to be found to turn the peasant world toward modernization. It was clear to recent college graduates that peasants would never do so on their own:

> Thoroughly isolated from the entire world by its illiteracy and age-old superstitions, the Russian village cannot change its prehistoric status by itself, at least under current circumstances. A force from outside is necessary.

> So far, the *intelligenty* and the specialists who have dedicated their efforts to the village can and should come to help the peasantry, to give them at least an opportunity to become familiar with rational management, to help with purchasing the necessary tools and seeds, to provide them with good sires. And all this can be done.[23]

The inertia of intelligentsia paternalism is very vivid in this assessment of the peasantry. On the other hand, the new understanding of the intelligentsia–peasant relationships broke with the nineteenth-century tradition formed under the influence of the cult populist thinker, Nicholas Mikhailovsky, and his romantic concept of "the hero and the crowd." In 1882, at the peak of the "Great Terror" of radical populists, Mikhailovsky taught that a few outstanding individuals can change the course of history, and the crowd should simply follow and obey them. He severely criticized the idea of "small deeds" as an improper application of the energy of the "heroes."[24] The new generation of Russian intelligentsia tacitly defied the sociological model of the still influential Mikhailovsky: in the worldview of Russian Progressivists, the role of the leaders was to trigger the initiative of the people themselves. As Doctor of Veterinary Medicine (University of Bern, 1906), Karl Sakovskii, urged in 1907, "For the sake of mass improvement of peasant husbandry, it is necessary to involve the agricultural population themselves in activity in that direction."[25] When in December 1909 Peter Struve stated that "The question of the economic revival of Russia is first of all a question of creating the new economic man,"[26] he merely summed up public discussions of the previous half decade. The new Progressivist-type program of social activism—a focused and more elaborated version of the "small deeds" theory—was a program of modernizing the peasant mentalité, rather than the public institutions. The role of educated activists was, then, the role of decisive "actors of modernization," who, working as rural professionals, attempted to trigger mechanisms of self-propelled modernization among the peasants. Any administrative and even legislative measures were perceived as futile by adherents to a new program of the "apolitical politics" of public self-modernization. As a prominent ideologist of the new movement put it, implicitly referring to the Stolypin reforms, "Essentially, no legislative form by itself can determine the direction of a cultural process. It is energized by creative forces that instill a certain social and cultural content into it. [This content] can be different, depending on the...tasks these forces choose."[27] It means that the subject of the politics of modernity was *obshchestvennost'*—the organized educated public. Peasants were expected to evolve from the initial object of this activity to its subject after acquiring initiative and critical thinking of their own. The classic formulation of this politics was produced in 1911 by a member of the 1910 class of the Moscow Agricultural Institute, Alexander Chaianov. In a speech delivered at the Moscow regional congress of rural professionals, he suggested that all of them should strive "[b]y means of impacting upon the mind and will of the economic people [*khoziaistvennykh ludei*], to awaken initiative in their milieu, and...to direct this initiative in a most rational manner. In a word, to change old ideas into new in the minds of the local population."[28]

In a few years, Chaianov would become a leader of the Organization-Production school in Russian rural studies, the scholarly current that shaped the methodological framework and theoretical premises of the public modernization campaign in the countryside. The name of this school of economic theory sounds as odd in Russian as it does in English, but it captures the very essence of the Progressivist approach to socioeconomic reality: Chaianov and his colleagues believed that the rational organization of agricultural production based on more accurate explanatory models and statistics could radically change the very structure of economy and society.[29]

Although the modernization project of the post-1905 agricultural specialists focused on the transformation of mental rather than institutional structures, it still needed certain institutional frameworks for its implementation. The ideal model was found initially in the Italian *Cattedra Ambulante di Agricoltura* (pl., *Cattedre Ambulanti*), or the mobile consulting bureaus staffed with two or three specialists in agriculture.[30] Such a mobile bureau of agriculture would stay in a place for a few years, establishing contact with the population and propagating rational techniques of agriculture. To adopt the advanced techniques and to purchase "the necessary tools and seeds, ... good sires," poor peasants needed money that they could not obtain through regular bank credit. Hence, mobile bureaus organized agricultural cooperatives among the propagated peasants, and taught them how to run those organizations. Cooperatives as registered corporations could guarantee the repayment of bank credit and thus accumulated much needed money at a modest rate. Furthermore, buying wholesale was cheaper, and the quality of goods was secured by official contracts. When the cycle of teaching–organizing–implementing was completed in the course of a few years, a mobile bureau moved to a new location, where the fame of its accomplishments had already prepared the way for a new magical transformation. This is how contemporary British observers in 1901 described the *Cattedre Ambulanti*, which they translated as the "Traveling Schools":

> [T]he Traveling Schools ... subsidized to some extent by Government, but founded by private initiative and chiefly supported by the Provincial Councils and private Savings Banks, are bringing a very practical kind of teaching to the peasant's door. Entirely the creation of the last ten years, they number thirty-nine, chiefly in the North, but including a few in the Center and South. The duties of the traveling teacher are multiform. He gives fifty or sixty lectures in the year in different centers; he has practical demonstrations; he supervises experimental plots; he sits in his office every market-day for oral consultation; he has classes in special subjects, such as grafting and pruning; he trains elementary teachers to lecture in their turn on agricultural subjects; sometimes he publishes an agricultural journal; he keeps an outlook for phylloxera and superintends the measures to stamp it out, if it appears; sometimes he has nurseries to supply American vine-stocks, or introduces bulls and rams of improved breeds; he organizes fruit shows; he introduces, where he finds it possible, Village Banks and Cooperative Dairies, or preaches the advantages of joining the local Syndicate. It is a work, that probably has no parallel either in France or England, and its practical usefulness is matched by its popularity. The cost of each "chair" varies between £184 and £750.[31]

Both the organization of "traveling schools" and its description by British commentators reveal the innate Progressivist discourse of modernization through self-organization and self-improvement. Contemporary statistics showed the remarkable effectiveness of *Cattedre Ambulanti*, hence Russian social activists found them to be the key to success in modernizing the peasantry.[32] Thanks to the mobile bureaus of agriculture, the knowledge of the few was able to change the lives of the many. The secret was in awakening the initiative of the masses, in mobilizing them by means of cooperative organizations. Alexander I. Chuprov, a Russian economist and a prominent public figure, for the first time introduced Russian *obshchestvennost'* to the phenomenon of mobile bureaus in 1900 in a series of articles published in the liberal newspaper *Russian News*.[33] At that time the idea of the Italian mobile bureaus in Russia could be regarded only as another beautiful daydream of the intelligentsia, which had no chances of even approaching realization. Nothing could move freely around the prerevolutionary Russian countryside, and certainly authorities would not allow *intelligenty* to propagate any ideas among the peasantry from the *cattedra*.

Yet the idea was not forgotten. In the spring of 1905 another distinguished rural scholar, D. N. Prianishnikov, decided that it was the right moment to bring up the issue of the Italian mobile bureaus. In a few issues of the special weekly *Messenger of Agriculture* he published a long article in which he drew direct parallels between the Italian experience and Russian reality.[34] As it turned out, it was badly timed, for the wave of agrarian disorders in the Russian countryside put the issue of legal initiatives among the peasantry on hold for a while. It took literally a new generation of rural scholars to put the idea of mobile bureaus of agriculture into practice. A pupil of Professor Prianishnikov at the Moscow Agricultural Institute, Alexander Chaianov, during his trip to Italy received firsthand experience of what he had read about in the writings of his tutors. In 1908, he published an article in which he put together a draft of the public modernization program and a description of the Italian mobile bureaus as a model of the institution that would carry out this program.[35] This line of argumentation received an authoritative endorsement from a champion of zemstvo agronomist assistance to the rural population, Kharkov province agronomist Viktor Brunst (1867–1932). In a 1909 article he surveyed the methods of agricultural assistance and education in all major European countries including Italy, as well as the experience of the United States, and came out with recommendations identical to those advanced by Chaianov.[36] Thus, after the Revolution of 1905–07, the professionally trained Russian public came out not only with a new project of modernization but also with an institutional vehicle for its implementation.[37] The scheme of modernization of the peasantry through the coordinated efforts of the corporation of rural professionals, modeled after the Italian mobile bureaus, became known in Russia as "public agronomy": a movement of the educated segment of society (*obshchestvennost'*) to assist other members of society, through self-organization and self-mobilization.

Localizing international experience: Precinct agronomy

Even after the partial democratization of the Russian political regime brought by the Revolution of 1905, the project of the mobile bureaus could not be implemented in Russia in its original version. First of all, the political climate in the countryside remained very tense long after the agrarian disturbances of 1905–06. It would take the combined efforts of the third and fourth elements and almost a decade to transform the configuration of rural politics into one more tolerant toward public initiative. Second, the vast Russian countryside could not be efficiently served with a few dozen mobile bureaus, which were apparently sufficient for relatively compact Italy. No independent public association could afford hundreds of mobile bureaus, while the state was not interested in supporting such an initiative. It was left to rural professionals to find in Russian conditions an adequate substitution for the Italian mobile bureaus.

One solution was proposed by Samara agricultural specialists, who took a familiar type of agricultural society as the basis for a "mobile bureau" on Russian soil. On November 7, 1907, the authorities registered the Samara Society for Improving the Peasant Economy (*Samarskoe obshchestvo uluchsheniia krest'ianskogo khoziaistva*, hereafter SOUKK). Unlike traditional agricultural societies, this Samara society was organized by rural professionals, rather than local landowners. In a few years, the number of members reached 50. There were three people on the Society's Board in 1910. A. V. Teitel, the chief zemstvo agronomist of the Samara province, was the initiator and chairman. The Samara district zemstvo agronomist, K. N. Mukhanov, served as secretary of the society. The attorney-at-law, P. I. Sakharov, was treasurer; he also provided legal counsel on the pages of the society's journal the *Samara Agriculturist* (*Samarskii zemledelets*). Besides agronomists, among the most active members were also a government specialist in husbandry; a State Bank inspector of credit cooperatives; and even a gentry landowner (who happened to be in charge of the Society's Bureau for peasant cooperatives).[38]

The members of the SOUKK met with peasants and explained to them the rational techniques of land cultivation, and even organized peasant excursions to the local experimental agricultural station.[39] Active propaganda for corn cultivation, sustained by the distribution of seeds on favorable terms, resulted in a significant spread of corn crops over the central districts of Samara province.[40] The members of the SOUKK also assisted in the foundation of a few agricultural cooperatives and in marketing for the local peasant creameries. From 1910, the society published the *Samara Agriculturist*, which soon received recognition all over Russia. The government allocated considerable funds in the form of special-purpose grants to support the activity of the SOUKK.[41] Driven by the enthusiasm of its members, mainly rural and town professionals, and supported by the state, SOUKK survived up to 1917 when its activity came to a grinding halt.

The success of the Samara Society for Improving the Peasant Economy proved that an analogue to the Italian mobile bureau of agriculture could appear on Russian soil. Yet, to accomplish significant results, it had to be staffed with dozens of professionals instead of two, and could not be really mobile while it worked with the same part of the peasant population for many years. A society such as SOUKK could be envisioned in any principal town with a considerable community of professionals, but what about the "deep backwaters?" It could not become the prototype for any nationwide organization.

Apparently, the enormous self-assigned task of modernizing the bulk of the Russian rural population by the intelligentsia could not be accomplished exclusively by means of public initiative. It was the "second element" of Russian modernized society, or the zemstvo, that provided the framework for an institution that would accommodate the raison d'être of Italian mobile bureaus to the specific Russian conditions. Soon after the 1905 Revolution, the most progressive zemstvos of Samara and Perm provinces proposed to establish an all-Russian network of precinct agronomists. Hitherto, in most zemstvos there was one provincial agronomist accompanied by district agronomists, some of whom had a deputy or a second district agronomist. The project of a precinct agronomy network planned to multiply the number of zemstvo agronomists by dividing each district (*uezd*) into a number of precincts. The precinct agronomist was supposed to follow the program of mobile bureaus, only staying in the same place and gradually expanding his or her influence over the neighboring villages. That was the most ambitious project of "small deeds" the Russian *obshchestvennost'* ever came out with, for it envisaged the cooperation of all "three elements," thus raising it to the level of an all-national task:

> It is clear, that under circumstances when the population and draught animals survive at the expense of food aid, and the zemstvo institutions—[survive] by efforts of the government, the latter, in its turn, often had to make unprofitable international loans. These, altogether, have compelled [us] to adopt measures toward increasing the productivity of the agricultural trade.
>
> In line with measures of rational land utilization, this goal can be achieved only by means of broad agronomic assistance to the rural population of the country. To make the aim of agronomy organization achievable,...it is necessary...to create it in the form of precinct organization, such as that of medicine or veterinary.[42]

As a matter of fact, the very idea of concentrating the work of agronomists over one or two counties (*volosti*) was expressed as early as 1888 at the agronomist conference of Perm province. In 1889, the province zemstvo board passed the appropriate resolution, but nothing happened for the next ten years, until Vladimir Vargin became the provincial agronomist. Even then, the system of district agronomist deputies, adopted by the Perm zemstvo, was not exactly an example of precinct agronomy either in principle of organization or in the quantity of agronomists.

During the last decade of Imperial Russia, the precinct agronomist became a synonym for the progressive representative of *obshchestvennost'*, an emerging new ideal type of the *intelligenty*: a professionally trained person, who applied the received knowledge to serve the people and who transcended the notorious "insurmountable gap" between educated society and "the common folk". The seemingly obscure profession of the precinct agronomist, who was supposed to live and work in the depth of rural Russia far away from the city centers that formed public opinion, became highly visible during the interrevolutionary period. To a great extent, that visibility was produced by the numbers of precinct agronomists, or, more precisely, by their rapid increase. Between the First Russian Revolution and the World War I, the number of precinct agronomists had grown by a factor of 64, from 27 in 1906 to 1,726 in 1913, almost doubling every year.[43]

Despite general agreement on the goals and basic principles of precinct agronomy, every provincial zemstvo had to decide many practical questions in accordance with local conditions. By 1912, observers distinguished three major types of precinct agronomy organizations. The agronomists of Ekaterinoslav province were known for their high public activism. They dedicated their major efforts to propagating rational techniques among the peasants and organizing peasant cooperatives. The Kharkov agronomists represented another type, focusing on work with individual farms or particular branches of agriculture. Finally, precinct agronomists in Moscow province were distinguished by their "communal" spirit: they worked on improving organization plans for entire peasant communes.[44] Doubtless, the ideological preferences of a local zemstvo and its third element had a great impact on which of these main models they chose. The case of Kazan province provides an insight into the medial approach toward precinct agronomy, characteristic of many zemstvos.

Precinct agronomy was introduced in Kazan province in 1910, but it was not until the summer of 1911 that the detailed guidelines for this institution were elaborated by the conference of provincial agronomists and representatives of the zemstvo board. It was decided that in the future Kazan province would be divided into 56 agronomy precincts, each precinct not to exceed 1,000 square versts (440 square miles) and to include no more than 100 villages. It should be staffed with an agronomist with no less than a mid-level agricultural college education, and a deputy agronomist. The precinct agronomist should assist all zemstvo taxpayers, without discrimination, including members of communes and those who parted from the commune under the provisions of the Stolypin land reform. Precinct agronomists were to help individuals seeking professional consultation, but to base their main educational activity on collectives of enterprising peasants, that is, cooperatives organized with their assistance. Agricultural cooperatives would become the means of agronomists' mass intervention into peasant affairs.[45]

We may assume that the three types of precinct agronomy distinguished by contemporaries corresponded to three stages of this institution's development: from propagating the basics of intensive agriculture, through technical assistance to individual peasants, to the modernization of masses of farms. As we have seen, the project of precinct agronomy organization in Kazan province did not particularly prefer any of the three available models. When in September of 1913 the Alatyr district zemstvo agronomist, D. S. Smirnov, assessed in retrospect the three years of precinct agronomy's existence, he pointed out that in 1910 the main task of precinct agronomists was to familiarize themselves with the precinct and to disseminate agricultural knowledge among the peasants.

> At that time, an agronomist-prophet was required, awakening the folk masses and instilling in their consciousness the idea of the necessity of a rapid transition to better economic techniques.... At present, the main task of the precinct agronomist is...to put into practice the principles that were previously expressed only in the form of conversations, lectures, and agricultural courses.[46]

Thus, in the course of three years a transition was made in the self-perception of an agricultural specialist from the prophetic figure of the precinct agronomist, the public activist, to the type of agronomist, the professional, who delivers on the earlier promises. These two positions differed dramatically in terms of the dominant modes of interaction of the rural professionals with the peasantry. While intelligentsia prophecies were "monological" by definition, professional services could be performed only as a part of the dialogical relationships between the professional and the client.

To transcend the spheres of public discourse and professional ideology and get insight into the everyday life of the early precinct agronomists, let us look at the experience of six precinct agronomists of the Samara district in 1910. The first agronomy precinct was established in the Samara district in 1907 simultaneously with the foundation of the SOUKK. For a while, it was unclear which type of professional assistance to the peasantry would become the model for other regions of Russia. However, in 1910 it was evident that the future belonged to precinct agronomy, as there were already six agronomy precincts in the Samara district (and 11 precincts by 1913). At least four out of six precinct agronomists in 1910 had received higher professional education (versus only half of Samara precinct agronomists in 1913). All four had graduated after the Revolution of 1905, thus sharing the formative experience of the new generation of Russian intelligentsia.[47]

Precinct agronomists lived in villages with the peasants, renting cottages or, sometimes, only part of a cottage. Mikhail Frankfurt, who happened to be one of the founding fathers of the SOUKK, enjoyed the privilege of living in the provincial capital Samara, as he was also a government specialist in husbandry for the province of Samara. Aleksei Kiselev moved from one village of his precinct to another, looking for decent accommodations, but failed to find any, and finally also settled in Samara. During wintertime, from November to February, agronomists lectured in the villages about the most urgent issues of local agriculture. For instance, agronomist Kiselev visited 20 villages of his precinct, where a total of 2,740 peasants listened to his explanations of how to minimize the impact of drought on their crops. Hence, during one winter he managed to visit 30 percent of the villages and speak to 26 percent of the homeowners of his precinct. He even managed to sell 187 books on agriculture, and distributed 70 more for free. This was a rather successful season for agronomist Kiselev, much of this success owing to the fact that his precinct was the central one in the district, neighboring the important commercial city of Samara. Peasants of this precinct were to a greater degree integrated into market relations than the peasants living in the southern part of the Samara district. They approached Kiselev, asking his assistance in changing their crop rotation scheme, and enthusiastically organized cooperatives.[48] In a word, this was almost an ideal materialization of the project of precinct agronomy.

The young agronomist Andrei Ryshkin was less lucky. The year 1910 was his first year as a precinct agronomist. His precinct was larger than Kiselev's and located on the eastern outskirts of the Samara district. His account of his first year contains invaluable psychological details of what a rural professional experienced at the beginning of his work among the peasants.

> From the very beginning, I had to meet that wall of peasant mistrust toward any new person dressed in civilian clothes, [the wall] that does not fall even after close acquaintance, and over which many initiatives have been broken.

> ... I visualize my precinct as a huge enemy territory, where every inhabitant is locked up in his own fortress. Each fortress has its own peculiarities, its weak points. The attacking agronomist must study all particularities of those fortresses and start with the weakest, and only then together with the defeated attack the stronger.

> ... I have experienced all these discomforts of a beginning agronomist, and I would not like to experience the role of newcomer in a new place once again.[49]

Ryshkin discovered that the peasant audience was not prepared to be lectured in the way he and his fellow agronomists had been during their school years. Peasants eagerly supported any conversation on a concrete topic, but his attempts to generalize, to present them his vision of a rationally organized farm, estranged the audience. Following recommendations of the city educators, Ryshkin tried to catch their attention by showing slides with a "magic lantern." As he found out, while the overly abstract slides were of little help in supporting his point, they attracted to the lectures a lot of children, for whom it was pure entertainment. Unlike Kiselev, Ryshkin encountered no interest in books in the peasant milieu.

Despite all of these handicaps, Ryshkin managed to deliver 21 lectures—the same number as Kiselev—which were attended by some 1,500 peasants. When in spring all precinct agronomists began distributing agricultural machines from zemstvo warehouses and rental centers, peasants from Ryshkin's precinct showed incredible demand for seed-drills and harvesters.[50] Hence, although deaf to certain prophecies or prophets, peasants learnt practical things very fast. As it turned out in summer, those peasants who plowed up their fallow lands following the recommendations of their agronomists, harvested two to four times more crops than the average.[51]

Concluding the story of the first Samara district's precinct agronomists, we must add that neither triumphant Kiselev nor frustrated Ryshkin kept their jobs for another full year. Kiselev was promoted to the position of Buzuluk district zemstvo agronomist in the same Samara province. Ryshkin, apparently, moved to Novocherkassk, where he became secretary of the Don Agricultural Society, a job that did not require direct communication with peasants.[52] By 1914, the number of Samara District agronomists had almost doubled, and their composition somewhat changed. As we have mentioned, their average educational level decreased, while their jobs began requiring fewer prophetic skills and more proper professional qualities.

Surviving internal conflicts

Notwithstanding different local conditions, and regardless of personal experience and skills, the project of precinct agronomy seemed to accommodate quite adequately the main principles of the Italian mobile bureaus, while adjusting a foreign model of a modernizing institution to Russian realities. Even though by the eve of the World War I Russian public agronomy had become a significant phenomenon in its own right, its success was still measured by comparison to the Italian system.[53] Precinct agronomy increased the demand for rural professionals, multiplying the ranks of the third element. This, in turn, led to serious social conflicts among the rural professionals drawn into the movement of "public agronomy." The right of recent college graduates representing a new generation of Russian intelligentsia to leadership in the community of agricultural specialists and the preeminence of the agronomists among other professions were not taken for granted.

A dramatic increase in the numbers of zemstvo agronomists after 1906–07 upset the traditional hierarchy of seniority. The massive opening of new positions prompted a rapid promotion of even recently hired specialists. When younger professionals obtained formal status, it did not necessarily guarantee senior colleagues' recognition of their expertise, and the stance on the peasant land commune became a decisive factor in the power struggle within the profession. Rural professionals sought zemstvo support in their resistance to government pressure to give priority treatment to peasants leaving the commune—a pressure that was especially strong in the early stages of the Stolypin reforms. Besides the complicated task of maneuvering between government and zemstvo interests, junior members of the agronomist profession had to deal with the cautious attitude of their elder colleagues while attempting to apply professional expertise and implement their vision of peasant modernization.

The question of the land commune was the central problem debated at agronomy conferences of the Russian provincial zemstvos in 1907, immediately following the decree of November 9, 1906, which inaugurated the Stolypin agrarian reforms. Among the first was the agronomy conference of the Olonets zemstvo in Karelia, convened on February 12, 1907, to discuss an agenda consisting of seven articles. From the very beginning, a conflict erupted between the participating representatives of the second and the third elements. Opening the conference was K. K. Veber (1855–1929), who got his degree back in 1878[54] and had been the provincial agronomist for almost ten years. Veber suggested that participants should stick to the preliminary agenda, that is, start with the problem of peasant hayfields, then proceed to the problem of stock-breeding, and so on.[55] His plan was immediately disputed by the young Lodeinopolskii district agronomist Ivan Luke, who had received his diploma of higher agricultural education as recently as 1905.[56] Luke argued that it would be preferable to start with the last, the seventh question: "How much does the existing [communal] system of land cultivation obstruct individual initiative, and consequently solid success in raising the peasant economy?"[57] His statement violated subordination and brought forward a politically painful issue, thus provoking the intervention of the provincial zemstvo board chairman, N. A. Rat'kov.[58] Provincial agronomist Veber supported the chairman, but his arguments were of a professional rather than an administrative character.[59]

The conference proceeded according to the original agenda, but on February 14 came the turn for the controversial seventh question. The Kargopol district agronomist K. Iu. Krylov, who had graduated from a mid-level agricultural college in 1889,[60] delivered a report based almost exclusively on traditional populist theoretical arguments stating that the land commune did not contradict agricultural progress.[61] His presentation caused a heated discussion. While younger agronomists I. D. Luke and G. P. Semenov did not see in the commune a serious obstacle to technical innovations,[62] the provincial agronomist Veber cautiously aligned himself with Rat'kov. The zemstvo board chairman Rat'kov was torn between practical considerations and ideological preferences: he supported the government anticommune policy, but he could not imagine how to mobilize peasants for public works such as road construction if all of them left the communes.[63] Finally, everybody agreed that in practice the land commune was an obstacle to agricultural progress. However, the preceding polemic revealed contradictions not only between zemstvo officials and agronomists but also within the ranks of the third element. This conflict seems to have had a generational dimension. Representatives of the new generation of rural professionals displayed an attitude different from that of their senior colleagues. While Krylov, a graduate of a mid-level agricultural college during the epoch of counterreforms, demonstrated loyalty to old populist beliefs and failed to provide analytical comments on the topic of his report,[64] the provincial agronomist Veber, educated at the Munich polytechnic some ten years earlier, preferred to avoid any politically charged issues. Younger professionals, regardless of their educational level, were thinking along more pragmatic lines, clearly distinguishing between their political priorities and professional expertise.

A similar situation occurred during the agronomy conference at the Kazan zemstvo board on August 17–24, 1907. The discussion initially was sluggish, revolving around petty issues, until the Spassk district agronomist, D. A. Il'inskii, posed the question "to what extent does the commune obstruct the intensification of the peasant economy?"[65] Dmitrii Il'inskii graduated from the Moscow Agricultural Institute just a year earlier, in 1906,[66] and demonstrated professional and social attitudes quite typical of the new cohort of agricultural specialists. In his opinion, it was the general ignorance of the peasantry that hampered agricultural progress, and not the commune itself. Agronomist Il'inskii supported his point with examples of Bavarian peasants in Germany and Volokolamsk peasants in Russia, who had achieved significant progress in the communes. Both arguments were dismissed by participating zemstvo officials. They also hinted to Il'inskii that the dissolution of communes would finally provide the agronomists with an opportunity to prove their usefulness by assisting the farmstead peasants. Il'inskii did not argue in support of the institution of land communes, as he had not argued against it before. Instead, in accordance with his technological approach to the agrarian question, he proposed that the conference develop a project of land cultivation for peasants who had left communes for farmsteads. To his surprise, none of the active critics of "economically reactionary" land communes had any idea how to approach this task. They asked agronomist Il'inskii himself to elaborate such a project for the next agronomy conference.[67] As was the case during the Olonetsk conference six months earlier, a senior representative of the third element at the Kazan conference—provincial agronomist V. I. Kotov, who had graduated from the

Moscow Agricultural Institute some ten years earlier than Il'inskii—did not want to speak on politically complicated topics. It would take a few years to bridge the generation gap within the community of agricultural specialists.

With growing homogenization within the most numerous group of agricultural specialists, the agronomists, a new professional conflict erupted. Zemstvo veterinarians, the oldest and once the largest fraction of rural professionals, in the early 1910s were slow to emulate new standards of professional autonomy and public-minded social activism. As late as 1913, the third all-Russian Congress of Veterinarians resembled an official meeting of the pre-1905 epoch, rather than a professional convention of the inter-revolutionary period. The congress was opened on December 29, 1913, in the Kharkov Opera House by the chief manager of State Horse Breeding, Prince N. B. Shcherbatov, in the presence of the reactionary Kharkov governor M. K. Katerinich. First of all, the Congress approved the text of a telegram to the minister of interior expressing the veterinarians' utter loyalty to the emperor. The audience listened to the telegram while standing. The first general session of the congress did not take place until January 2, four days after the opening of the congress, which ended on January 6. This scenario was simply unimaginable at an agronomist or cooperative congress.[68]

The veterinarians' conformism can be explained in part by the origins of their profession in government service, while the agronomists had set the standards of their profession as zemstvo employees enjoying relative professional autonomy, long before the appearance of government agronomists. The community of veterinarians was extremely diverse, and often only a common educational background united those who worked at city slaughterhouses and those who served in the army. It was extremely difficult for such a diverse corporation to develop a distinctive common identity, especially on terms acceptable for the most publicly concerned, but not the largest, group of zemstvo veterinarians (who comprised just over a quarter of the entire corporation).[69] Furthermore, the nature of the veterinarian service differed greatly from that of the agronomists; while the latter were oriented toward social and economic engineering, the former was associated with sanitary, if not police, tasks often requiring the killing of animals suspected to be infected.[70] Still, the professional pride of village veterinarians and their feeling of seniority were hurt, and in the early 1910s they revolted against the agronomists' "usurpation" of the modernization movement.

Already in 1907, when agronomists were still a minority among the rural professionals, veterinarians Krasnopol'skii and Sakovskii called upon their colleagues to shift their focus of activity from sanitary efforts to assisting the peasants with stock breeding.[71] However, for various reasons mentioned earlier, nothing really changed in the status and activities of the zemstvo veterinarians during the following years. Five years of the Stolypin reforms and the emerging "public agronomy" project made the veterinarian a marginal figure among the rural professionals, the majority of whom were engaged in reforming the countryside. In 1912, veterinarians rushed to recapture their lost positions and to claim their share in the skyrocketing prestige of the rural professionals.

In December 1911, the veterinarian A. Orlovskii complained that while veterinarians should have been the champions of stock breeding, "the entire leading role has been occupied by the agronomists—a noisy role—while the practical work is done by veterinarians."[72] In February 1912, his more militant colleague expressed confidence that agronomists would inevitably fail to secure the positions they had occupied and speculated about the future of their confrontation:

> Battle with agronomists, even if ever happens, is not particularly dangerous. Besides their inadequate... training for stock breeding... agronomists are inferior to us in their number (I mean agronomists with complete [i.e., higher] education)....

> The only advantage of agronomists under the present circumstances—is their better organization and solidarity.

> ...We must go public and speak about ourselves and our profession. One must be surprised how little about "veterinary" gets into the general press!...In this respect, agronomists have left us behind, and not only have informed everyone about themselves, but sometime even advertise themselves like American agricultural machines, and thus reach their goal—people talk about them.[73]

This militant veterinary practitioner proposed to publish compromising materials about agronomists and stage a public campaign to promote the image of the veterinary profession, but veterinarians, indeed, lacked the necessary solidarity and dedication. In Kazan province, where annual conferences of agronomists had became the norm long before, the conference of veterinarians convened in April 1912 was separated from the previous one by an interval of ten years. The conference voted for a resolution recommending that veterinarians work for the improvement of peasant stock breeding "in line with agronomists,"[74] but it turned out that even the structure of veterinary organization in the province had not been clarified until that conference.[75]

While the Kazan veterinarian Simagin complained in a professional periodical that local agronomists protested when he attempted to become active in zemstvo-run stock-breeding initiatives,[76] Kazan's chief inspector of husbandry, Sabinin, argued at the veterinary conference that veterinarians deserved such mistrust. He said that veterinarians ignored "the creative work that demanded complex discussion and solid knowledge" initiated on a regular basis by agronomists, and called veterinarians to work alongside agronomists to improve husbandry, leaving the question of leadership to be decided by the future.[77]

Agronomists, apparently, did not take seriously the militant aspirations of veterinarians,[78] which only reinforced the veterinarians' sense of inferiority. At the end of 1914, some veterinarians still complained bitterly about the unfair competition of agronomists, only now they did it without their former passion and optimism.[79] The insecurity of veterinarians who wanted to lead the general movement instead of being junior partners instilled in them an apocalyptic vision of the history of their profession. Veterinarian K. F. Pavpertov presented the veterinarian as the ultimate victim of the people, *obshchestvennost'* and "even" the government. He joined the profession in the late 1880s, when "every recently graduated veterinarian did not know himself, who he was, and in public life administrators looked askance at him and said that he was in the bestial line."[80] According to Pavpertov, it took about ten years to establish veterinary in public opinion as a profession. At the beginning of the twentieth century, veterinarians were enjoying a certain recognition, dominating the rural professions, when suddenly public agronomy took off.[81] A representative of the older generation of rural professionals, Pavpertov could explain the tremendous success of the agronomists only by citing the joint conspiracy of the government and part of the *obshchestvennost'*. Quite typically for the nineteenth-century semiprofessional, semiservice stratum, his ideal of professional success was limited to acquiring recognition and "a certain acknowledged narrow track in life." The aspirations of the new generation of Russian intelligentsia were beyond his understanding (while the Kazan veterinarian Simagin, who graduated from a veterinary institute around 1910, actually strove to join the new social movement). Envious of agronomists enjoying government support and subsidies, the old veterinarian Pavpertov did not ask for the same support for his fellow veterinarians; instead, he called on the government to stop supporting agronomy, hoping thereby to reach a status quo ante[82]— apparently, a kind of status quo Russian rural professionals enjoyed before the Revolution of 1905.

Those veterinarians who joined the modernization movement of the rural professionals were not in the mood to lament about the lost veterinary paradise. In 1913, veterinarians of the Stavropol district (Samara province) for the first time "went to the people" in the manner typical of public agronomists. They participated in a district husbandry exhibition, the kind of enterprise usually run by agronomists. The success of the veterinary section featuring photographs, instruments (including a microscope), and 215 demonstrative materials exceeded their expectations. About 4,000 people, "not just intelligentsia but also common folk," showed interest in the veterinary section.[83] This recognition by the public was coupled with a small grant from the Stavropol district zemstvo to support a local veterinary museum.[84] It is characteristic that such very different representatives of the veterinary community—a zemstvo veterinarian, two government veterinarians, and two veterinary assistants—joined their efforts for the exhibition.[85] The internal professional boundaries were overcome by the common goal of achieving the professional status enjoyed by other rural specialists.

Of course, not everyone agreed with the Progressivist scenario of combining professional duties with a certain public agenda. Many professionals just wanted to make a career. Among the zemstvo employees, such people were more likely to belong to the older generation of professionals, for young people looking for careers joined the government service, which provided many more benefits. Those younger professionals, who in the century's first decade consciously joined the ranks of the third element, more often than not did so for ideological reasons, forming the core of the new generation of Russian intelligentsia. This became the major driving force of the new movement at the beginning, and the main threat to its sustainability later. The very fusion of ideological zeal with pragmatism that characterized the new public modernization campaign in the Russian countryside took shape through and as a result of dramatic inter- and intraprofessional conflicts. The new social movement became possible only because of the successful homogenization of the initially disintegrated community of rural professionals and the negotiation of a common professional ethos and public stance.

Part II
Dynamics of Modernization

4
The Ambivalent Role of the State: A Conservative Patron and a "Progressive" Rival

From repressions to self-restrained control

The rising stratum of agricultural specialists and the program of social reformism they advanced faced a complicated political situation in the countryside, with cautious zemstvo boards and suspicious government officers strongly opposing any initiatives. Initially, the number of rural professionals was too insignificant to regard them as a serious force: in 1906, in the empire as a whole there were 55 government agricultural specialists and instructors, 387 zemstvo agronomists and specialists, and about 3,220 veterinarians, of whom some 33 percent were in zemstvo service and about 50 percent in state (including military) service.[1] While the agronomists and agricultural instructors were a relatively new phenomenon, the veterinarians comprised the bulk of rural professionals, and they were always regarded as a politically unreliable element of rural society by the government. They attracted the most police attention as a major group of rural specialists, which contributed to their belated joining of the ranks of rural professionals-modernizers only in the 1910s. Along with rural teachers, veterinarians became the main target of the government attack on revolutionary intelligentsia in the countryside. In 1906–07, many veterinarians became victims of political repression. Every issue of the monthly *Veterinarian Review* named people who were arrested or lost their jobs on political grounds. These reports give insight into the insane working conditions with which Russian rural professionals coped at the beginning of the twentieth century.[2] Indeed, some of the reports sound tragicomic. As late as 1912, the Kharkov governor prohibited the sale of veterinary books at zemstvo veterinary precincts.[3] The veterinarians in state service were especially thoroughly scrutinized by the authorities. There was no way a socially active veterinarian could be regarded as a loyal employee.[4] It seemed that political reaction gave to rural professionals little opportunity to step forward with a modernization program of their own.

However, while one branch of the government used traditional police measures to try to eliminate any possibility of unregulated public activity, the new agrarian policy of its other branch was unleashing social forces on an unexpected and unprecedented scale. A comparison of the history of agricultural modernization in Russia and its sociopolitical antithesis, the United States, reveals a striking similarity. In 1926, the Progressivist educator and agricultural expert, William Bizzell, distinguished five stages of American society's involvement in the modernization of the countryside from the eighteenth century to the twentieth century:

1. The organization of agricultural societies;
2. interest in rural and community fairs;
3. the establishment of the agricultural press;
4. the opening of agricultural schools;
5. and the establishment of state and federal agencies for the promotion of agriculture.[5]

Despite the actual and imagined omnipotence of the Russian state, it intervened in the process of rural reformism only at the very last stage, when *obshchestvennost'* and zemstvos had substantially succeeded in leading Russia through the previous stages mentioned by Bizzell, and in exactly the same order. This does not mean that the role of the state was of secondary importance.[6] Between 1895 and 1913, the annual expenses of the Department of Agriculture had risen by more than a factor of 12 (although during the same period, the 34 oldest zemstvos increased their spending for various agricultural measures by a factor of almost 18). The number of agricultural schools under the auspices of the Department of Agriculture had increased by a factor of 4.5, and so had the number of pupils.[7]

The most immediate result of the government's "turn to agriculture" was the opening of new jobs available to the third element. The rate of annual increase of agronomists in the state and zemstvo service was 31.5 percent in 1910, 55 percent in 1911, and 40 percent in 1912, fluctuating between 25 percent and 95 percent during the period between 1907 and 1914.[8] Still, new positions opened at an even faster pace. As we have mentioned, in 1910, the Moscow Agricultural Institute was 400 graduates short of meeting the requests of local agronomy agencies.[9] Although the government employed only one-third of all agronomists, it played an important role in stimulating zemstvos to hire more rural specialists. Every year from 1910 on, the Department of Agriculture doubled the sum of money granted to zemstvos and agricultural societies to cover the salaries of rural specialists. But there was a catch: these grants were given only on the condition that the recipient institution matched the expenses for agronomy personnel ruble for ruble.[10] Although quite aware of the potential trap in the government offer, even the most conservative zemstvo boards could not avoid the temptation to receive extra money. As a result, by the beginning of the World War I, there were altogether 4,402 agronomists, agricultural specialists, and instructors in zemstvo service and 2,369 in state service (not counting 1,982 agricultural elders who had only primary education), 1,374 zemstvo veterinarians, 2,900 state veterinarians, and 2,638 zemstvo veterinarian assistants.[11]

Zemstvo and public associations were the major employers of agricultural specialists, leaving government agencies far behind. Besides direct subsidies to the third element, government intervention in the sphere of modernization of rural Russia had another, more profound result. It was the new political climate in the countryside, the new mode of "political correctness" that elevated the status of the third element and its initiatives in the eyes of the local authorities and zemstvo bosses. The case of remote Buinsk district in Simbirsk province provides a vivid example of how such factors affected the prospects for modernization of the peasantry.

The Buinsk zemstvo board represented that type of conservative second element that later would make many historians believe in the "myth of 'zemstvo reaction'" (as Kimitaka Matsuzato put it).[12] It is all but appalling to read minutes of the zemstvo board's sessions in 1908; every initiative of the district agronomist Vladimir Ivanov was blocked by the local second element. On September 29, 1908, the board declined his suggestion to establish the position of second district agronomist: they did not perceive the activity of one agronomist (i.e., Ivanov himself) as useful enough to warrant hiring a second.[13] A month earlier, during the session of the Buinsk zemstvo economic council (a joint organ of the second and third elements), agronomist Ivanov was confronted by the influential zemstvo deputy and wealthy landowner, Mikhail Fedorovich Bonch-Osmolovskii. Bonch-Osmolovskii embodied the image of an ultraconservative gentry representative and a typical *krepostnik* in terms of the Leninist sociology of the period (if not of the more refined theory of the "gentry's turn to the land" of Roberta Manning).[14] Vladimir Ivanovich Ivanov, on the contrary, demonstrated the inferior status of a member of the third element.[15]

The maltreatment of agronomist Ivanov by the local "second element" cannot be explained simply by the prevailing "zemstvo reaction," insofar as precisely at this time the new phenomenon of precinct agronomy was taking off in the more liberal and richer zemstvos of neighboring Samara and Perm provinces. Furthermore, the very definition of "reaction" or "radicalism" was changing as a result of the interaction of the "three elements" of Russian rural society, a point missed by structuralist-inclined historiography.[16] The second element of the Buinsk district remained predominantly conservative during the following years, promoting the notorious Bonch-Osmolovskii to higher levels in the zemstvo hierarchy.[17] Nevertheless, this did not mean that the Buinsk zemstvo vegetated in the shade of "gentry reaction" until the Revolution of 1917. Government support for agronomy assistance to the peasantry came to the aid of the third element in the Buinsk zemstvo. On February 12, 1910, the Buinsk zemstvo board, lured by the promise of government funding, decided to introduce precinct agronomy in the district.[18] And so, four agronomy precincts and an experimental field were founded—more than the district agronomist Ivanov had dared even to dream about only one and half years earlier.

One can imagine the difficulties encountered by new precinct agronomists in a place as patriarchal as the district of Buinsk. According to their reports, in 1910 local peasants heard for the first time in their lives about grass cultivation and the sorting of grain at sorting stations; some villages would not even allow a precinct agronomist in.[19] In a few years the situation changed. Peasants became more responsive to agronomist persuasion and organizational activity. From 1913 on, rural cooperatives began mushrooming in the Buinsk district,[20] which indicated the awakening of peasant initiative, a key prerequisite for public self-modernization. All of this was achieved by four young precinct agronomists who graduated from mid-level agricultural colleges in Kazan, Samara, and Viatka about 1910,[21] thus representing the same generation of new Russian intelligentsia. The technology of social engineering that they applied worked even in the severe conditions of a remote district with a hostile second element and ignorant peasantry. Last but not least, despite the official request that agronomy should privilege "Stolypin" peasants (i.e., those who separated out individual farmsteads from land communes),[22] the state played an important, if ambivalent, role in launching the project of "public agronomy."

The double role of the state as the greatest conservative force and at the same time an important factor of modernization was represented by the two major state agencies involved in rural affairs: the Ministry of Interior, especially its Department of Police, and the State Administration of Land Settlement and Agriculture (GUZiZ), especially its Department of Agriculture. While the GUZiZ (as of October 26, 1915—the Ministry of Agriculture) in general was associated with such unpopular measures as the forceful dissolution of peasant communes, the Department of Agriculture was perceived as a progressive agency with a purely professional agenda. Its vice-director, and from 1914 its director, D. Ia. Slobodchikov, had the reputation of a "red director" among his colleagues.[23] The Department of Agriculture supported initiatives of rural professionals and agricultural cooperatives, and directed research efforts in different aspects of agriculture.[24]

The daily confrontation of these forces, the police and the agricultural agency, took the shape of conflicts over the political loyalty of individuals and whole organizations. The Department of Police rarely intervened directly in the affairs of the Department of Agriculture if there were no grounds for arrests and convictions. However, it carefully collected compromising materials that could be used by local authorities as a pretext for banning certain public initiatives. For instance, there were compromising dossiers on the Moscow and Kharkov agricultural societies,[25] and the Kharkov governor referred to those dossiers in December 1911 in a letter to the Department of Agriculture objecting to the project of short-term courses for agronomists designed by the Kharkov Agricultural Society. Although he claimed that the courses "doubtless will be used for agitation purposes,"[26] he could only postpone their opening from February to October, as if those ten months could have changed the political profile of 17 suspicious Society members out of the total number of 44.[27] Hence, police surveillance, a serious weapon in the hands of a conservative official, could not stop public initiative, but only temporarily hamper it. This weapon was absolutely powerless in cases where officials looked favorably upon the modernization activity of rural professionals. For example, in September 1909, the Moscow Town Governor (*gradonachal'nik*) asked the governors of neighboring provinces for their opinion on the prospective regional congress of agronomists in Moscow. The governor of Vladimir province was against such a congress. In his view, all agronomists were socialists, and the task of improving agriculture was the exclusive concern of the GUZiZ.[28] The governor of Kostroma province, on the contrary, was enthusiastic about such a congress. The assumed socialist sympathies of agronomists did not intimidate him:

> ... for Kostroma province, where agriculture is at a low stage of evolution, a congress of agronomists could be very useful in assisting the intensification of agriculture, especially during the current transition to farmstead ownership. Because of the socialist complexion of the participants in the congress, it should be surrounded by special conditions, which would limit the competence of the congress to purely agronomic issues.[29]

The regional congress of agricultural specialists finally took place in Moscow on February 21–28, 1911, despite the fierce resistance of some officials. This congress that made Alexander Chaianov famous in professional circles became a landmark in forming the self-perception of rural professionals as a particular group of public modernizers.

After the Revolution of 1905, while it was difficult to halt the activity of entire organizations or congresses, it was still possible to control individuals. Any public gathering immediately attracted the attention of the police, even if it was organized by a state agency. All participants were screened in case any of them had a previous criminal record.[30] The system of political control over individuals found its most complete implementation in the phenomenon of references of political reliability (*spravki o blagonadezhnosti*). All state institutions making personnel-related decisions were obliged first to check with the local governor's chancellery to learn whether there were any objections to a particular candidate; the chancellery then requested a reference of reliability from the police. The Department of Police held

2.5 million records on 2 million Russians, every year some 150,000 new dossiers were added.[31] Given the very small size of the Russian secret service cadres and a limited number of people engaged in any publically visible activities (legal or criminal), these numbers are shocking as indicators of the efficiency of the "undermodernized" old regime. The emerging public sphere in Russia developed in the gaps left within the carcass of the administrative system, gradually expanding at the latter's expense. Therefore, virtually any major step in the domain of public relations was associated with crossing the boundary of the state-regulated sphere and hence with obtaining a reference of reliability.[32]

In 1910, the Ekaterinoslav governor did not confirm the nobleman Dmitrii Korolevtsev for the position of Bakhmut district zemstvo agronomist because of his negative political credentials. Unaware of the exact charges, agronomist Korolevtsev appealed directly to the Ministry of Interior asking to be cleared before the Ekaterinoslav governor. According to his police dossier, Korolevtsev participated in student unrest at the Riga Polytechnic in 1894, and "propagandized among peasants" in 1906–08, when he was a district agronomist in Kharkov province. The charges were serious, and the Ministry of Interior left the final decision to the Ekaterinoslav governor, who had already indicated his negative attitude to Korolevtsev.[33] We do not know where Dmitrii Korolevtsev worked for the following few years, as his name disappeared from the lists of agricultural specialists. We may assume that government pressure compelled him to leave the sphere of public agronomy. However, late in 1913 or the beginning of 1914 he got a job in Saratov province on the Lower Volga, as the Tsaritsyn district zemstvo agronomist.[34] He moved from Ukraine to a climatically similar part of the Russian empire, where his compromising political background apparently did not bother the authorities.

The troubled story of agronomist Korolevtsev testifies to the absence of a sustained state policy of political repression. It was not the police who initiated Korolevtsev's persecution, and even with his negative record of political reliability, it was the personal choice of the Ekaterinoslav governor to fire or keep him (and certainly of the Saratov governor, Prince A. A. Shirinskii-Shikhmatov, a dynamic administrator and a man of many interests). A booming job market and a diversification of employing agencies minimized the effect of individual cases of political persecutions. In the fall of 1912, the Stavropol governor ordered the firing of Stavropol city agronomist and director of the local agricultural experimental station, N. K. Pokhodnia, who supported an opposition candidate during the elections to the state Duma.[35] Just a few months later, agronomist Pokhodnia became head of the entire network of experimental stations and laboratories of the Kiev Agricultural Society, which was obviously a more important position.[36] Gradually, the entire procedure of political screening became simply routine paperwork.[37]

While the system of political repression was potentially available to the government, the priority was given to cautious support of modernization activities over sustaining a state of political and economic stagnation. This attitude of the most reactionary government agency, the Ministry of Interior, was expressed in a circular memo to provincial governors, no. 64251 of April 5, 1909, signed by the deputy minister, notorious police general Pavel Kurlov:

> ... [there is] an undoubted aspiration of antigovernment parties to use the professional movement....

> The Ministry of Interior acknowledges that trade unions and cooperatives by their nature are intended to benefit the economy and laboring masses, and takes all possible measures to support those institutions in their normal condition, however, it cannot allow the antigovernment forces to realize their plans.

Therefore, it was recommended to the governors to collect information on those institutions, thoroughly investigate their activities, screen individuals for political reliability, and authorize the opening of new organs of public initiative only after adequate investigation.[38] Gradually, the practice of cautious toleration lowered the threshold of police anxiety and vastly expanded the sphere of social activism considered "legal." By 1913, it was already sufficient just to inform the police about tremendous enterprises, such as the launching of a special agronomy train by the Vladikavkaz railways, without asking for preliminary consent.[39]

As the available archival sources suggest, in the absence of any coherent government politics aimed at impeding the rising public modernization movement, when certain rural professionals got into trouble with the police, it was the result of professional or personal rivalry rather than an orchestrated police showdown. At the same time, individual decisions were often informed by the logic of institutional solidarity or rivalry. On September 28, 1913, a clerk at the Arsk agricultural zemstvo warehouse (Kazan province), Emel'ianov, informed to the police on the manager of the warehouse's workshop, Amosov. In June, Amosov had supposedly allowed himself offensive comments about the emperor's sexual life. Since the allegation was brought forward three months after the alleged crime of lèse majesté took place, it was obviously an act of personal revenge. Nevertheless, on October 3, a formal investigation began. Despite the ridiculous nature of the accusation, the deputy head of the Kazan Gendarme Administration, Lieutenant-Colonel Subbotin, and the deputy provincial prosecutor, Samartsev, conducted a month-long investigation. The governor revealed a personal interest in the case. Despite such an impressive reaction to the charge, both Emel'ianov and his victim Amosov were punished only by being placed under "special surveillance" by the local police. The reaction of the Kazan district zemstvo officials in this complicated situation was really crucial for the outcome of the investigation and quite telling. The chairman of the zemstvo board Novitsky and board member Serebriakov gave an excellent reference to Amosov, while Emel'ianov was characterized as a troublemaker and a drunkard. Moreover, in the middle of the investigation, telltale Emel'ianov was fired by the zemstvo with a dreadful reference, while "unreliable" Amosov was subsequently promoted to a position as agricultural warehouse manager.[40] Thus, a low-ranked representative of the third element was defended by local second-element representatives, while the police demonstrated their zeal to fight any case of subversion but not to the extent of disrupting local economic organizations.

If one can speak about the persistence of "old absolutist methods" of political repression in interrevolutionary Russia, it is probably more true of the political culture of the individuals belonging to the generation that matured before 1905, than of the government agencies, which were usually following their instructions and regulations. For example, some agricultural specialists in state service regarded professional competition from the third element as a threat to their status, which previously had been secured simply by their place in the service hierarchy. These representatives of the pre-1905 type of semiprofessional civil servant would not see anything wrong in seeking the authorities' intervention in a professional dispute, something absolutely impossible for the new breed of professionals, who shared an *obshchestvennost'* ethos. The authorities, in their turn, did not look with enthusiasm at the prospect of settling professional conflicts masked by political invectives, as the controversy among Kazan specialists in apiculture had shown in 1915.

The Kazan Society of Apiculture, led by a professor of the Veterinary Institute, V. I. Loginov, was one of the first and most active apiculture societies in Russia. Enjoying recognition outside and within Kazan province, this voluntary association of agricultural specialists and amateur beekeepers played a key role in propagating apiculture and educating beekeepers.[41] At the same time, the position of senior instructor of apiculture of the State Administration of Land Settlement and Agriculture (GUZiZ) in Kazan province was occupied by Vladimir Merkulov, a 48-year old hereditary "honorary citizen" with a prerevolutionary service record. He was deeply upset by the necessity to act via the structures of the influential Society of Apiculture, which was subsidized and hence recognized by the GUZiZ, thus making his position one of secondary importance. He did not share the professional ethos of his younger colleague in Samara province, who pledged to serve the beekeepers' cooperatives of the province as a public-minded specialist rather than as a state servant.[42] Merkulov wanted professional recognition equal to his official status, yet he did not want to work on a team and could not compete with the nationwide fame of Professor Loginov.

Frustrated, Merkulov at first decided to refer the matter to a higher authority, though in a strange manner. In the fall of 1915, he wrote a letter on behalf of some fictitious anonymous "peasants-beekeepers" and sent it to his superior, the inspector of agriculture of Kazan province, A. P. Gortalov, the chief representative of the GUZiZ in the province. In that letter, the "peasants-beekeepers" claimed in pseudo-folk language that Professor Loginov had disrupted apiculture in the province and mismanaged government funding, and that he should be replaced by a really learned man, Merkulov. Inspector Gortalov conducted a formal investigation and did not find any evidence supporting these claims.[43] Disillusioned, Merkulov decided to transfer to his native Poltava province, but before leaving he made a last attempt to take over the Society of Apiculture. On November 27, 1915, he filed an official report to the Kazan chief of police repeating the allegations of the "peasants-beekeepers," and adding that Professor Loginov had referred negatively to the Orthodox Church and the emperor. The police opened an official investigation on December 1, which continued for more than a year and was closed "for lack of evidence" of the alleged crime. During interrogation, members of the Society of Apiculture unanimously claimed Loginov's innocence, while accusing inspector Merkulov of careerism and insufficient professionalism. The GUZiZ took the principled position of supporting the head of a public organization, Loginov, against its own employee Merkulov. Both acting inspector of agriculture Gortalov and his predecessor Markovnikov issued statements in support of Professor Loginov.[44]

This incident suggests two possible and certainly not mutually exclusive interpretations. First, it probably testifies to the preeminence of local political loyalties and interests over the central and more abstract ideological dogmas: both Gortalov and Markovnikov, the acting and the former top GUZiZ officials in the province, were local rich landowners and members of Kazan "high society," and Professor Loginov, noted all over Russia, was also a prominent member of that society. While police did not have any particular incentives to go after Loginov, Kazan notables and public activists used all their influence to support his case. On the other hand, it is important that the police did their job ignoring an opportunity to curb a public initiative by using the case of Professor Loginov as a pretext for taming an overly independent and influential organization. The second conclusion we may draw from this case is that the GUZiZ, in its turn, chose to support a public figure and rural professional rather than a government employee, thus putting professional and state interests above the petty concerns of a department's prestige. This conjunction of attitudes of the major state agencies involved in rural affairs and the successful mobilization of public interests and solidarities at the local level explains the astonishing scale of success of the self-mobilization of Russian society involved in reforming the countryside. By sponsoring public initiatives in the field of countryside modernization and lacking resources and incentives to curb any particular activities (with the exception of proved subversion), the government agencies could not be regarded as an obstacle to the public modernization movement. To symbolize its endorsement of the rural professionals' efforts, the government even established a special Romanov badge of honor for outstanding accomplishments in the sphere of agriculture.[45] While it was still controlling, and by no means liberal, government policy did not consciously block the initiative of rural professionals and was more responsive to the arguments of economic rationality than to the appeals of narrow-minded reactionaries. Even the predominantly anti-statist "public modernizers" had to admit that government agricultural policy was becoming "in some aspects close to the actual interests and needs of agriculture."[46]

Cloning "public agronomy," government style

The positive role of the state in promoting the project of societal self-organization by the community of agricultural agronomists and *obshchestvennost'* resulted more from important structural factors than from any particular benevolence and liberalism on the part of the authorities. By introducing the complex of measures known as the Stolypin reforms, the government had endorsed a course of economic modernization that was based on essentially Progressivist premises in its outline and formed under the influence of the expert community within the *obshchestvennost'*. In a nutshell, it was a program of radical social engineering in the modern understanding of the term, which means that it required a systematic social policy rather than a one-time radical intervention by the legislator. The Imperial state was capable of changing the legal status and even the name of entire social groups overnight, but it was only learning to institutionalize the implementation of social policies within the existing system of administration.[47] Only able to prosecute or tolerate social activists, but not to direct them, the police eventually gave up attempts to differentiate between pro-government and potentially subversive types of professional expertise in the countryside. In its attempts to go beyond purely administrative measures, the government had to rely on the human resources and intellectual potential of the *obshchestvennost'*.

Despite the old tradition of mutual mistrust and suspicions, the government and the *obshchestvennost'*-based cohort of rural professionals gradually drifted toward closer cooperation. The new generation of Russian intelligentsia was formed by the very same "perceptual revolution" that, according to David Macey, in governmental spheres formed "a new generation of enlightened or liberal bureaucrats"[48] as a social cohort with distinctive educational and career backgrounds that brought about the Stolypin reforms.[49] With cautious support from a new generation of Russian bureaucrats, the new generation of Russian intelligentsia staged the first large-scale social engineering project based on coherent social policy that was not confined to juridical novelties and administrative measures.[50] The similarities in the genealogy of government and *obshchestvennost'* reformism should not be overshadowed by the image of the Stolypin reforms as a ruthless implementation of the administrative fantasies of a well-ordered countryside, so widespread in historical studies. In both cases, the ultimate goal of modernizing efforts (by rural professionals or top bureaucrats) was seen in helping peasants to set free their economic and intellectual potential (only bureaucrats could use the powerful tool of legislation, while professionals had to concentrate on "small deeds"). In both cases it was important to find somewhere a precedent for successful implementation of the proposed modernizing strategy. While the professionals operated in the transnational context of the Progressivist culture of modernization, borrowing different aspects of their program from Italy, Denmark, or Germany, it seems that the government officials were more concerned with finding "indigenous" solutions that would fit the peasant realities.[51]

It was exactly the discovery of the potential for evolution and the adaptability of the traditional peasant household that influenced the actual outline of the government reforms. The future chief ideologist and executor of the Stolypin land reform, the Danish emigrant C. A. Koefoed, was serving as assessor to the State Noble Land Bank in 1901, when he took a business trip to Mogilev province (in present-day Belarus) and discovered a village that on its own initiative had replaced communal landholding with individual farmsteads.[52] Koefoed wrote a memorandum and later a book advocating peasant land consolidation and the dissolution of the peasant commune as a short path toward more efficient agriculture and a better peasantry. Eventually, he was put in charge of government legislation on the agrarian question. It was very important for Koefoed (Koffod in the Russian spelling and in the writings of some modern historians) to stress that his plan offered nothing new or radically alien to the peasant routine. On the contrary, the proposed measures just fostered and facilitated processes already under way in countryside. He (and most of his associates in the government) did not deny the peasantry's rationality or claim that peasants could not learn and adapt to new ways of life. He just questioned the speed at which innovations spread among the peasantry. Judging from his case study in Mogilev province, we may estimate this speed at about a mile per year. Given the size of the Russian empire, this progress was not very encouraging. Thus, the architects of the Stolypin reforms appeared to be even more concerned about legitimating their actions based on the peasants' natural predisposition toward the proposed measures than their critics, who were dreaming about Italian mobile bureaus and Danish dairy technologies.

The actual implementation of the government reform was also affected by the program and practices of the rival "public agronomy" project, with the only ideological element that could be claimed by government modernizers as their own "know-how": a paranoid fixation on the peasant land commune. The attitude toward the land commune was an important test for the new thinking in general,[53] regardless of one's political affiliation. Those who reduced the whole complex problem of radical socioeconomic transformation to the question of the commune's future were thinking in the archaic categories of naive laissez-faire liberalism or populist radicalism.[54] At the same time, the very concern of the government with the commune is understandable: the abolition of the commune's artificial legal bonds was the only possible administrative element of social engineering at the disposal of Stolypinists. All other measures simply mirrored, and therefore eventually endorsed, the project of the "apolitical politics" of public agronomy: Stolypin's legal reforms were accompanied by lavish government subsidies to zemstvo agronomy and also by the creation of the state system of agronomy service—virtually from scratch.

The very parallelism of the thinking of the new bureaucrats and new rural professionals and the understanding by the government reformers of their dependence on public activists reinforced rivalry and stimulated attempts to marginalize the initiative of public activists. In 1907–11, until Stolypin's death, the most dynamic faction of Russian officialdom attempted to overrun the *obshchestvennost'* in its modernization activity. A state strategy of cloning public initiatives in order to hijack them and put them under government control, so typical of the politics of political demobilization, was employed by the Stolypin government. Both the network of zemstvo-based precinct agronomy and the system of social mobilization through public campaigns became targets of emulation and manipulation by the government and conservative politicians associated with government circles.

Despite government pressure on zemstvos and agricultural specialists to give preferential treatment to peasants leaving the land commune, the principle of equality of all zemstvo taxpayers did not allow even conservative zemstvos to insist on a more favorable attitude of rural specialists to farmstead peasants. Besides, in regions with strong communal traditions, zemstvo boards were clearly procommune (e.g., in the Central provinces populated by ethnic Russians, or in Kazan and Ufa provinces populated by very communal-minded minorities).[55] Indeed, the intent of Stolypinists to separate "the sheep from the goats," or the farmstead farmers from the communal peasants, equally contradicted the populist legacy and the modern professional ethos of the zemstvos's "third element." To offset this resistance and to undermine the project of "public agronomy," the government founded its own network of agronomists and instructors at land settlement commissions that were in charge of breaking up communes.

The formal split between the zemstvo and government agronomists occurred during the fateful regional congress of rural specialists in Moscow in February 1911,[56] but tensions between the two had began long before. The case of Kazan province reveals the roots of alienation between the two groups of rural professionals. To begin with, the very climate of the Special Agronomy Conference of the Kazan provincial land settlement commission in December 1909 differed substantially from that of a typical zemstvo agronomy conference. With almost no agricultural specialists present, that conference was dominated by leaders of the provincial first and second elements. The conference spent much of its time expressing faithfulness to the emperor and elaborating the text of telegrams to the prime minister and the chief manager of the GUZiZ. Instead of debating practical agronomy measures, the participants extensively criticized the zemstvo and its "futile" agronomy organizations.[57] On December 9, 1909, the conference decided to introduce a separate government agronomy organization staffed by 12 agronomists, with salaries set about 30 percent higher than zemstvo agronomists received at the time (apparently, to secure the superiority of the government agronomy organization and its cadres). They would have to establish 30 sample farms with incredible subsidies of 330 rubles per farm, introduce potatoes to the peasant, and propagate the same five-field intensive scheme of crop rotation throughout the entire vast Kazan province.[58] In essence, this was a borrowing from the blueprint of the public agronomy program, only abridged and remodeled to fit the administrative approach to modernization. While "public agronomists" insisted on the study of local conditions prior to giving any particular recommendations, government modernizers already had universal recipes for any occasion.[59] While the former began their modernization activities with "sermons," the latter were encouraged to use coercion.

The zemstvo and the government agronomists differed dramatically even in their dress. Zemstvo agronomists were free to dress casually and tried to look as natural as possible in the peasant environment. The dress of government agronomists was standardized by law. While on the job, they had to wear a particular uniform. It included a dark green woolen coat or a double-breasted frock-coat, each with two rows of six buttons, and embroidery on the collar and cuffs. The uniform trousers matched the coat in color. There was also a black woolen (or a white pique) vest with six small buttons, and a narrow black necktie, knotted with a small bow. On their heads, government agronomists wore a dark green cap with visor and a cap-badge. In cold weather, they wore a double-breasted overcoat with 12 buttons or a greatcoat with hood, also dark green. Gloves of kid or suede completed the uniform of the government agronomists, who in peasant eyes must have looked like land captains. In summer, the government employees were allowed to wear white linen double-breasted frock-coats and trousers, which made them resemble rich summer residents or landowners—also not favorite figures among the peasants.[60]

It is therefore no surprise that the zemstvo became outraged by the arrogance of the government agronomy organization in interfering with the zemstvo's own policies and activities. In turn, the zemstvo's third element could not but be hostile to the "impostors"—the government land settlement agronomists who compromised the "true" rural professionals and their program, while being put into a preferable position. As a result, a top official of the Kazan provincial land settlement commission complained in November 1911 that district zemstvos blocked the land settlement officials from participation in zemstvo agronomy efforts.[61] Thus a political and ideological conflict acquired an institutional dimension. In the situation of conflict between the zemstvo and the land settlement authorities, the GUZiZ chief official in Kazan province, the inspector of agriculture Gortalov, supported the professional zemstvo efforts rather than the administrative policy of his own department (probably for the same reasons that he would support Professor Loginov four years later).[62]

Hostility between the third element and the government land settlement commissions' rural specialists culminated in 1912, when it took the form of a direct political confrontation. An incident at the Moscow provincial zemstvo agronomy conference in early 1912 became a nationwide scandal. The Bogorodskii district agronomist Stanislav Freitag simultaneously held the positions of district zemstvo agronomist and district land settlement commission agronomist. When he arrived at the conference, he was confronted by the Bronnitsk district zemstvo agronomist, A. Zalog, who was supported by other agronomists. Demonstrating their protest against the "opportunist" position of Freitag, all agronomists left the conference after submitting a written protest to the conference chair. This politically charged incident drew the attention of the Moscow governor, who ordered the zemstvo to fire Zalog as a politically unreliable person.[63]

Adherents to the public modernization program denounced the land settlement agronomists as betrayers of the third element's professional ethos, which necessarily included a strong public service component:

> . . . agronomists in the role of servicemen not only lose any energy in the bureaucratic atmosphere, but under the influence of circulars and guiding directives forget the sense of their activity. Many of them use orders on paper, authority intonations, and threatening shouts in disseminating agronomic knowledge, and often lower the prestige of the agronomist as a public activist in the eyes of the local population.[64]

As a matter of fact, ideologists of the public modernization campaign did not dispute the necessity of the rational land settlement of peasant landholding. Alexander Minin, a colleague of Alexander Chaianov in the Organization-Production school in rural studies, the Chernigov provincial zemstvo agronomist, and a graduate of the Moscow Agricultural Institute in the class of 1911, addressed this problem in a special article. He argued that rural professionals faced a conflict between their public temperament and professional expertise: they were torn between a desire to give preferential treatment to the poorest peasant and a rational understanding that work with well-to-do farmers was more efficient. From a professional point of view, rationalization of the land tenure regime was an absolute necessity, which would benefit, however, only midsize households. To fulfill the public duty of the rural professionals, Minin recommended that the land settlement be turned to the advantage of poor peasants by urging them to consolidate their small plots together and found collective farms.[65] In conclusion, Minin appealed to practitioners:

> I realize how risky it is for an agronomist to associate his name with land settlement, but what is to be done? One must take into account that land settlement is a fact of tremendous importance, while the role of Pilate washing his hands is not always an expression of civic virtue.[66]

It was rather the local practitioners and some representatives of the second element who perceived and presented the professional and institutional rivalry as a major programmatic conflict between the "bureaucracy" and the "public initiative."[67] Eventually, the government acknowledged the absurdity of the coexistence of two parallel agronomy organizations and the futility of attempts to overrun the booming zemstvo-based public agronomy. Around 1913, GUZiZ began transferring its land settlement agronomy organizations to the local zemstvos under the condition that the latter continue the land settlement activity.[68] Some unambitious district zemstvos even tried to avoid appropriating the government agronomy organization with all its responsibilities, and the consequent further expansion and strengthening of the local third element.[69] The government's withdrawal from the competition with "public agronomy" on the eve of the World War I meant defeat in the rivalry with the *obshchestvennost'* over the mobilization of peasants, and the triumph of the *obshchestvennost'*-backed "third element." This decision also made the business of land settlement less politically charged, and changed the attitude of the representatives of the third element toward taking government-sponsored positions.

A bureaucratic imitation of public initiative

As the GUZiZ was launching an ambitious program to create an alternative network of agricultural assistance to the population, a group of public figures associated with government circles put forward their own version of the popular yet completely controlled movement for modernization of the countryside. In 1908, a new initiative was announced that was expected to gain support for the unpopular Stolypin land reforms by utilizing some techniques and rhetoric of the *obshchestvennost'* modernizers. To be sure, the democratic potential of any popular movement jeopardizes the goals and ideals of conservative and loyalist politicians who ventured to rely on controlled mass mobilization.[70] The innate "Russianness" of the peasant (understood as loyalism, patriotism, and nationalism) and the "Slav" nature of the Russian Empire were the two pillars of the conservative political discourse in early twentieth-century Russia.[71] Therefore, mobilization of the population along a nationalist agenda under the supervision of reliable individuals and institutions was seen as a lesser evil and an opportunity to hamper or even hijack the growing public modernization movement. On September 2, 1908, the Society for the Assistance of Revival of Agriculture and Popular Ability to Work, the Russian Grain, was registered in St. Petersburg.[72] Alexander Stolypin, the conservative journalist writing for the newspaper *Novoe vremia* (New Time) and brother of the Prime Minister Petr Stolypin, became its chairman. His close associate in the Russian Grain was Dmitrii Vergun, editor of the magazine *Slavianskii vek* (the Slav Century) and émigré from the Habsburg Empire, where he was prosecuted as a leader of the Russophile Ruthenian movement. Among the board members and close associates of the new society were spouses of such politicians as the former director of the Department of Agriculture and chairman of the Third Duma, son of the famous Slavophile, Nikolai Khomiakov, or the Duma deputy and leader of Russian nationalists, Count Vladimir Bobrinskii.[73] Russian-enlightened bureaucracy and the rising nationalist movement composed the dual social base of the new association. The goal of the Russian Grain was formulated as assisting the peasants in the improvement of their farming skills by getting firsthand experience on the most advanced farms in Russia and abroad, mainly in Slavic countries.[74] Peasants would thus be easily persuaded of the advantages of intensive farming on private property, and without the potentially subversive mediation of Russian agricultural specialists.

The very idea of this society was conceived in May 1908 during consultations of the delegation of Czech leaders of the Neo-Slavist movement headed by Karel Kramář with Russian politicians in St. Petersburg, primarily members of the interparty Association of Public Figures (*Klub obshchestvennykh deiatelei*).[75] Working on the preparation of the all-Slav Congress in Prague, Czech Neo-Slavists and their Russian counterparts (mainly affiliated with the Octobrist Party) drafted its agenda carefully, ignoring politically charged issues and concentrating on questions of cultural and economic cooperation.[76] Apparently, the idea of sending Russian peasants to much more advanced Czech farms was a direct product of these conversations, as the composition of the first Russian Grain initiative group's meeting suggests. Probably, the very idea belonged to Evgraf Kovalevsky, an Octobrist and Duma deputy with expertise in schooling and education exchanges.[77] The first "founding" meeting took place in his apartment, where he made a formal presentation of the project to the invited guests, which suggested his authorship.[78] However, he was never mentioned in connection with the Russian Grain's activities again after the society's formal registration in

September 1908, which may be seen as an indication of the society's shift toward a more rightist and nationalist orientation.

The Slav Congress that took place in Prague on July 13–17, 1908, had a major impact upon the shaping of the Russian Grain: the only practical outcome of the congress was its endorsement of a variety of cultural and educational collaborative initiatives, with a special emphasis on the desirability of peasant educational trips to farms in advanced Slavic lands.[79] Resolutions of the congress and the firsthand observation of the superior state of Czech agriculture[80] compelled members of the Russian delegation to the congress, dominated by conservative and nationalist politicians such as Count Bobrinsky, to find the amorphous program of the Russian Grain particularly attractive. Their support and connections in the top echelons of the first and second elements had propelled yet another philanthropic society to the status of an all-imperial public actor.

From the very beginning, the Russian Grain intended to get control over the public initiative and to distribute government and zemstvo funds in accordance with its own vision, as no provisions were made to secure starter capital for the organization: there was no endowment grant and the wealthy founders and members of the new association were required to pay a symbolic fee of one ruble (although donations were welcomed).[81] However, within two years the Russian Grain's budget had reached 30,000 rubles[82]— an amount comparable to the budget of a zemstvo agronomist organization in a midsize province. The chairman, Alexander Stolypin, used one of the largest newspapers in the country, *Novoe Vremia*, as the mouthpiece of the new society, and its affiliates in the government and provincial administrations mobilized resources that they could control.[83] The cases of three local chapters of the Russian Grain provide a close-up portrait of this peculiar version of modernization politics.

The first chapter of the Russian Grain was opened in Perm province, often regarded as cradle of the public agronomy movement, one year after the registration of the St. Petersburg society. Probably sometime in the spring of 1909, the provincial governor, Alexander Bolotov, joined the Russian Grain, which became a decisive factor in the society's success in Perm. The official provincial newspaper *Permskie vedomosti* (the Perm News) was used as a free advertisement resource, which also placed an automatic stamp of official approval and sanction on all its publications.[84] When the Perm chapter was opened on October 22, 1909, the entire provincial administration joined it following their governor, including the vice-governor, member of the Provincial Office for Peasant Affairs (*po krest'ianskim delam prisutstviia*), manager of the Peasant Land Bank, manager of the State Property Administration, and others.[85] These people guaranteed the "proper behavior" of the new public body but could not secure its bold goals: to establish throughout Russia "sample schools, farms, moving agricultural exhibitions…, the distribution of agricultural publications in hundreds of million copies…, presentation of comprehensible lectures, and so on…. Today [the Society] has just a thousand members, while tens and hundreds of thousands of members are needed!"[86] The society's proponents presented the initiative of the Russian Grain as the only response of Russian educated society to the government modernization campaign, thus totally ignoring the rising public agronomy initiative.[87] To replicate or appropriate the program of "public agronomy" and to become a broad movement, the new society needed the organizational and financial support of the zemstvos and the expertise of rural professionals. Quite predictably, the peasant-dominated zemstvo and the third element of Perm province were reluctant to support a rival project at their own expense.[88] Luckily for the Perm Russian Grain, Governor Bolotov was a typical representative of the "new generation" of Russian bureaucrats, in terms of both age and service ethos,[89] and his sincere enthusiasm won the cautious support of some district zemstvos: chairmen of three district zemstvos (out of 12) became members of the society, and their zemstvos combined contributed a few hundred rubles to its budget.[90] As a result, the Perm chapter of the Russian Grain not only assisted four local peasants to travel to Moravia at the expense of the central organization but also financed the trip of four other peasants from its own budget.[91]

With eight peasants sent to study advanced farming techniques between 1909 and 1914, Perm province occupied tenth place among 44 provinces that participated in the initiatives of the Russian Grain.[92] By contrast, Samara province sent only two peasants abroad over the same period. This under-representation is easily explained by the profile of the local chapter of the Russian Grain society that was established in Samara in April 1911. This was a purely bureaucratic endeavor: the first 63 members of the society included virtually the entire top stratum of the Samara "first element," featuring such officials as the provincial factories inspector, head of the Samara Post and Telegraph District, and the manager of the Samara branch of the State Bank. Governor Nikolai Protasiev was, of course, the chairman. District and provincial Marshals of Nobility played a prominent role in the society, but not a single representative of the zemstvo, which had been so active in modernizing the countryside since the establishment of SOUKK in 1907, joined the Samara chapter of the Russian Grain. It comes as no surprise then that the society with 107 members by 1912 had a very tiny budget and could only afford itself to send small groups of peasants for one-day excursions to the local Bezenchuk agricultural experimental station. The two peasants who actually went abroad were funded by the St. Petersburg office of the Russian Grain, and the sole responsibility of the local chapter was to select the right candidates. It failed on both occasions: one peasant was sent to Denmark but apparently could not overcome the language barrier and cultural shock and returned home in less than two months. The second peasant spent the whole term of 11 months in Moravia, but having come from a family that owned 200 hectares of land he got hardly any relevant experience in that region dominated by small-scale farms.[93]

The cases of both Perm and Samara represented an attempt to arrange for a broader public initiative by purely bureaucratic means. The relative success of the former and complete failure of the latter could be explained by differences between key personalities. Governor Bolotov, 43 years old when he initiated the establishment of the Russian Grain chapter in Perm, had a genuine personal interest in its success. When he retired from the governorship soon afterward, Alexander Bolotov became a full-time executive in the St. Petersburg office of the Russian Grain (he was vice-chairman of the society and head of its Peasant Commission).[94] Governor Protasiev represented a different generation or another type of Russian bureaucrat. He was 56 by the time he opened the Samara chapter of the Russian Grain, and he probably did so just because it was the right thing for an active governor to do. Protasiev was a "professional governor": he had served as governor of Olonets in 1902–10 before coming to Samara, and in 1915 he was transferred to the position of Kharkov governor.

While Samara was at the very bottom of the list of provinces sending peasants abroad with Russian Grain assistance, and Perm somewhere in the middle, Voronezh province was second only to Tver in terms of its activity: 21 peasants from the province participated in the Russian Grain programs. This number still seems insignificant given the size of the empire and even of the province alone. Obviously, with all its local chapters, the Russian Grain still evolved along the same model as Samara SOUKK established in 1907, or any other philanthropic or educational society. It could not be used as the basis for a truly massive modernization campaign. However, it is hardly accidental that the share of Voronezh in the Russian Grain's activities was several times larger than the majority of the other 43 involved regions could boast. Unlike Perm and Samara, Voronezh district zemstvos supported the Russian Grain both organizationally and financially: two-thirds of the peasants participating in its programs received grants from local district zemstvos. St. Petersburg paid for only four Voronezh peasants. The unprecedented responsiveness of the Voronezh zemstvo can be explained by the same human factor that accounted for the hostility in Samara and elsewhere: the founding father of the Russian Grain, Evgraf Kovalevsky, had served for 20 years as district, and then provincial zemstvo deputy, and was elected to the state Duma from Voronezh province.[95] Apparently, his personal and political ties with the local second element secured support for the new St. Petersburg initiative.

The local Russian Grain chapter also differed from a typical, bureaucracy-dominated provincial branch. It was founded in 1911 by Voronezh ultra-nationalist leader and head of the local division of the All-Russian National Union, Vladimir Bernov.[96] That was only one of the initiatives he staged before the election campaign to the fourth state Duma, and apparently not the most important in his eyes.[97] The budget of the Voronezh Russian Grain was probably even more modest than that of its Samara counterpart: it did not allocate any subsidies to peasants and could serve only as an intermediary between the central office and local zemstvos. However, even though Bernov himself was a medium-level official in the provincial administration, the opening of the Voronezh Russian Grain was not a formal bureaucratic act. The combined efforts of the nationalist-dominated local Russian Grain and Octobrist zemstvo turned out to be quite efficient. And again the role of personal contacts was crucial in mobilizing public initiative in Russian imperial society: obviously, it was an active member of the All-Russian National Union in St. Petersburg and a close affiliate of Russian Grain, Count Bobrinsky, who gave his Voronezh colleague the idea if not a request to open a local branch there. This makes the Voronezh case similar to the otherwise quite different cases of Perm, Samara, and even St. Petersburg itself. An attempt to build a broad public initiative to foster modernization in the countryside, parallel to both government-sponsored reforms and the zemstvo-based public agronomy project, was confined to the niche of pre–mass-politics public initiatives. Without an institutional framework of its own, the Russian Grain depended on the benevolence, enthusiasm, and competence of provincial governors. It was sustained by and spread through personal networks, which were also very important in promoting the public agronomy project. But whereas the personal dimension of the latter acquired a scale of mass phenomena, and can be described in terms of professional solidarity, social cohort affinity, and generation bonds, the growth of the Russian Grain was provided by contacts and alliances among selected individuals.

Equally unsuccessful were attempts by the Russian Grain leaders to position themselves ideologically between the fierce anticommunal propaganda of the government and the prophecy of joint *obshchestvennost'* efforts by the proponents of public agronomy. This can be seen even in the story of the society's slogan, which was expected to express the essence of its program. Initially, it was formulated as "You Should Know Labor and Persistence" (*Znaite trud i stoikost'*)[98]—a formal and meaningless motto and purely technical as were the initial goals of the Russian Grain.[99] After the Prague Neo-Slavist Congress, leaders of the Russian Grain realized that their society should be sending peasants abroad to study farming techniques in advanced Slav lands with individual landholding, which resulted in an attempt to fuse moderate Slavophile rhetoric, nonaggressive anticommune preaching, and enlightening pathos. The official slogan of the Russian Grain, at least as of 1911, read "Not by Telling but by Showing" (*Ne rasskazom—a pokazom*),[100] although for some reason back in 1909, Russian Grain supporters in Perm believed that it was "Not by Order but by Showing" (*Ne prikazom—a pokazom*").[101] The latter version, of course, sounded rather revolutionary in the context of forceful government measures aimed at dissolution of the land commune and could not remain the society's official slogan for any significant period. While some 250 peasants sent abroad by the Russian Grain and also sent to advanced farms in the Russian empire between 1909 and 1914 had had a truly remarkable experience and documented it in their letters to the Society (as will be discussed below), the Russian Grain was just a well-funded educational enterprise without a clear agenda. Even Russian nationalism and the Slavophilism of some of its founders remained a low-key issue, as the Russian Grain sent many peasants to non-Slavic regions such as the Russian Baltic provinces or Denmark, and was establishing contacts with Argentina and the United States.[102]

Despite its close affinity to Russian officialdom, the Russian Grain shared many elements of its loose ideology with "public agronomy" and the Progressivist culture of modernization. When the 1900 World Exhibition in Paris introduced a new model of piecemeal social reformism as an integral part of the general image of triumphant modernity in its "Social Economy" pavilion, the projects of practical social amelioration were presented by people such as the author of the term "social engineer," William Tolman, or the renowned French reformer, Benoît-Lévy. Evgraf Kovalevsky, who in 1912 authored the first Russian universal schooling law, was a typical intermediary between Western Progressivism and Russian educated society. Kovalevsky not only attended the exhibition himself (perhaps not missing the Social Economy pavilion) but also brought several groups of Russian schoolteachers, expecting them to spread a new culture of modernization among their pupils. His initial project for the Russian Grain was very much in line with the Progressivist focus on rationalization and societal improvement through the enhancement of human capital, and akin to *obshchestvennost'* efforts at self-modernization. Those people who came to dominate the Russian Grain attempted to use it as a Trojan horse against rising public self-mobilization. They failed, just as the government failed to overrun the zemstvo-based network of public agronomy with the well-paid cohort of agronomists-servicemen. This shows that the Russian public sphere of self-organized and self-mobilized *obshchestvennost'* was more efficient than the state. The government depended on the rising stratum of public-minded agricultural specialists both intellectually and in terms of practical implementation of its policies. With the gradual softening of anticommunal pressure and the dismantling of a separate agronomist network by the government, rural professionals became less suspicious of the state and less adamant in their adherence to the intelligentsia's anti-statist "political correctness."

5
The Economic Foundations of Social Mobilization

Even though the majority of agricultural specialists who joined the post-1905 public modernization movement in the Russian countryside believed in its apolitical nature, we may still characterize this campaign as an important political factor. Indeed, the tense and ambivalent relationships with the state and zemstvo institutions alone testify to the political potency of the "public agronomy" project. The interaction of rural professionals, cooperative activists, and educators with the peasants made the part of Russian educated society that was concerned with agriculture a de facto important political actor.

Not unlike conventional party politics, the Progressivist-type "apolitical politics" had to rely on some self-regulated "machine" of human resources mobilization and financial accumulation that would efficiently convert political capital into status and wealth. In the absence of formal institutionalization, it still had to find mechanisms to sustain the determination and enthusiasm of the followers above and beyond any ideological allegiances. Let us turn to the "ground zero" of the process of political mobilization, societal self-organization, to see how the personal agendas and interests of individuals interplayed and contributed to the formation of broader social group interests.

Structurally, the most fundamental prerequisites for the development of a new public initiative after the Revolution of 1905–06 included the partial withdrawal of police control coupled with the state's active socioeconomic policy and rising zemstvo activity. These institutional changes affected the status of the "third element" as a group, or the prospects of implementing the project of "public agronomy," but individuals rarely perceive reality at this abstract level. Hundreds of people entered the ranks of agricultural specialists after 1905, not necessarily for idealistic reasons. The attraction of a progressive social cause coincided with the advantages of a particular career track associated with certain financial benefits and social status. Once "inside," in the course of education and later socializing with their peers, newcomers had to adjust to a certain professional ethos and corporate code of behavior. The social networks of the emerging new generation of Russian intelligentsia and the reformist program of public agronomy depended on such mundane factors as state of the job market and the prospects of a salary increase with a certain employer. Even the ideological zeal of agricultural specialists could be partially explained by nonpolitical considerations, for the sense of a particular mission helped many of them to cope with routine life in Russian villages. Money could tempt people to go there, but often was not enough of an incentive to make them stay.

In March 1908, the Kazan police intercepted a letter from a former revolutionary, alias Lel'ka, from Vologda to one Seregei Golovin in Kazan. Lel'ka wrote that only a few years earlier his entire life had been dedicated to the revolution and to protest against the existing regime. Now the situation had changed, and so had his life:

> I enrolled in the courses of land-surveyors that are open from March 1 in Vologda.... They will pay [during summer practice] 40 rub. per month, and then will give 50 rub. in resettlement allowance, plus travel allowance.... There are 50 of us. One can enroll with the educational background of a town primary school.
>
> I would like to finish these courses, for they give [me] the right to a piece of bread, and even a substantial one, especially during the summertime.[1]

This transition of Lel'ka corresponded to the major turn of Russian educated society from political confrontation with the regime to a more productive involvement in the national economy, especially its agricultural sector.[2] Even the technical occupation of land-surveyor offered a decent paycheck and something even more important—a new form of solidarity and comradeship, which found its expression in the ethos of professionalism. The Society of Russian Land Surveyors was among the most active trade unions in interrevolutionary Russia.[3]

The first and most important characteristics of a new job market for agricultural specialists after 1905 were its diversification and inclusiveness. Even the state turned out to be a quite nondiscriminative employer. According to the *Statutes of the Government Civilian Service*, 1896 edition, neither religion nor ethnic origin could be used to bar an individual from joining state service.[4] With the advent of women's vocational education, it became a common practice for GUZiZ to hire female specialists in agriculture. The graduates of agricultural institutes with diplomas of the first and second degrees became servicemen of the tenth and twelfth classes, respectively, upon joining state service,[5] and in 1907, GUZiZ recognized private agricultural courses as having equal rights with state institutions of higher learning.[6] The skyrocketing dynamics of enrollments in private courses can be seen as dual evidence of both increasing public concern with agriculture

Table 5.1   Extracts from the register of salaries of the Chistopol' district zemstvo employees in May 1910.[7]

Position in the Chistopol' zemstvo	Annual salary, in rubles
Chairman of the board	2,300
Two members of the board, each	1,200
Secretary of the board	1,200
Senior physician	1,800
Six precinct physicians, each	1,200
34 medical assistants, depending on education and seniority	360–600
Two veterinarians, each	1,200
District agronomist	600
Gardener	600
169 primary-school teachers, average	420
Agricultural elder (assistant)	300

and growing possibilities for those wishing to earn their living as agricultural specialists.

At the same time, zemstvos had almost a half-century-old history of hiring rural professionals. The case of the Chistopol' district zemstvo (Kazan province) provides insight into the hierarchy of zemstvo employees, in terms of their salaries, on the eve of the big takeoff of agricultural specialists (Table 5.1).

After the zemstvo board members and clerks, physicians and veterinarians constituted the highest paid group among the 274 employees on the Chistopol' zemstvo's payroll. Agronomists earned no more than some of the medical assistants. The salaries of the most numerous group of zemstvo employees, teachers stood between those of rural professionals and unqualified "blue-collar" employees. While salaries in other zemstvos varied slightly, 35 rubles a month (or 420 a year) remained the lowest rate for employees with at least several years of high-school education.[8]

As we saw in the introduction of the government agronomy organization in Kazan province in late 1909, the salaries of government agronomists were deliberately fixed some 30 percent above the average zemstvo agronomist salary. Meeting the challenge of the government agronomy organization and being caught up in the general turn to agriculture, zemstvos after 1910 significantly raised the salaries of agricultural specialists. On the eve of the World War I, the minimum salary of an agronomist with a higher education was about 1,800 rubles a year, while the majority of physicians still received 1,500.[9] In fact, each zemstvo had its own financial policy, which contributed to the "fluctuation" of agronomist cadres.[10] Usually, precinct agronomists were paid between 1,500 and 2,500 rubles, district agronomists between 2,000 and 3,500, and provincial agronomists between 3,000 and 5,000 (these figures include salary and travel allowance).[11] Thus, agronomists and agricultural instructors became the highest paid group of rural professionals.

At this point it is necessary to put these figures into some historical perspective.[12] The salaries of Russian rural professionals were high, even by international standards. As we mentioned, in 1914 Russian agronomists noted with satisfaction that they were paid more than their Italian colleagues and role models. In the United States, the county agricultural agent was the closest analogue to the Russian zemstvo district agronomist. In 1914, the average salary of 1,436 county agents was $1,200 a year.[13] At the current exchange rate of approximately 2 rubles per dollar,[14] the county agent's salary translated into 2,400 rubles, a quite typical annual salary for district agronomists. In contemporary money, that $1,200 converts to some $26,000 per annum.[15] Unfortunately, direct comparisons can be used only as a superficial illustrative device that cannot claim any degree of accuracy on methodologically sound grounds. It would take special research to translate the ruble value of the early 1910s into modern currency using the method of index numbers and comparing not just wholesale prices but the analogous "consumer baskets."[16] Besides, over the span of the past century, living standards have changed dramatically, and even by expressing the value of 1913 Russian currency in contemporary money would not give the full picture. While, say, that $26,000 (in rural areas) as an annual salary does not look particularly impressive today, the analogue of this sum in 1913 currency can be viewed differently compared with the earnings of other categories of working people. Let us attempt to evaluate the earnings of agricultural specialists by comparing them with the income of certain reference groups.

In the early twentieth century, before serious changes in the peasant economy began taking place, Russian economists believed that the peasant consumption pattern was stable and could be expressed in monetary terms as a of about 50 rubles a year.[17] Consequent studies proved that 50 rubles a year provided only the physiological minimum necessary to survive, and that peasant consumption tended to rise under favorable market conditions and with higher productivity of the peasant farm.[18] For us, the important thing is to establish a minimum income required to survive physically in the countryside, and we can accept a statistically constructed 50-ruble benchmark as such a minimum. In comparison with the peasants, even a poor village teacher earned considerable money.

The lowest social strata in the cities, workers and individual craftsmen, earned quite a bit more than the peasants, but no more than village schoolteachers.[19] The average wage of an industrial worker was only 264 rubles a year in 1913.[20] Wages varied significantly from region to region, and from one type of work to another.[21] Hence, the minimum standard of living in the cities was very diverse. Those agronomists and instructors, who resided in the cities and whose salaries began from 1,500 rubles a year, were as much elevated over the typical minimum city income as rural teachers were over the minimum peasant consumption standard.

In comparison with city professionals, agricultural specialists appeared to be somewhere in the middle, earning more than secondary-school teachers, about the same as engineers, but less than some lawyers or physicians with an established private practice.[22] Agronomists actually complained that while their salaries were higher than those of zemstvo physicians, the latter had many advantages over agronomists in terms of free and better housing, periodical salary increases, and the possibility of private practice.[23] Indeed, unlike other rural professionals, many precinct agronomists paid for housing from their salaries, about 10 percent of their annual income.[24] As we have seen in the case of the Samara precinct agronomist Aleksei Kiselev, sometimes it was impossible to find a decent dwelling in a village for any amount of money. According to a 1913 survey, the typical apartment of a precinct agronomist in Ufa province had one or two poorly furnished rooms, with a total area of some 56 square meters (607 square feet).[25] In fact, zemstvos constantly built and furnished new housing for agronomists, but they could not catch up with the pace of the agronomist network's expansion in the early 1910s, and for a while many agronomists were left to their own devices.

The precinct agronomists were most desperate and eager to accept any type of zemstvo housing, while the agricultural instructors enjoyed the variety offered by the urban real estate market, and the zemstvo physicians or teachers at the agricultural vocational schools had a longer experience of living in public housing provided by the zemstvos. The latter had formed a quite detailed vision of the ideal housing for a rural professional residing in the countryside. One such model envisaged a small one-floor cottage on a brick foundation, with all the basic utilities (including running water, toilet, and bath), with a total area of approximately 115 square meters (1,215 square feet), or two times the size of the average Ufa agronomist's dwelling. The assumption was that the rural professional (in this case, a schoolteacher) is married,

> he has a "normal" family—2 to 3 children, but the old folks do not live with them. In such a case, in the house there should be: a study room, a bedroom, a dining room, a children's room, then—an anteroom, a kitchen, a porch, a closet, and a lavatory.[26]

In March 1913, the twelfth agronomist conference of Ufa province attended by 62 agricultural specialists and zemstvo officials formed a special commission to outline the budget and detailed project of a precinct agronomist's residence, which should include the agronomist's house, the hut of the agricultural elder to be shared with a laborer, a shed, a barn, and other facilities. The commission was headed by the provincial agronomist Rezantsev, who received his bachelor degree in 1906, and included five district and two precinct agronomists, who got their degrees between 1880 and 1909 but held positions with seemingly no correlation between seniority of status and age.[27] They not only produced a detailed plan of the precinct agronomist house—almost identical to the abovementioned teacher's house, only more spacious[28]—but also calculated the cost of the project to the last nail (including rulers and inkpots in the agronomist's study room). It seems that Ufa agronomists were quite confident about zemstvo financial backing: the house alone should cost 2,400 rubles,[29] and by allocating 2 rubles for a special sign "Zemstvo Agronomist,"[30] commission members clearly indicated their intention to spend zemstvo money at their discretion. The total ticket for a precinct agronomist residence reached 6,700 rubles, and with 39 precincts in the province the program asked for the fantastic sum of over a quarter million rubles. That was some 70 percent of the budget of a provincial zemstvo agronomist organization in 1913, but the provincial zemstvo board supported the project expecting the Department of Agriculture to chip in on a fifty-fifty basis.[31] The almost fourfold increase in the budget of the agronomist organization over the course of six years (1907–13)[32] was the primary factor fostering a feeling of security and rising expectations on behalf of Ufa agricultural specialists.

The appearance of such plans testified to the fact that rural professionals wanted to enjoy the comfort of civic life in the countryside, and to serve as experts deserving a certain material status, rather than prophets suffering for and with their flock. Although some aspects of the agricultural specialists' financial situation required further improvements (for instance, the system of periodic salary raises),[33] in the 1910s, many zemstvos demonstrated eagerness to move mountains for agronomists, often at the expense of other professions.[34] Doubtless, in a few years, zemstvos would have provided the necessary adjustments to the agricultural specialists' salaries and benefits. Besides, the agronomists' claim that they could not benefit from private practice was only partially true: numerous short-term agricultural courses for peasants provided lecturing agronomists with additional income.[35]

While the agricultural specialists in the government service enjoyed financial benefits approximately equal to those of the zemstvo agronomists, veterinarians lagged somewhat behind the more popular agronomist profession. Although their salaries had already risen by 1906–07,[36] before the big surge in the agronomists' financial status, the average veterinarian salary was then frozen for a while at 1,200 rubles a year.[37] Those living in towns sometimes managed to multiply their basic income almost fourfold thanks to private practice or the holding of several positions at a time.[38] The zemstvo precinct veterinarians rarely had such opportunities, and envied the more prosperous agronomists to the point of plotting a "rebellion" against their alleged domination.

Veterinarian and agronomist assistants earned even less, from 360 to 600 rubles a year, depending on their educational background and their employer's enthusiasm.[39] Still, their salary was comparable to the income of village teachers, while the education of many "agricultural elders" consisted of one or two classes of the parish school.[40] It is no surprise then that some teachers were thinking about joining the more profitable occupation of agricultural specialists.[41] The increasing number of agricultural elders (over four years, from 1909 to 1913, their number had grown tenfold) indicated that the agricultural specialists' profession was maturing.[42] The expansion of a stratum of agricultural elders completed the differentiation of functions within the profession, and strengthened the link between the trained professionals and their uneducated clients.

The massive "turn to agriculture" after the First Russian Revolution contributed to the threefold increase of the pedagogical staff of 288 educational institutions of the Department of Agriculture.[43] In 1914, there were 1,774 full-time pedagogues, and 692 part-time (many of whom were priests who taught the Bible).[44] Some 365 teachers of "special subjects" (i.e., agriculture and stock breeding) in the mid-level colleges and vocational schools of various types were paid on average twice as much as teachers of "general subjects," usually from 500 rubles a year in practical schools to 2,000 rubles in mid-level colleges.[45] About 50 of the agricultural-school directors earned between 2,000 and 2,500 rubles. For some agronomists without a higher education, the position of director of a primary agricultural school could be seen as a quite appealing option, even though it offered just 1,500 rubles of annual salary along with free housing in a more or less "civilized" township.[46] The demand for schoolteachers of agriculture was so high that the State Council and the State Duma sanctioned the establishment of a special Agricultural Teachers Institute.[47] Thus, the professional track of agricultural education not only attracted additional participants in the countryside modernization movement but was even capable of redistributing human resources within the movement itself.

In the 1910s, however, a new agriculture-related profession appeared, which seemed to be even more attractive, that of cooperative manager. Their ranks were constantly growing in numbers, while no specific training was necessary to get a prominent and well-paid position in this occupation-turning-profession. It seems that the emerging cooperative structures emulated the zemstvo hierarchy. Thus, the second-level associations of individual cooperatives (i.e., which usually united all the cooperatives of a certain district) offered their board members salaries identical to those of district zemstvo board members. For instance, in 1915, three board members of the Birsk Union of Credit Cooperatives received salaries of 1,200 rubles a year each, plus significant travel allowances. The chairman of the Rzhevsk Intermediate Association of Cooperatives was paid 2,400 rubles, while board members received from 600 to 1,800 rubles annually, depending on the importance of matters under their supervision.[48]

Finally, a limited number of positions were open to rural professionals in the sphere of academia. For many, it was less attractive in financial terms than fieldwork. During their graduate school years, students could receive a monthly institution stipend of 50 rubles or a ministerial stipend of 100 rubles at best, while their former classmates started their field careers as precinct or district agronomists with salaries of 150 rubles. After years of study, having defended extensive *magister* theses, assistant professors, who were no longer so young, would receive the basic salary of a district agronomist, while only a handful of full professors at the apex of their careers received 4,500 rubles annually.[49] True, many professors earned additional income by lecturing elsewhere, but in general only a devotion to scholarship drew young professionals to the sphere of academia, alongside an ambitious desire to become the intellectual and ideological leaders of their profession.

The agricultural specialists' booming job market together with the most "progressive" ideological climate in this relatively new occupational niche secured an unprecedented influx of women, contrary to the usual assumption that women were historically admitted first to the less prestigious and worst paid professions.[50] During the historic Moscow regional congress of agricultural specialists in February 1911, a resolution was adopted stating that female agronomists should be granted equal employment rights with men in the zemstvo agronomy organizations.[51] The prerequisites for this attitude within the profession had already been set in the 1860s, when the Petrovskaia Agricultural Academy, for the first time in the history of Russian higher education, admitted women to study along with men. The mother of Alexander Chaianov was among the first female graduates of the academy.[52] Together with Alexander Chaianov, in the ranks of the class of 1910, eight women (6.4 percent of all students) graduated from the Moscow Agricultural Institute.[53] In general, by 1914 women constituted 30.5 percent of all students receiving higher education in Russia, while in France their share was about 10 percent, and in Germany only 7 percent of all students.[54] In 1914, a total of 1,026 women were studying in four higher agricultural courses in Moscow, St. Petersburg, and Saratov.[55] By January 1, 1915, 3,560 women were studying in 62 (out of 341) agricultural schools of all types, except for mid-level colleges. Their share varied from 35 percent in agricultural institutions of higher learning, to some 13 percent in secondary vocational schools, to 42 percent in the permanent special courses targeting mostly peasants.[56]

The St. Petersburg Stebut women's higher agricultural courses were the oldest public institution of this type. In 1909, the budget of the Society for Assistance to Women's Agricultural Education, the founder and sponsor of the Stebut courses, exceeded 60,000 rubles—twice the budget of the pro-government Russian Grain society during the first years of its activity.[57] The courses' graduates were granted the same rights as graduates of state institutions of higher learning. During the first decade of their existence (1904–15), the Stebut courses registered 1,214 students, 514 of whom completed the full program of studies. Although only 33 of them received formal diplomas as agronomists, those who chose not to take qualifying exams went to work in agriculture-related professions.[58] The employment tracks of those Stebut graduates who provided information about their jobs reveal a remarkably high share of women occupying positions that granted significant professional authority and independence. Almost half of the respondents worked as agronomists and agricultural instructors (29 and 19 percent). Over 8 percent became teachers. The rest were employed in laboratories, experimental stations, or research centers, apparently as assistants.[59]

In March 1909, the founders of the Stebut courses complained that the recently founded educational institution "does not yet enjoy the broad popularity so necessary for fundraising."[60] Just a few years later the atmosphere around the courses was characterized by the professors themselves as a boom, demonstrated by the "intensified enrollments of the past two years" (i.e., 1912–13).[61] In 1914–15, only 330 out of 1,000 applicants were admitted to the Stebut courses, which suggests the courses' incredible popularity.[62] There was also a considerable demand in the society for women who had successfully completed agricultural courses even without getting formal diplomas.[63] The typical starting salary was at least 1,200 rubles a year,[64] which was almost unthinkable in any other occupation open to women.[65] In fact, as agronomists or employees of experimental agricultural stations, female specialists earned much more, and in some districts women accounted for 30 percent of all agricultural specialists.[66] They were paid the same salaries as men in the same positions, and there were no restrictions on female agricultural specialists' marital status.[67] In return, female rural professionals were expected to work on equal terms with men. One can imagine how intensive the job of female field specialists was from the account of the instructor of credit cooperatives, Zinaida Zhukova. From January 1 to September 30, 1913, she attended 35 agricultural cooperatives and in six of them conducted full-scale audits. During this period, she spent 75 days (27 percent of all the time) on the road traveling some 2,890 versts (1,916 miles).[68]

To finish the discussion of agricultural professions as a particular career path, let us look at the opportunities for creating "careers" that these occupations offered. Many contemporaries explained the painful problem of the "fluctuation" of the agronomist cadres in the 1910s by the desire of rural professionals to find better conditions of work.[69] Indeed, with the existing demand for agricultural specialists and the inequality of payment offered by different zemstvos in the early 1910s, it was only natural that many specialists looked for a significant improvement in their finances. That horizontal mobility was complemented by incredible opportunities for upward mobility. Young professionals, only one or two years after graduation from an institute, often took positions as heads of agronomist organizations of entire provinces, supervising hundreds of employees.[70] By 1914, out of 150 top agricultural specialists who directed government or zemstvo agricultural assistance at the provincial level, one-third graduated after 1905 (32 percent). Almost 40 percent graduated during the pre-1905 decade, and 29 percent had more than 20 years of working experience.[71]

In their twenties to early thirties, post-1905 graduates achieved the highest positions in their profession, which in the academic sphere would take at least ten more years. It is the existence of the phenomenon of the new generation of Russian intelligentsia and the generational character of society's turn to agriculture after the First Russian Revolution that explain the disproportional representation of young agricultural specialists at the very top of their profession. The professional solidarity of agricultural specialists reinforced that feeling of a distinctive community, which the demoralized traditional intelligentsia was rapidly losing after the defeat of the 1905 Revolution. Society's "turn to agriculture" in the years after 1905 vastly increased the social and economic status of rural professionals. Those who joined the ranks of agricultural specialists for idealistic reasons were rather unexpectedly rewarded by a broad social recognition of the values and deeds of the third element. This recognition closed the old gap between the "conscious intelligentsia" and the upper-class "qualified [*tsenz*] society" of the second and the first elements, and confined the idealistic intelligentsia's zeal for serving the common folk to purely legal professional activities.

The increased social and financial status of agricultural specialists and fine prospects for career growth changed the usual disdain of the intelligentsia for "lay success." For the first time in Russian history they were promoted and rewarded for serving the people, and more skillful specialists succeeded better than amateur agitators both in helping peasants and in improving their own situation. Thus the fast growing job market of agricultural specialists not only served as a powerful "attractor," "sucking in" the most active and capable elements of Russian educated society, it was also reconfiguring traditional boundaries within that society. While such episodes as the conflict between zemstvo and government agronomists or a veterinarian "conspiracy" against the agronomists indicated that the traditional intelligentsia ideological ethos did not disappear completely, the impact of professionalization on the agricultural specialists was immense. It compelled the intelligentsia-turning-professionals to think along more pragmatic lines and taught them to discipline their social thinking. At this level, society's self-organization took the form of redistribution of human capital in accordance with new patterns provided by both the new job market and new ideological preferences. Broadening public support for reforming the countryside through the professional expertise of agricultural specialists in zemstvo service succeeded in uniting and mobilizing traditionally conflicting elements of Russian society: populist-minded social activists, middle-class professionals, and wealthy zemstvo bosses.

The system of societal self-mobilization that emerged based on economic interests as much as on idealistic goals was a political force in a very broad sense, perhaps even broader than a tradition of new political history would suggest. It was political not so much by its *action* as in its *function*. This distinction is important in the context of Russian imperial society, where political action beyond subversive revolutionary activity was institutionalized only in 1906. Even then in the minds of many government officials there was little difference between the opposition parliamentary parties and radical underground revolutionary groups.[72] Under these institutional circumstances and within this mental framework, politics found other venues. In a society where belles-lettres was a sphere of politics par excellence,[73] everything had a political dimension and a symbolic political meaning: one's dress, haircut, or choice of profession. This type of political *gesture* could have little impact on the political situation in the short-run perspective, but at the other pole, equally insignificant factors affected the fates of the whole empire: a clique of favorites, an imposter visionary, or just a sovereign's whim. Between these two extremes lay an imperial society that was arranged into estates and orders by law but was constantly reshaping and rearranging itself to accommodate the new economic and political realities. The repertoire of political *actions* available to major social actors was limited, but the very constellation of social forces and their rearrangement directly influenced the distribution of authority and thus acquired a political *function*.

This point is best illustrated by the attempts of the government and conservative public figures to secure the regime's leadership in reforming the countryside vis-à-vis the public modernization campaign, as described in the previous chapter. The Progressivist ideology of the new generation of Russian intelligentsia as a loose yet distinctive social group, and the project of public agronomy as a vehicle for countryside modernization were explicitly apolitical and nonpartisan. The mobilization of a significant part of Russian educated society around this program, and the production of a certain intellectual know-how resulted in a virtual surrender of the government to the public initiative. The emerging civil society, or the *obshchestvennost'* in the language of the epoch, gained control over reforming the largest part of Russia as well as over the redistribution of resources. That was done mainly outside the sphere of politics proper, but it had tremendous political significance and thus performed a political *function*.

6
From Knowledge to Influence: Building a Bridge to the New Peasant

"The peasants of today" as future partners

The project of "public agronomy" aspired to no less than changing the techniques of land cultivation and the very mode of economic thinking of peasants, which implied that rural professionals should somehow be in a position to persuade peasants to change their ways. The authority of agricultural specialists, particularly of those in zemstvo service, was based primarily on their special knowledge and professional expertise. The ambivalent formula "knowledge is power" in translation into Russian requires a clarification of meaning and a choice of wording: "power" can be translated as either "authority" or "force." Similarly, in the Russian historical context of the early twentieth century, the most accurate way to apply the Baconian formula would be a somewhat different phrase: "knowledge grants influence."[1] This significantly limits the applicability of Foucauldian interpretations that suggest the almost unchallenged and unrestricted power of rural professionals to manipulate both public discourse and culturally discriminated against peasants.[2] While their special expertise did allow Russian rural professionals to have a certain political weight as a group and to interfere with the peasants' routine, they did not have any unquestionable institutional or even symbolic "power," that is "authority." Their "power as influence" was a result of social consensus involving many parties, and had to be negotiated on every occasion. The outcome of the attempts to close the gap between "educated society" and the peasantry by drawing the latter into the process of society's self-organization and self-modernization rested entirely on the ability of the agricultural specialists to generate a broad response to their initiatives from the peasants.

On November 15, 1909, a pilot issue of the magazine the *Peasant Cause* (Krest'ianskoe delo) was published in Moscow. A private enterprise, this non-partisan magazine "for peasants and rural intelligentsia" featured among its editorial board members all of the prominent liberal Moscow professors, including almost the entire staff of the Moscow Agricultural Institute.

These people united to address the peasants not as potential rebels (as radical populists did), or as humble and industrious supporters of the regime (as conservatives imagined them), but as an economic group with interests of its own. In that pilot issue, Alexander Minin, a young ideologist of the public modernization campaign and a typical representative of the new generation of Russian intelligentsia, attempted to explain in pseudo-folk style the essence of their modernization program to peasants:

> What matters is, therefore, not only wealth, but something else. This "something else" is knowledge.
>
> …And so, when it became clear that it was impossible to improve the peasant economy without knowledge, it was decided to send to the peasant this man, who would help him [the peasant] to improve his farm, who would give him knowledge, teach him how to do it. Such a man, a mobile teacher who knows agriculture well, is called an *agronomist*.[3]

This quotation captures the moment of an attempted translation to the peasantry of the intellectuals' own project of public agronomy, fashioned after the Italian mobile bureaus (cf. the notion of a "mobile teacher"). Yet, speaking to someone does not guarantee getting the message across. In this chapter, an attempt will be made to examine the interaction between the rural professionals and the peasants, and the possibility of establishing a dialogue between them.

The infamous cholera riots of the 1890s, when peasants murdered physicians suspected of causing the disease, and the refusal of peasants to let agronomists into their villages that we mentioned earlier—all contributed to creating the image of the dark and inert peasantry that dominated public discourse at the beginning of the twentieth century.[4] What, then, gave the rural professionals hope that peasants were able to understand their modernization sermon and to respond adequately? One way to deal with this question would be simply to dismiss the intellectuals' pessimistic characterization of the peasantry as unscrupulous discursive projections by manipulative literati, as some historians do (apparently implying that peasants were neither uneducated nor poor or conservative).[5] Alternatively, we should go beyond holistic categories and differentiate "the peasantry," this time not in terms of economic status or regional variation, but according to the degree of the peasants' integration into a larger society. The decades preceding the "agrarian turn" in Russia after 1905 and the experience of the abortive revolution and subsequent political demobilization resulted not only in the formation of a new generation of enlightened or liberal bureaucrats and the appearance of the new generation of Russian intelligentsia but also in the rise of a new generation of peasants:

… in contrast to medical services or public education, economic policies required a certain intellectual level on the part of peasants—they had to understand the advice given to them and be ready to assume the risk of innovations. At the beginning of the twentieth century, a new generation of peasants educated in zemstvo schools took central stage in Russian rural life, and zemstvo activists who organized these peasants into rural credit cooperatives found it possible "for the first time" to oppose an organized conscious nucleus to a benighted (*temnye*) unorganized mass.[6]

Schooling alone could not produce "a new peasant." According to the 1911 Zemstvo Survey, just 16.5 percent of rural teachers thought that schooling made their pupils socially conscious. Among those respondents, 216 teachers specifically stated that educated peasants had closer relations with the local intelligentsia.[7] We can estimate the number of representatives of "a new generation of peasants," who not only adequately responded to the modernization efforts of the agricultural specialists, but actively participated in this work.[8] These peasants composed the majority of voluntary correspondents of the zemstvo statistical bureaus.

In 1909, when the spiral of countryside modernization was just unwinding, the 14 oldest provincial zemstvos alone had almost 17,000 full-time village correspondents,[9] some 64 percent of whom were peasants.[10] Their task was to report a few times a year on the prospective and actual harvest, the prices of land, grain, its transportation, and, in some provinces, even the dynamic of local markets.[11] This alone made the voluntary correspondents the most economically conscious part of the peasantry, thinking in terms of market conjunctures and regarding agriculture as a phenomenon of "production" rather than an element of the traditional peasant way of life. The common zemstvo practice of rewarding the voluntary correspondents with agricultural periodicals and popular brochures only reinforced the position of village correspondents as the "outpost" of rural modernization. Those few tens of thousands of peasants, the voluntary correspondents, constituted the vanguard of the "new peasants" that actually had already been involved in a dialogue with agricultural specialists when the broad modernization campaign took off. They also constituted the bulk of the first peasant readers of agricultural periodicals.[12] It was they, whom the magazine the *Peasant Cause*, and dozens of other popular periodicals, addressed in the first place.

Indeed, it was very important to rural modernizers to sustain rapport with their readers, and especially with peasants, to be assured that the people, whom they were about to assist, were actually interested in modernization. Sections like "Questions and Answers" in periodicals regularly checked the "pulse" of intelligentsia–peasant communication, while special, usually annual, surveys of subscribers provided a complex sociological map of the readers. We shall take a close look at the audience of the *Samara Agriculturist* and the *Peasant Cause*, which represented two major types of agricultural periodical for peasants and rural intelligentsia: a local publication with a run of 1,000 to 1,500 copies, and a central (usually Moscow-based) periodical, publishing prominent agricultural experts, with a run of over 2,500 copies.

The *Samara Agriculturist* was published by the SOUKK with an average run of 1,300 copies. Subscriptions covered only half of the publishers' expenses,[13] so the magazine was a noncommercial publication with the primary goal of enlightening its readers. It regularly published replies to the peasants, who actually read the magazine and even wrote to the editorial board: 99 replies in 1911, 247 in 1912, and 415 in 1913.[14] The majority of subscribers lived in Samara province, but the magazine also circulated in neighboring Saratov, Simbirsk, Ufa, Kazan, Orenburg, and Astrakhan provinces. According to surveys conducted by the editorial board in 1911 and 1913, the majority of their readers (at least, their most active ones) were peasants. In 1911, 39 percent of those who returned the questionnaires were peasants, while in the 1913 survey their share increased to 54 percent.[15]

The Moscow *Peasant Cause* was a private enterprise, and hence could not afford a charitable attitude toward its readers, but with an average run of 3,000 copies it could sustain the same low subscription fee as the *Samara Agriculturist*.[16] Unlike the regional Samara magazine, the *Peasant Cause* was read all over European Russia. The map of its subscribers published in August 1911 showed that the magazine had the same number of readers in the central Moscow province as in distant Perm province (together accounting for 30 percent of all subscribers). There were equal percentages of subscribers in neighboring Kaluga province and in Viatka province, some 700 miles away from Moscow.[17] Among the readers who answered the magazine's questionnaire in 1911, 56 percent were peasants (which is close to the 54 percent of the *Samara Agriculturist*'s survey in 1913).[18]

The rural modernizers were very concerned about how comprehensible their publications were. Throughout the period under consideration, intellectuals issued recommendations on how to write for a peasant audience,[19] educational literature was thoroughly reviewed prior to being recommended for mass distribution among peasants,[20] and special lists of popular books on various aspects of agriculture were distributed.[21] As readers' replies indicated, the peasants well understood the intentions of the intelligentsia and appreciated their efforts:

> I am sorry—writes a peasant from Kherson province—that I am completely uneducated and cannot express what I think, and only thank and rejoice at those people who want equality and rights.[22]

While the Moscow *Peasant Cause* was viewed by the peasants as a kind of general sociopolitical publication, a source of information about the state Duma and current agricultural policy (despite many special articles on agricultural topics),[23] the regional *Samara Agriculturist* was read for its practical advice.[24] These were two different types of enlightening narrative, for peasants accepted the general character of reports on recent Duma decisions, but demanded utter specificity from the publications on how actually to improve the efficiency of their farms.[25] Many peasants even suggested that the *Samara Agriculturist* should publish articles written by other peasants who had implemented certain improvements because they did not think that examples from large estates were of much use to them.[26]

In general, despite the snobbish discursive projections of the rural intelligentsia and even of some better educated peasants,[27] the peasant readers of the agricultural press were able to react adequately to the appeal of rural professionals even in its most abstract written form. In fact, the actual number of peasant readers of the agricultural press was many times higher than the mere quantity of peasants-subscribers, because every issue was usually read by all of the literate and interested neighbors of the actual subscriber.[28] The agricultural press stimulated the "new peasants" to stick together, forming the nucleus of a peasant branch of the emerging rural public sphere.[29]

The modernization project of "public agronomy" with its stake on the network of zemstvo precinct agronomists reserved an auxiliary role for the agricultural press, while putting major stress on the personal educational efforts of rural professionals. Those efforts were institutionalized in two major forms: one-day village lecturing by a precinct agronomist or an agricultural specialist on a particular subject, and short-term (usually fortnight) courses taught by a number of specialists. Between the First Russian Revolution and the World War I, the funding for such educational activities increased almost 40-fold.[30] In 1913, some 1,580,782 peasants attended 43,763 one-day lectures in 11,762 villages.[31] During the same year, almost 100,000 peasants studied in 1,657 short-term courses, and in 1914, 2,500 courses were planned (because of the war, only half of them actually took place).[32] These impressive figures are important as an indication of the rural professionals' role in the mobilization of peasants, and as evidence that in a single year, some 12,000 agricultural specialists and their assistants were able to reach a very significant part of the peasantry.

Let us now go behind the screen of statistics accumulating the experience of thousands of rural professionals and millions of peasants, to look more closely at the phenomenon of short-term agricultural courses as a situation of their direct and sustained interaction. The week-long courses held in November 1912 in the village of Nadushite (Soroki district, Bessarabia province) demonstrated the ritual character of the first meeting of professionals-modernizers and peasants. By noon on November 15, about 30 peasants had gathered at the county court building—the only public place in the village suitable to host the courses. The village priest held a service, and then delivered a short speech about the usefulness of knowledge. He reassured the peasant audience that agronomists wished them well, and the courses were formally opened. Then came the turn of two agronomists, a veterinarian, and a zemstvo instructor of apiculture to introduce themselves as pedagogues. They began their educational work by explaining the necessity for agricultural knowledge, supporting their case by examples from the United States and other foreign countries. Over the course of a week, four specialists lectured for 23 hours on various topics, in Moldavian and Russian (depending on the audience), using slides and simple experiments (e.g., demonstration of a sprouted seed). Every day 50–60 peasants were present, but only half of this number attended all the lectures.[33] Thus, despite deep cultural (and often linguistic) differences, they finally met voluntarily in the local courthouse: rural professionals inspired by their science and foreign examples, and peasants who were encouraged by their priest. Nowhere in this scene were the state or zemstvo authorities visible, for the county court was more a peasant estate institution adapted to the needs of peasants' interaction with the larger society[34] than a locus of state power.

The importance of the voluntary character of peasant participation in the modernization efforts of the agricultural specialists cannot be overemphasized. Probably for the first time in modern Russian history, peasants eagerly reacted to the initiative of the educated elite to change their ways, which made their interaction a dialogue, and not the usual "discursive dictate" of modernizers over an "objectified class." In fact, the demand for short-term courses was greater than the agricultural specialists could accommodate.[35] Moreover, peasants displayed incredible enthusiasm about such abstract and complicated subjects as agricultural economics or cooperative theory.[36] This attitude testified to the broadening isthmus between the agricultural specialists and the peasants. Potentially, this could lead to the transfer of initiative in their dialogue from the educated elite to the peasant masses.

A well-documented story of the fortnight-long agricultural courses that took place in Elets (Orlov province) in February 1912 highlighted the complexity of the dialogue between agricultural specialists and peasants. Announcing the forthcoming courses, agricultural specialist Nikolaev explicitly stated their goal was to make the graduates "actual assistants to agronomists, and pioneers in agriculture," who would contribute to the economic development of Elets district, and Russia in general.[37] These fortnight-long courses cost the zemstvo 707 rubles, about half of a precinct agronomist's annual salary.[38] That was enough to teach, accommodate, and feed 55 students from all corners of Elets district for two weeks. The available statistics portray a typical student: in his early twenties, he was most likely to be already the head of a household. Graduated from a zemstvo primary school, he was a well-to-do member of a land commune living some 24.5 miles (39.5 km) from the site of the courses.[39] Yet, this relatively prosperous commune-minded peasant left his farm and made a several-hour-long trip to learn from city educators.

The sponsors of the Elets courses secured a number of written accounts from their former students. These peasant accounts can be deconstructed endlessly down to their syntax in a search for "real" peasant voices behind the layers of agronomists' propaganda, dominant political discourse, and the zemstvo education system. What is important here, however, is the peasants' willingness to enter a sphere that could not be described in their traditional mythopoetical language. They borrowed clichés and expressions from the modernized elite, but only those clichés and expressions that most adequately expressed their nonrationalized feelings.[40] The following are the most characteristic passages from actual peasant accounts (perhaps stylistically edited by the publishers):

> Upon our arrival at the courses, none of us students understood anything, but, like children, looked to everything in nature, and could not answer questions that seem so simple to us now.... It took a lot of effort from the gentlemen lecturers to overcome our devotion to the old methods of land cultivation. Nevertheless, they have reached their goal, and, by means of their knowledge supported by experiments, have proved to us the flimsiness of the old ways.... These courses are a new foundation of agriculture.... Lecturers with their knowledge work as masons; the zemstvo—as a foreman, and we—students—are the bricks, from which a new building will be built.[41]

> All of the lectures that I attended produced a very good impression on me.... It does not matter, if we forget all those sophisticated terms, they are not so important. The main thing, the basis of all notions, will not evaporate and dissolve in space, but yield a rich harvest.[42]

> Agricultural knowledge is necessary and relevant for the village. With the increase in the cost of living, old grandfathers' [ways of] tilling the land become awfully unprofitable these days.... The countryside is still a virgin land, which requires much work—and personally I say, rewarding work, despite our ignorance and loyalty to our grandfathers' customs.... True, there were few of us, only 50 people from the entire district. But all these people were really people of the land, who went to the courses not for entertainment, but to learn new methods of land cultivation, to get some knowledge of how to do agriculture.[43]

True, peasant reports mirrored the modernization discourse of the agronomists. For a while, some peasants were eager to perceive themselves as simple bricks in a magnificent building of intelligentsia modernizers, but they also had a rationale of their own to participate in the modernization movement.[44] The story of the Elets fortnight-long courses had its conclusion, which, at the same time, marked the beginning of a new process. Some eight months later, peasant Vlasov, who in his comments on the courses earlier that year had expressed a belief in "the rich harvest" of the received seeds of knowledge, wrote a letter to the Elets agricultural magazine about his experience in farming after the courses:

> When I returned home from the courses, everything appeared to me in a new light, as if had I just put the received knowledge into practice, I would become richer. But all this was hampered by a wave of the people's ignorance, and here we, with our knowledge, get into a dead end. Take Easter, for example. Very often, it coincides with spring sowing, as was the case in 1912. According to the old traditions, there can be no fieldwork during the Easter holidays. Meanwhile time is passing, the sun is burning fiercely and drying the soil…. Some of the peasants, including myself, began to work from the fourth day of Easter, under a storm of mockery. Had this happened 10–15 years ago, we, the cultural workers, would have been [severely] punished…. But everything in the world changes. We were just reproached that we did not believe in God, and for this, God would yield us nothing. But the results were unexpected: those who sowed during the Easter holidays, received 90 poods of oats from a *desiatina* on average, while those who waited until the end of the holidays received only 48 poods from a *desiatina*…. If one asks directly: where is the exit out of this situation, and how to teach the peasant population good farming? The answer will come by itself, namely—to educate them. It is also necessary to prove experimentally the advantages of scientific methods over archaic ones. Examples and experiments are very important. Peasants often distrust the lectures of agronomists in the villages and the books on agriculture. But when instead of their three haycocks from a *desiatina* they see 10 haycocks harvested by a neighbor, then mockery will yield to a serious attitude toward business.[45]

In this letter, we do not find many echoes of the intelligentsia modernizers. Every point made by peasant Vlasov is rooted in his own experience, reflections, and calculations. He even elaborated his own frame of reference, arguing against conservative traditions, supporting his ideas by quoting from a publication by the peasant Usachev, and referring to the agricultural specialists only indirectly, in instructing them how to improve their modernization activity. At this point, a former passive partner in the dialogue, the peasant, became equal with the modernizer. By violating communal traditions, the new peasants put at risk not only their reputations but also their welfare and even freedom.[46] Hence, the peasant's call "It's your turn, gentlemen *intelligenty*. Come help us peasants"[47] transcended the sphere of imitative discourse as it acquired quite a material response in the form of real professional assistance and actual improvement in crop yields.[48]

A partial decade of the public modernization campaign in the countryside is too short an interval to measure its success in terms of statistics, although the major trajectories of the changes in Russian agriculture corresponded to the plan elaborated by the architects of the agrotechnical reform.[49] While agricultural specialists often took credit for improvements in the productivity of peasant farms, their chief concern was to establish partnership relationships with peasants.[50] Obviously, the vision of modernization as a process of conquering enemy territory was not the unique fantasy of the young agronomist Ryshkin that was discussed earlier. That is why every report of a district or provincial agronomist organization to the annual zemstvo session had extensive attachments consisting of complete lists of peasants who had adopted this or that improvement measure, like registers of captured enemies. Thus, the widespread conviction among agricultural specialists on the eve of the war that "the peasantry is being swayed"[51] did not require statistical proof; every agronomist knew the names of all the peasants in his or her precinct who had introduced a multifield system of crop rotation or experimented with fertilizers.[52]

For agricultural specialists, the immediate goal was to mobilize the "new peasants" and to agitate as many tradition-observing peasants as possible. However, that would establish just an "outpost" in the midst of "enemy territory." To get complete control over the countryside, agricultural specialists needed more time and more peasants who were already educated in the spirit of modernization. That is why in the 1910s so much stress was put on the vocational schooling of peasant children. The system of education required a major institutional overhaul to suit a new modernization paradigm, including the revision of basic textbooks. Even "new peasants" educated in zemstvo and state schools were influenced by a "hidden curriculum" (Ben Eklof), which in many aspects contradicted the ideology of public modernization. Consider, for instance, a math textbook published in 1901, which featured typical schoolbook problems. A wine trader bought a few buckets of wine, diluted it with 9 buckets of water, and sold it at a profit of 51.5 rubles. A gentry landowner bought land for 27,000 rubles. Another fellow left 26,000 rubles to his heirs; one invested in agriculture and lost, while the other put his money in the bank and profited, which indicated that agriculture had no prospects. Problem no. 34 informed the pupil that in industry men earned five times more than women, while problem no. 222 discussed usury calculations from a lender's point of view.[53] In 1915, a special textbook was published, which put all sums in the realistic context of contemporary agriculture, both large-scale estates and small peasant farms. In the group of problems concerning life insurance, three people purchased insurance: a 37-year-old individual farmer; a 32-year-old village teacher; and a 25-year-old priest. The textbook also provided information about actual premium rates for other categories. When calculations involved crop yields, the textbook referred to the experiments of the Bezenchuk agricultural station, or Professor D. N. Prianishnikov of the Moscow Agricultural Institute. In other problems, individuals A, B, and C joined a creamery cooperative; a peasant purchased land through the Peasant Bank; while the State Noble Land Bank acquired a nobleman's estate.[54]

Peasants, even in remote districts, demonstrated enthusiasm and interest in education.[55] On the eve of the World War I, annual growth in the number of agricultural schools reached 5 percent.[56] In 1912, the Statute of the Popular Agricultural Schools laid the foundation for a truly mass agricultural education. It oriented the curriculum toward physics and agricultural studies, while leaving only one lesson a week for "religious-moral talks," which replaced traditional Bible studies.[57] Another goal of the new type of agricultural school was to attract girls along with boys.[58] Quite surprisingly, the decades-long attempts of modernizers to attract peasant women to education were met by an overwhelming response in the 1910s from peasants of all ages.[59] Ben Eklof hesitated to attribute this phenomenon clearly either to the success of the school system or to changes in the peasant mentality.[60] For us it is important to stress that the dramatic increase in the education of peasant girls and women testified to the success of the dialogue between rural professionals and peasants, which constantly involved more and more categories of villagers.[61]

The early 1910s witnessed a radical change in the pattern of employment of peasants graduating from the agricultural schools. Early reports sent alarming messages, later repeated by historians, that educated peasants tended to abandon agriculture.[62] However, the data for 1913 show the opposite trend. Depending on the type of agricultural school, only 1–9 percent of the graduates left the sphere of agriculture.[63] A case study of the Belebei (Ufa province) primary agricultural school confirmed and clarified the picture drawn by the general statistics: while education expanded one's social and material status, the majority of agricultural-school graduates worked in the village as qualified assistants to the agricultural specialists.[64]

The phenomenon that we describe as the "new peasantry" was known to the villagers of the interrevolutionary period as the peasants "of today" (*iz nyneshnikh*). Here is a firsthand description of the peasants of today by the peasant S. Matveev: "They do not drink vodka. Good, ambitious managers, they are close-fisted, tough people, and they assess all phenomena in the world by their economic value. They look at the world with hungry-curious eyes. They are much involved in public activity, so to speak—'big shots.' They like to reason very much.... They do not know how to sing songs, for it does not suit them to sing songs, and they do not have songs of their own yet."[65] If Russian Progressivist-minded reformers of the countryside sought a new economic man in the village, they should have looked no further, as they actually found a partner for their modernization activity. Though hardly an easy object of ideological manipulation, a new generation of peasants eagerly utilized everything truly relevant and practical that agrciultural specialists could offer them. They were modern both in their cultural out-look and farm organization. A young precinct agronomist, Nikolai Pospelov (Ardatov district, Nizhegorod province), who graduated from an agricultural institute in 1912, was excited to discover that all six farmstead owners in his precinct had introduced intensive systems of crop rotation and regularly read agricultural literature, and one of them even kept a diary.[66]

In 1914, the government inspector of agriculture of Kharkov province, Sergei Teitel, an 1893 graduate of the Petrovskaia Agricultural Academy, for-mulated his vision of the current state of relationships between the rural specialists and the new peasantry:

> Even not so long ago, it was the agronomist who was seeking the farmer…while at present, we witness the opposite trend: the farmer is seeking the agronomist to consult about various economic problems, but mostly to settle doubt: "should he break with the old habitual order of farming?" Such requests are addressed by individual farmers…and commune members, who have also realized the impossibility of keep-ing with old ways…. This decisive economic change can be called a critical moment in agronomist activity…. The psychology of the farmer and the agronomist is changing: previously, the farmer was in doubt, and the agronomist generously and firmly gave general useful directions, and nowadays it is just the opposite.[67]

This optimistic assessment of the situation was supported by a general trend of intensification of peasant–professional contacts. In 1902, the Ufa provincial zemstvo distributed a questionnaire among the voluntary peasant correspondents of the statistical bureau to find out how much they knew about the activity of zemstvo agronomists. From 43 to 66 percent of the respondents, depending on the district (52 percent on average for six dis-tricts), knew nothing about the existence of zemstvo agronomists and the nature of their activity. In 1910, when Ufa agronomists were just beginning their wide-scale activity, 815 respondents answered the same questionnaire. This time, only 13–29 percent (18 percent on average) had never heard about zemstvo agronomist activity. Only 17 percent of the respondents in 1910 mentioned cases of peasants' consulting with agronomists on their own initiative; 88 percent of those who received an agronomist's advice were satisfied with the results achieved.[68] From what we know about the Ufa agronomist organization's activity, in 1913, the same questionnaire would have revealed a greater degree of peasant awareness about agricul-tural specialists and their services. In 1913, voluntary correspondents of Samara province answered a similar questionnaire distributed by the provin-cial zemstvo. On average, 92 percent of the respondents knew about the existence of an agronomist in their locality. Only 22 percent of those who answered the questionnaire had never seen an agronomist working in their village. In 82 percent of the cases, an agronomist assisted a respondent upon request, and 84 percent of all respondents were satisfied with that assistance.[69]

The Samara survey not only testified that peasants had become partners in dialogue with rural professionals but also indicated that peasants were an active part of this dialogue, initiating four out of every five contacts. The rising activity of peasants imposed on the agricultural specialists a burden of responsibility for the results of their expertise. Previously, only the state could evaluate the level of specialists' training, while their professional performance was controlled by the zemstvo or other employing agency. Therefore, it was a genuine shock to some rural professionals when peasants began to demand effective and reliable services from them instead of simply being grateful for receiving pieces of scholarly wisdom. Also in 1913, the peasant Nal'chenko sued a young zemstvo veterinarian, S. Mikhailov, for 200 rubles to compensate for a stallion who died during an unsuccessful castration by a veterinarian assistant in the presence of Mikhailov. This, and a few similar incidents, made the veterinarian Mikhailov very uneasy about performing his duties and retaining his authority as an educated modernizer in the countryside: "In performing a castration after that, you tremble in the hope of a positive outcome." Veterinarian Mikhailov learned the reciprocal nature of dialogue the hard way.[70] The next stage in the peasant–professional dialogue would involve peasants uniting in associations to implement or challenge the recommendations of countryside modernizers.

Rural modernizers and the politics of peasant cooperation

While the progress of the changing economic rationality of peasants cannot be estimated quantitatively directly,[71] the mass cooperative movement that sprang up after the 1905–07 Revolution may be viewed as a direct result of the growing socioeconomic initiative of the peasantry and as an indicator of such progress. For almost half a century after the Great Reforms of the 1860s, the Russian intelligentsia attempted to introduce the idea of cooperative economic associations to the laboring classes. With few exceptions, all those attempts were dictated by ideological concerns, not by pragmatic economic considerations.[72] In fact, a quite naive moral economy was the only type of economic discourse mastered by the traditional intelligentsia. Hence socialist utopianism was their way of treating both micro- and macroeconomic problems, from high interest rates charged by village usurers to the land hunger of the peasantry.[73] All of these early experiments with rural cooperatives failed, because peasants did not understand the ideological message of the educated dreamers and could not grasp the economic benefits of the cooperative business to their predominantly self-sufficient family production units.[74] It would take an absolutely different socioeconomic situation in the countryside and new types of economic thinking in all quarters of Russian society to make the cooperative idea work.

The spread of all types of cooperatives after the First Russian Revolution was as intensive as it was unexpected. As late as the winter of 1910, a prominent cooperative activist, Vasilii Khizhniakov, did not expect the cooperative movement to take off in the near future without state and zemstvo assistance.[75] For a while, seasoned veterans of the cooperative movement thought it was just another short-term fashion that would disappear soon without a trace, as had often happened before.[76] However, this time the cooperative movement was not relying entirely on the enthusiasm of intelligentsia activists, as in the nineteenth century. While the proliferation of credit associations could be explained by government financial and political intervention,[77] the skyrocketing rise of consumer cooperatives testified to the emerging economic initiative of the masses. The statistics of consumer cooperative associations registered from 1865 to 1911 point to the countryside as the main driving force of the cooperative movement after 1905. Almost 84 percent of all rural consumer societies emerged during the six postrevolutionary years. Of all existing town cooperatives, only 48 percent were founded after 1905, while consumer societies uniting industrial workers or civil servants experienced a more or less sharp decline in the tempo of growth.[78] In absolute figures, in 1906–11, 4,807 rural consumer societies were registered, 6.3 times more than all other types of consumer cooperatives combined.[79] Other categories of rural cooperatives also mushroomed during this period.[80] Cooperatives were the basic interface of peasants' active (conscious and voluntary) integration into the national economy and society. By 1917 the cooperative movement had established itself as a powerful socioeconomic factor,[81] a distinctive segment of socioeconomic reality and a frame of reference of its own. In 1915, the agricultural machines and tools plant of A. L. Shklovskii's heirs in Elisavetgrad found it appropriate to add the following slogan to its advertisements: "The supplier of Cooperative Unions, Moscow People's Bank, and Zemstvos" (as the producers of Smirnoff vodka and other goods proudly announced that they were "suppliers to the royal court").[82]

The dream of several generations of educated Russians seemed to become reality: the people finally "woke up" from their passivity, which had been interrupted periodically by equally unintelligible outbursts of violence, to enter a conscious dialogue with their socioeconomic environment. But were the Russian intelligentsia actually ready to recognize the initiative of peasants and workers as their equal partners in the business of peaceful modernization of Russia? The new Progressivist stake on the inducement of public initiative by rational managers-organizers opposed the still influential paradigm of the classical radical intelligentsia of the older generation: "the heroes and the crowd." In the sphere of party politics this opposition has been interpreted as a major programmatic divide between the Menshevik and the Bolshevik factions in the Russian Social Democratic Party.[83] In the economic sphere, the situation was more complicated, as the leadership in the cooperative movement was not secured automatically, even for the professionalized intelligentsia, by the mere fact of their role in the production of meanings and interpretations of reality. As intellectuals discovered to their surprise, their role in the proliferation of agricultural cooperatives after the First Russian Revolution was niggling:

> There was a time, when all progressive layers of Russian society were enthusiastic about the idea of the *artel'*—our popular form of cooperatives. Yet, it is remarkable that exactly at that time...the cooperative movement did not take root in Russian soil.... Then, the intelligentsia cooled toward cooperatives and almost forgot about them.... However, while educated society lost any interest in cooperatives, it was awakened in the masses.... [The p]opular mind was working, and the more the consciousness of the masses grew, the clearer the necessity for self-assistance became by means of...cooperatives.[84]

Contemporary observers attributed the sudden success of the agricultural cooperative movement to the fact that after the defeat of the 1905 Revolution, peasants realized that it was the only way to improve their economic situation.[85] The actual development of peasant cooperatives contradicted many theoretical premises of the intelligentsia ideologists. For instance, specialized cooperatives focusing on a particular operation had proved to be more vital and effective than the universal associations envisaged by the early ideologists of cooperation.[86] Rural modernizers were ready to sacrifice their image of infallible experts rather than condemn the "mistakes" of the long-awaited peasant economic initiative. In fact, all the blame for the poor performance of the first cooperative associations was now put on their intelligentsia organizers:

> The organization of cooperatives in the [18]60s had the character of a pure intelligentsia enterprise. This means too broad a way of posing questions, the substitution of initiative with charity, the ignoring of details, which constitute the essence of [any] institution. Finally, everything was decided at the top, and not by the members of cooperatives themselves.[87]

Thus, the desire to keep up with the rising popular cooperative movement compelled part of the intelligentsia to distance themselves from the tradition of ideological patronizing of the people. In fact, those who preferred to associate themselves with the new type of social and economic thinking attempted to rewrite the traditional history of the intelligentsia in order to find in it the roots of the new attitudes.[88] At the same time, the everyday practice of the cooperative movement was proving that the traditional intelligentsia dilettantism was mortally dangerous to the movement and to its amateur intelligentsia leaders.[89] The ascent of a new generation of Russian intelligentsia was greeted as the answer to the chronic inability of the intelligentsia to lead the nation's economic renaissance: "That 'versatile worker,' that 'talented Russian nature,' who can do everything, is expelled now, and the European-type specialist, aka the public activist-citizen, has come to replace him."[90]

In the early 1910s, M. L. Kheisin and P. A. Sadyrin represented the most sober group of intelligentsia ideologists of cooperatives, who stressed the priority of pragmatic economic goals for the cooperative movement over ideological principles, and the role of the intelligentsia as qualified specialists rather than ideological masterminds.[91] At the same time, there were many cooperative ideologists who saw no need to adjust the attitudes of modernizing intelligentsia toward the people. These intellectuals perceived the rise of the mass cooperative movement as a splendid opportunity to use it for their own ideological purposes. V. V[orontsov] and especially M. Tugan-Baranovskii were among the most influential leaders of the "ideological" wing in the Russian cooperative movement. Neither age nor educational background distinguished these people dramatically from the pragmatic cooperative ideologists, but rather a very peculiar type of thinking, so characteristic of the Russian radical intelligentsia.[92] They preached the cooperative movement as a "third path" of social evolution, an alternative to both the capitalist economy and the socialist project.[93] This vision of a future "cooperative republic" was very much influenced by the writings of Charles Gide, the leader of the so-called Nime, or cooperative economic school, and a prominent figure of the French reformist movement. In his books, Charles Gide described a future democratic society as a world of cooperative producers and consumer associations acting in a rationally regulated economy, which virtually eliminated the figure of the middleman. While Gide himself put a major stress on proving the vitality of a small-scale business in the age of monopolies, his Russian readers were more interested in the theory of a distinctive "cooperative" socioeconomic system.[94] Virtually all intellectuals conceptualizing the cooperative movement paid tribute to the ideology of "the third path," but only the most fanatical representatives of the traditional, ideologically minded intelligentsia subordinated economic rationality to theoretical daydreams. Those ideologists attributed to the cooperative movement the militant goal of undermining the existing "capitalist" system and winning over the socialists.

Cooperative practitioners did not want to be pawns in the game of the intelligentsia dreamers. Having found themselves under increasing pressure from the ideological "generals," many local cooperative leaders staged a propaganda campaign of their own, discrediting the role of the intelligentsia in the cooperative movement. The reaction of the broad cooperative public to the intelligentsia was, in general, favorable, yet scarcely flattering.

A contributor to the Petersburg magazine *Splotchina* compared the intelligentsia to a poisonous medication: "one organism can sustain this dose, another one—that dose. The same is true of cooperatives: in some cases, the intelligentsia is much needed, in others it is even harmful."[95] With certain reservations, the local press tended to advocate the usefulness of the intelligentsia.[96] A very subtle sociological observation was provided by a certain Travinov in the pages of the magazine *Ural Cooperator*:

> There is no "intelligentsia" as a single phenomenon. There are different groups of intelligentsia with different attitudes toward cooperation, and there are individuals of various social standing energetically contributing to the development of cooperatives.

> ...With the advance of the cooperative cause, with the growth of the movement, one has to revise values that have been taken for granted. Rural cooperatives are becoming a broad popular movement, and the question of influence over cooperatives in the countryside becomes a question of influence over the toiling village population.

> It is certainly true that the intelligentsia, in general, is an organic part of the popular organism, but some groups of the intelligentsia...became an absolutely alien body to it.... [The cooperative movement] must produce its own toiling intelligentsia.... Initiative and self-assistance of the popular masses are the basis of cooperatives, and there must be no place for patronizing by alien elements, hence the intelligentsia who do not originate within the laboring masses can be only incidental to the cooperative movement.[97]

These ideas expressed by Travinov were more than a mere echo of the contemporary debates among the Social Democrats about the problem of a genuine worker intelligentsia. The rise of the cooperative movement opened the way for many members of the lowest social strata to important positions in cooperative business, which taught them to think in general socioeconomic categories and see things in a broad perspective. A vast group of cooperative managers attempted to counterbalance the cultural and political hegemony of the intelligentsia ideologists. This emerging "fourth element" of Russian rural society was suspicious of the other three "elements," including the agricultural specialists of the "third element," to the dismay of the latter. Cooperative activists could rely on the ever-growing cooperative press for expressing their own position, but they still remained within the larger public sphere structured by different types of cooperative ideologists. The duality and ambivalence of the cooperative movement was reflected in the booming cooperative press.[98] Unlike professional periodicals for agricultural specialists, and similar to the "thick journals" of the classical intelligentsia, all periodicals dedicated to general and theoretical aspects of the cooperative movement established by the ideologists of cooperatives were located in the capitals. At the same time, the burgeoning of magazines and newspapers dedicated to the cooperative movement was sustained by the periodicals published by local cooperative associations in the provinces, which discussed practical issues of everyday cooperative life. For every periodical title published in Moscow or St. Petersburg, there were about ten titles published elsewhere. This discrepancy reveals two different streams in the cooperative movement: a broad movement driven by the economic initiative of the population assisted and directed on a local scale by self-propelled activists and rural professionals, and a more elitist ideological current among the intelligentsia, which found in cooperatives a vehicle for political domination. It is scarcely accidental that no major journal dedicated to cooperative theory was published in the provinces, where the everyday "reality check" made the highbrow discourse of cooperative ideologists inapropriate. Unlike agricultural professional periodicals, provincial cooperative publications were devised not so much to educate their readers as to express their actual thoughts and feelings. Usually unprofitable and unable to pay their authors, these periodicals were published by local cooperative associations, and entirely depended on reports written by ordinary members of cooperatives. Rather than instructing the intelligentsia about how to write "popularly," cooperative periodicals encouraged common cooperative members to write, "do not be afraid of clumsiness and inexperience, what is important is your information, not the beauty of your language."[99] And they wrote,

> Timidly I write this message; but I cannot keep silent because of the feeling of joy that I experience while reading my own magazine, which can very much benefit me and my association; timidly I write this message, because I cannot put on paper what I have in my head, for I am a common peasant, my education is a popular school, but it was my lot to organize the Sviatotroitsk credit association, Anan'ev district, which began its operations in 1905.... In conclusion of my simple letter, I will say a sincere thank you to our teachers in cooperation, gentlemen Inspectors of Credit Cooperatives.[100]

Another man expressed his understanding of the cooperative movement, which, in his mind, quite comfortably accommodated democratic procedures, popular economic initiative, and the autocratic political regime:

> We sent our representatives to regional congresses, so after deliberating together they could intercede with the proper persons to raise our economic well-being. At least, we see the results of its [sic] efforts. We have what is most important for us—our own magazine, by means of which we can unite for further work benefiting our agriculturist-toiler. Gentlemen

Comrades! Do not forget that our Father Tsar has permitted us to unite and establish associations, the Government gives money from the State Bank, and our kind gentlemen inspectors stop at no difficulties to point out and direct us on the [right?] track—how to organize, run [cooperatives?] and use the rights given to us. Gentlemen Comrades! I am writing an article for the first time, and ask you to excuse if anything is not smooth; do write yourself also about everything that troubles your soul with the burden, while I write and rejoice—hurrah![101]

Quite in line with the general scheme of "public agronomy" modeled after the project of Italian mobile bureaus, rural cooperatives further emancipated peasant subjectivity, providing peasants with the social framework and economic grounds for self-organization and self-expression. This emerging new peasant subjectivity differed from the ideal schemes of educated modernizers, as it naturally relied on the familiar repertoire of social practices. The new "cooperative" culture was a hybrid phenomenon in more than one sense, bridging the cultural and economic isolation of different social groups and making elements of trans-national modernity "indigenous."

This hybridity is revealed in such symbolic events as local cooperative congresses and festivals. In the early 1910s, a tradition of annual cooperative festivals was developed in many regions of Russia. One of the first cooperative festivals was organized in Ekaterinburg on November 26, 1912, to celebrate the fifth anniversary of the first store opened by the Ekaterinburg Consumer Society. The festival took the form of a typical corporate celebration: the Consumer Society rented the city Commercial Club, which could accommodate 700–800 people (about half the society's members); entry was free but reserved for those who had obtained guest tickets far in advance. A usual program (songs, dances, and literature readings) was interspersed with speeches on various aspects of cooperative movement. To stress the special cooperative nature of the event, the club was decorated with the symbolic figure of a young worker holding a torch in one hand, and a sheet of paper in the other. The paper read, "Love the future, strive for it, work for it, bring it closer, take it into the present day as much as it is possible." Exciting for the audience, these words did not carry any particular political message.[102] This festival with its five official speeches was partly an attempt of the top cooperative managers to compensate for insufficient ideological activity in the past, and thus to reaffirm their status as "conscious activists" and not mere businessmen.[103] Still, the intensive ideological pressure during the festival was not so much politically charged as it was implicitly commercially oriented. Even a lecture about the role of women in cooperatives, while advocating the abstract principle of equality, carried a practical message: housewives should neglect the clumsiness, distant location, and long lines in cooperative stores, and shop only there.[104]

Even less formal were the festivities accompanying the opening of the first store of the Central Agricultural Society in Kostroma on February 3, 1914. Cooperative solidarity was expressed not so much through official speeches or visual decor as in keeping everything "cooperative": "'cooperative' priest, 'cooperative' deacon, 'cooperative' choir, and so on. Peasants did not want city priests, alien to the cooperative movement, to perform the service."[105] Unlike the Ekaterinburg cooperative festival, the Kostroma celebrations were not confined to one designated locality. After the dedication of the new store, all participants in the ceremony went down the street to the "sober" teahouse, where they had reserved a special room to continue the celebration. "'People' and 'intelligentsia' joined here in a single whole, united by the same feeling." Cooperative leaders delivered speeches on the history of their association, the importance of the sober way of life, and the prospects of the cooperative movement that were rewarded with ovations. Then, the church choir performed the Orthodox hymn "How Glorious" (*Kol' slaven*), and repeated it once again upon requests from the audience.[106] Thus, ordinary members of cooperatives celebrated their communality, which was based on economic interests. The identity of those "new people, people-comrades"[107] (aka "gentlemen comrades") was rooted not so much in ideology, which was rhetorically addressed by their leaders as in the plebeian culture of socialization (of town commoners, in Ekaterinburg; of peasants, in Kostroma).

On the eve of the World War I, the second congress of cooperatives of the Ekaterinburg district (Perm province) gathered the majority of local cooperative managers, including those who organized the cooperative festival described earlier. This gathering gives us a snapshot of the social composition of the district's most prominent "people-comrades." The available statistics demonstrate that, despite overrepresentation of the city cooperatives due to the significance of Ekaterinburg as a cultural and industrial center, almost 87 percent of the cooperative managers attending the congress were of peasant background, and 57 percent resided permanently in the countryside. In terms of occupation, about 37 percent tilled the land, 19 percent were craftsmen, 15 percent served in private firms, 14 percent were professional cooperative functionaries, and 12 percent represented "rural intelligentsia" (teachers, agronomists, and priests). Most of them (41 percent) were in their thirties. This social and demographic composition suggests that more likely than not, an Ekaterinburg cooperative manager was a young man of peasant background and living in the village, who earned his living by manual labor.[108]

Of course, the cooperative rank and file were not completely immune to the intelligentsia's ideological propaganda that penetrated even into reference materials such as statistical surveys or calendars.[109] They were proud of being a part of something more important than merely a grocery store that saved them a few rubles a month.[110] Ideology mattered little in everyday practice, which was shaped mostly by sober economic rationality. Even the members of town consumer cooperatives, who were more exposed to the propaganda of cooperative ideologists, did not reveal much ideological enthusiasm on a daily basis.[111] Usually, it was the competition with petty village traders that provoked managers of rural cooperatives to raise issues of cooperative orthodoxy.[112] Among cooperative managers and rural professionals, there was no general consensus about cooperative policy toward village traders and whether a conflict with a competitor qualified as a sincere expression of ideological zeal.[113] In 1912, a certain Vystavkin asked the Elets cooperative magazine:

> A peasant commune allocated a land plot for a consumer association to build a store next to a merchant's store. Is it possible to construct a store on the allocated plot adjacent to the merchant's store, or is a certain distance required? If it is required, how big a distance is necessary from the merchant's store?[114]

This was a clear case of confusion about cooperative "political correctness" caused by intensive anticapitalist propaganda by cooperative ideologists. The editorial board did not understand Vystavkin's concern, answering that there was no special law regulating such a situation, and advising him to contact a local insurance agent about fire safety regulations.[115]

The All-Russian Cooperative Congress that took place in St. Petersburg in March 1912 recognized that specific cooperative consciousness and ideology emerged only at the level of provincial or even all-Russian unions of cooperatives.[116] The intermediate stratum of cooperative managers displayed more concern about the ideological purity of their enterprises than ordinary cooperative members, trying to emulate the great theoreticians of the cooperative movement (especially in towns).

The ideology of "Cooperativism" united people of very diverse political views (from naive monarchists to socialists) and did not correlate with any particular party program.[117] Radical parties declared their intentions of taking over the cooperative movement, but to the best knowledge of the police, socialist infiltration into cooperatives was insignificant. According to the voluminous dossier of the Kazan Provincial Gendarme Administration, in 1909–10 only about 0.1 percent of all cooperative members in Kazan province could be suspected of any kind of illegal activity. Usually, that meant opposition declarations in conversations or reading (but rarely distribution) of illegal literature.[118] Like the "third element," the cooperative movement revealed itself as a particular political force only in its interaction with other major social actors on all levels of the social hierarchy.

First of all, rural cooperatives became a new factor in local village structures of power relations. Although this subject has not been studied by historians, there is evidence that cooperative leadership was often a subject of conflict within the local village elite. In 1912, the inspector of credit cooperatives Zviagintsev complained that the Trubetchinsk credit association (Lebedian district, Tambov province) had fallen under complete control of the local county board. It was even located in the county administration building, which naturally scared off peasants from the cooperative (instead of 9,000–10,000 rubles in deposits typical of other associations, the Trubetchinsk association had only 1,000 rubles). It seemed that the most powerful figure in this cooperative was a member of its board, who had served as a county elder for 28 years.[119] This strange combination of the influence of a county elder with the unpopularity of the county administration among peasants may have had a logical explanation. The power structure of the Russian village was traditionally organized along interfamily relationships. The only hierarchical structure unquestionably recognized by the peasants themselves was the clan, or large family.[120] Some clans were considered more "noble" than others. It was this peasant nobility that determined one's election as county elder, rather than wealth or economic efficiency.[121] That former county elder had likely become a key figure in the Trubetchinsk credit association for the same reason he received a high village public office: he belonged to the local village nobility. Another case supports this hypothesis. In 1912, a consumer society was organized in the village of Cherkasy. E. Butov was elected chairman, M. Butov was a bookkeeper and a local wine trader, F. Butov was a member of the revision commission, and I. Butov was hired as a salesperson. Obviously, the Butov clan gained control over this cooperative, probably over the resistance of less powerful families.[122] On the level of intravillage politics, the "new peasants" had to struggle with communal traditions not only in the sphere of economic innovations but also for control over the emerging structures of self-modernization.

The initial reaction of the zemstvo to the rise of the cooperative movement was ambiguous. While greeting cooperatives for their role in the improvement of the peasants' economic situation, some representatives of the second element were concerned about the potential competition of the cooperatives with zemstvo-run institutions, such as small credit funds.[123] The third element constantly pressured zemstvos to recognize cooperatives as a kindred social institution.[124] Ever since the Kazan regional zemstvo congress in 1909, it was often argued that because the district zemstvo did not have direct contact with the local population, cooperatives substituted for the nonexistent county-level "small zemstvo unit."[125] Hence, it was argued, the zemstvo must have changed the established pattern of economic assistance to the population: "cultural-economic measures should be the business of cooperatives under the direction of zemstvos and with zemstvo subsidies; economic measures should be the business of cooperatives based on [zemstvo] credit and resources of the farmers interested in those measures."[126] The cooperative leadership did not want to follow anyone's directions, though they expected help from the state and the zemstvo.[127] By the mid-1910s, agricultural cooperatives and zemstvos had reached a state of parity, which expressed itself in a partnership when their relationships were peaceful,[128] or a bitter rivalry when cooperatives competed with zemstvos over resources, particularly during the war.[129]

On the level of everyday contacts, cooperatives interacted with the "village intelligentsia" and rural professionals. The First All-Russian Congress of Cooperative Activists in St. Petersburg (March 11–16, 1912) adopted a resolution acknowledging that "successful development and correct functioning of cooperatives is impossible without the free participation of local intelligentsia—rural teachers, agronomists, instructors, physicians, medical assistants, et al."[130] It took time to make the rural clergy an organic part of the cooperative movement and to transform them into "cooperative priests" and "cooperative deacons." At first, Russian religious authorities were against village clergymen's participation in the cooperative movement. For instance, on April 13 and May 23, 1907, the Holy Synod issued rulings prohibiting clergymen from leading credit associations. In mid-1909, these bans were lifted.[131] Cooperative propaganda often appealed to Christian ethical values and symbolism, which facilitated the acceptance of the cooperative movement by the Church. A telling incident took place in 1916 in Ufa province. A low-rank clergyman (psalmist) organized a village cooperative. Local traders signed a complaint to the Ufa bishop, Andrei, arguing that a clergyman had no right to combine his church duties with cooperative activities. Bishop Andrei resolved the complaint as follows: "I thank the psalmist for his efforts to help his suffering neighbor; it is our common Christian duty to help our neighbor in need…. Hence, let help to the poor be blessed by our Lord, while I call God's blessing upon the psalmist."[132]

For agricultural specialists, cooperatives were an important part of the project of modernization of the peasantry,[133] modeled after the Italian mobile bureaus. However, it was not until the early 1910s that agronomists found themselves confident enough to begin supporting the cooperative movement on a regular basis.[134] The role of agricultural specialists in the cooperative movement and the place of cooperative activity in the regular duties of agronomists were often discussed at provincial and district agronomist conferences and cooperative congresses. These discussions revealed a differentiation within the ranks of the agricultural specialists, though not so dramatic as in the case of cooperative-minded intelligentsia. The minority agitated for transferring all agronomist activities to cooperatives, while the majority insisted that they were specialists in agriculture, not cooperative instructors, and hence should work with cooperatives while not forgetting their original tasks.[135]

Agricultural specialists valued the high status that the third element had achieved too much to trade it for the role of the "fourth element," that is, the rising stratum of cooperative activists (instructors of credit cooperatives in state service, cooperative instructors in zemstvo and, later, cooperative service, and cooperative managers). The approach of the rural professionals-modernizers to the role of agricultural cooperatives was well summarized by the Chistopol' precinct agronomist (Kazan province) Konstantin Pyzhenkov, who graduated from a mid-level agricultural college in 1911. At a district cooperative conference in June 1914, he argued that while it was unwise to resist the wave of land settlement that was sweeping over the country, the duty of agricultural specialists was to keep the village from falling apart, by encouraging the cooperative movement, which was to replace in part the vanishing land commune. The conference adopted his theses in their original version.[136] Available statistics indicate that agronomists dedicated their efforts to those cooperatives that were of primary importance for the economic progress of the peasantry.[137] There was another side to the relationships between agricultural specialists and cooperatives: as of 1912, large cooperative unions began hiring agronomists on their own, thus multiplying employment opportunities for rural professionals and giving them a sense of working directly for the people. The phenomenon of "cooperative agronomists" became a reality. Apparently, the first position of a cooperative agronomist was opened in 1912, in the Eisk division of the Kuban region, offering a salary of 1,800 rubles a year.[138] In 1915, two agronomists of the Dmitrov district zemstvo (Moscow province) switched to the Dmitrov Cooperative Union, which itself was established with the assistance of the zemstvo.[139] In 1916, the Tersk cooperative union followed suit by establishing an entire agronomy department.[140] Peasants were becoming not only partners, but employers of agricultural specialists.

Just as the "public agronomy" project, the cooperative movement offered representatives of the traditional radical intelligentsia an opportunity to change old ways, only without years of professional training. The story of A. Gusakov, a Petersburg librarian turned Viatka inspector of credit cooperatives, is very characteristic in this respect. Until early 1913 he was socializing exclusively within the community of socialist parties' members, who always spoke with contempt about cooperatives. Out of curiosity, he decided to be trained as a credit cooperative inspector for two months. He was impressed by the atmosphere in the Administration of Small Credit. In his opinion, this agency was completely apolitical and oriented toward actual economic assistance to the population. In June 1913, he was sent to Viatka province as inspector of credit associations and he worked in this capacity until at least the outbreak of the civil war.[141]

Paradoxically, a more intensive involvement of agricultural experts in the cooperative movement took place not in the areas with more passive peasants and weaker cooperative networks, but in the regions of most intensive peasant mobilization and highest initiative. A 1916 survey distinguished four modes of interaction between the zemstvo agronomists and rural cooperatives, ranging from close integration to occasional contacts. The closer ties characterized provinces with the most active agricultural specialists and market-integrated peasants (Poltava and Moscow provinces), while low integration was typical of Kazan and Simbirsk provinces, with weaker networks of "public agronomy" and less intensive peasant economy.[142] The trend toward closer ties with rural cooperatives became particularly visible with the outbreak of the World War I and the increased mobilization of the peasantry and intensification of peasant economy. For young agronomists just joining the profession after 1914, the centrality of rural cooperatives in their activity was taken for granted.[143] It seems that the project of self-modernization through "public agronomy" in Russia was quite realistic in its ambitious design. Despite its explicitly progressive quality characteristic of "high modernism,"[144] "seeing like a civil society" made this scheme to improve the human condition the one to succeed in the short run.

7
At the Crossroads: Coping with Modernization as Routine

A conflict of the double legacy: Intelligentsia vs. professionals

Paradoxically, at the peak of their success on the eve of the World War I, the community of reform-minded agricultural specialists declared a crisis of their cause. It seemed that the wildest dreams of the ideologists of the movement were becoming reality: the corporation of rural professionals had been recognized by the state and society; the project of public agronomy as a vehicle for modernization in the countryside had evolved into an established and ramified institution; and a large proportion of the peasantry had been drawn into the process of socioeconomic mobilization and integration into the larger society. The more successful the implementation of all the points of their initial program, the more frustrating it was to realize: they did everything right, but no miracle happened. The program of "small deeds," even when undertaken on a large scale, did not result in a "big leap," particularly after only several years of its implementation.

The agronomist Vasilii Onufriev was probably the first to publically ackowledge the crisis during the 1913 All-Russian Agricultural Congress in Kiev. He distinguished four "external" and five "internal" components of the crisis, starting with contemporary government policy, still repressive in some areas, and ending with the unregulated relationships among various groups of rural professionals (agronomists, veterinarians, cooperative activists, etc.).[1] Onufriev was a typical member of a new cohort of agricultural specialists: he studied at the Moscow Agricultural Institute in 1908 but left before obtaining a degree and worked as a precinct agronomist in Kharkov province. It was not until 1910 that he submitted his graduation thesis and was certified as an agronomist of the first class, together with Ekaterina Sakharova, Alexander Chaianov, Ivan Barkhatov, and other members of the class of 1910. In 1913 he worked as a district agronomist in Ekaterinoslav province.[2] To him, there should be nothing new in the problems that he outlined, and certainly none of them had been worsening over the five years of his professional practice. What really changed was a new understanding that a more complex implementation of the modernizers' ideals would take more than one decade, and that the career of agronomist was a lifetime occupation in the village, rather than a short-term heroic mission to the "enemy territory." It was said that this realization divided all agricultural specialists into pessimists and optimists. The pessimists could not adjust their high expectations and ideals to the prosaic reality of village life, while the optimists, presumably, were willing to adapt themselves to the conditions of professional work in the countryside at any cost, even by sacrificing some of their ideals.[3] Some observers calmly concluded that the "agronomist crisis" was the inevitable result of too fast an expansion of the system of public agronomy. Hence, the crisis was viewed as a moment for changing gears in rural modernization, from the exciting but superficial propagandizing activity to the routine but more fundamental work of actual improvement of the peasant economy.[4] Others saw the crisis as a result of certain errors in the initial program of public modernization.[5] In either case, a new scenario of actions, new goals, and a new normative model of the rural professional were needed:

> It is necessary to change the atmosphere of exaggerated expectations, which has emerged around agronomy, formulating its tasks in a different way and elaborating a new order [of actions] and plans.[6]

Apparently, the "heroic period" of the public modernization campaign was over, half a decade of self-mobilization had exhausted the enthusiasm of the self-appointed modernizers, and the question of sustainability of the movement became urgent. This was a quintessential problem of the Progressivist-type reformism based on public campaigns for a certain cause. In fact, the fundamental task consisted in finding a new balance between the two major components of the public modernization project: the intelligentsia's ideological zeal and professional technical expertise.[7] The first option relied on the holistic humanist tradition of the Russian intelligentsia that was obsessed with the fear of any type of human "incompleteness" and "alienation," whether a "differentiation" within society or a "narrow specialization" of the individual. Regardless of party affiliation, social activism based on this worldview acquired a form of body politics, be it individual terror or an educational "going to the people" campaign: the conscious personality was seen as both the means and the end of social interaction, and no formalized routine (technology) was tolerated in the process. The second option relied precisely on rationalized routine as a way to compensate for the erosion of genuine enthusiasm as the main source of public mobilization.

The new generation of Russian intelligentsia combined traditional intelligentsia ethical ideals and a new appreciation for professionalism and technology. This fundamentally binary worldview is not just a sociological construct attributed to a group retrospectively, but a basic characteristic of individual personalities, as can be seen in the personal writings of Ekaterina Sakharova or Alexander Chaianov.[8] For a while, this dualism was reconciled by the escalating success of the movement and the inertia of the original enthusiasm. The perceived "crisis" probably indicated a moment when, in order to sustain its activity, the social group of rural modernizers faced a necessity to become less a group of "intelligentsia" and more one of "professionals."

Those who stuck to the orthodox version of public agronomy as merely a legal cover for intelligentsia service were ready to give up their recently acquired positions rather than to sacrifice the classical ideal of the agronomist, who "must be the ideological leader of the economic life of the population, he must be…first of all, an agitator [and only] then a technician, cooperative activist, and so on."[9] It was difficult to resist this holistic image dating back to the populist tradition, which many agricultural specialists inherited in their alma mater, the Moscow Agricultural Institute, from the founding father of Russian "public agronomy," Professor Aleksei Fortunatov. As late as the fall of 1913, Fortunatov opened his course on agricultural economy at the Agricultural Institute with the following verse:

> When a student for the first time
> Comes to a certain higher school,
> He must remember that in his new lot
> His typical characteristic will be—the *intelligent*.
> He should not strive for getting useful knowledge
> As his priority.[10]

No less clumsy in Russian than it sounds in a literal English translation, this verse reflected the conscious position of populist ideologists of "small deeds," who did not want the intelligentsia to be transformed completely into professional intellectuals and continued educating generations of students in this spirit.[11] As late as January 1914, the Congress on Agricultural Education in Moscow downplayed the importance of practical training in agricultural schools.[12] Following suit, even some pedagogues of provincial vocational schools tried to emphasize the priority of holistic intelligentsia education over professional training.[13] It was this determination of the patriarchs and some practitioners of public agronomy to cultivate universal agricultural specialist/intelligentsia activists that brought out the "agronomist crisis." Scores of young professionals found themselves upon graduation equipped with little more than naive enthusiasm and an array of theoretical concepts, which were of little help in confronting everyday practical problems. Some of them, such as Sergei Fridolin, experienced years of humiliation when they failed time and again in performing the role of universal "scientifically educated preacher." Others, such as Arkadii Zonov, who graduated from an agricultural institute two years later than Fridolin, were luckier or more talented, and for a while succeeded in playing the role of universal rural modernizer. (In 1913, the Viatka district zemstvo agronomist Zonov wanted to combine the functions of agronomist, teacher in an agricultural school, and informal leader of local cooperatives).[14] By the early 1910s, Sergei Fridolin had found his calling as a narrow specialist in animal husbandry, blaming his alma mater and its professors for not providing him with the necessary practical knowledge. For him, the agronomist crisis was a logical outcome of the faulty educational and professional program. People like Zonov or the agronomist Onufriev blamed dozens of "internal" and "external" factors for slowing down the triumphal march of public agronomy, admitting very reluctantly the inadequacy of the universalistic model of the agricultural specialist under the new socioeconomic circumstances.

The problem of insufficient practical training of agricultural specialists was by no means new. Already on the eve of the public modernization campaign in the countryside, in 1909, the influential conservative government specialist in agriculture, Konstantin Golovin (1842–1913), sarcastically concluded,

> The pupils of higher agronomy schools, mainly of the Petrovskii [Moscow Agricultural] Institute, are of such quality as to be suitable chiefly for the writing of semischolarly articles, and they scarcely promise profits for a landowner, who would ask them [for advice].[15]

Naturally, Moscow ideologists of "public agronomy" dismissed this criticism: Golovin represented an alien ideological camp, and the students were not supposed to work for the "profits of a landowner." At the same time, provincial educators, who were less affected by ideological considerations than their Moscow colleagues, argued in favor of more practically orientated agricultural training. The director of the Kazan middle-level agricultural college, Nikolai Il'inskii, advocated certain changes in the curriculum. Quite in line with the dominant intelligentsia sentiments, he was against the reduction of courses in history and literature, for they were indispensable in cultivating "a broad view on the surrounding reality, moral values, and flexible language," and compensated for the routine of special courses. However, he wanted a considerable expansion of the time reserved for practical training, at the expense of classes in math and German language.[16] A subordinate of Il'inskii and his opponent, the manager of the college agricultural farm, Nikolai Utekhin, went even farther. He argued that it was necessary to create a network of primary agricultural vocational schools, each specializing in a certain aspect of agriculture, eliminate the mid-level stage of agricultural education, and reform the Moscow Agricultural Institute into a pure academic institution. The primary schools would produce practitioners, while the institutes would concentrate on research.[17] I. Dolgikh, who began teaching at the Pskov agricultural college in 1909, a few years after his graduation from the University of Iuriev, changed the curriculum at the college on his own initiative. In 1910, he introduced special lessons on public agronomy for senior students. He also obliged them to read a number of professional periodicals, the proceedings of all agricultural congresses, and the materials published by the Moscow Agricultural Society and the Department of Agriculture. In terms of practical training, Dolgikh first introduced excursions to peasant farms. In 1913, amid the talks of the "agronomist crisis," he realized that there was a better way to give students real-life experience. As of 1913, college seniors participated in the annual budget studies. Each student had to examine two farms, which took from 40 to 80 hours. This work allowed students to learn the structure and functioning of a peasant farm in detail, while their attitude toward the peasant became more realistic and sober, as Dolgikh put it.[18] There were recurrent attempts to encourage agronomists to begin tilling the land as farmers, thus getting necessary practical skills and earning recognition as authorities in agriculture among the peasants.[19]

Besides the resistance of the old intelligentsia-minded educators, some students opposed the idea of enhancing the professionalization of agricultural training, wishing instead to retain the legacy of radical studenthood (*studenchestvo*),[20] often against their own practical interests. For instance, in 1912 a group of students of the St. Petersburg Agricultural Courses wrote an open letter protesting the transformation of the courses into an analogue of state institutions of higher learning. This prospective transformation, granting the rights of a state institute to the courses, would have benefited the graduates in terms of their chances on the job market, while imposing certain restrictions over admission requirements. Together with tightening the discipline of studies, those measures were supposed to raise the professional qualities of the students. The authors of the open letter did not want any of those benefits:

> In exchange for lentil soup from the government, the courses will be deprived of the freedom of creative initiative, which is so important in the interests of *obshchestvennost'*. Instead of [being] public activists, as the courses' students should be, they will become the civil servants of agronomy.... The ideological level of the courses will be significantly decreased under these conditions, and instead of friends and sincere... assistants of the people, the courses will produce individuals armed with diplomas but deprived of ideological stimuli.[21]

While being quite in line with the traditional ideal of "conscious" students' behavior, this demonstration of protest by no means represented the dominant attitude of students during the epoch of the "agronomist crisis."

On October 29, 1913, at a session of the student circle "Agronomy and Life" at the very same St. Petersburg Agricultural Courses, Aleksei Gern made a presentation on the topic "What Should an Agronomist Look Like?" The main thesis of his presentation resonated with the ideological component of the project of "public agronomy": the agronomist should be a public activist par excellence, and not a narrow-minded specialist, like other rural professionals.[22] In the immediate aftermath of the 1905 Revolution this statement would have been unanimously welcomed by the audience. In 1913, however, Aleksei Gern was confronted by all the members of the circle "Agronomy and Life," who were supposed to represent the most conscious students, preparing for their mission in the countryside. The majority argued that "now specialists are needed, not 'legendary heroes [*bogatyri*] of thought.'"[23] Finally, even the students at the Moscow Agricultural Institute, the stronghold of the opponents of professionalization, began expressing concerns over the necessity to improve their practical training.[24]

Thus, students faced the same problems as their elder colleagues, but unlike the agricultural specialists, who joined the ranks of rural modernizers during the big takeoff in 1907–10, agricultural students in the mid-1910s were free of many intelligentsia ideological stereotypes and knew precisely what skills they would need upon taking jobs in the countryside. The very role of student associations in the form of specialized circles and study groups suggested a higher degree of professionalization of the agricultural students in the 1910s in comparison with their predecessors. Before the First Russian Revolution, even professional student circles scarcely differed from clandestine political unions.[25] The postrevolutionary student body demonstrated a quite different attitude. Student professional circles proliferated and their membership skyrocketed,[26] while student periodicals reaffirmed the pure professional goals of the circles and their members.[27]

A similar transformation occurred with another important element of the intelligentsia public sphere—professional congresses. A contemporary observer wrote with surprise that neither of the two major agricultural congresses of 1913 displayed any signs of political agitation or ideological biases so characteristic of previous intelligentsia forums.[28] According to the *Agronomy Journal*, 1913 witnessed a stream of agricultural congresses at all levels:

> Apparently, we have here some more or less constant social current that has special goals, socioeconomic meaning, and content of its own.... Unlike in past times, we observe an increased differentiation of the public elements, social concerns, and demands of certain groups, which are gathered at different congresses. [T]he very proceedings of congresses now more often have a so-called "business" character...and appeals toward the organized [public] initiative become almost the major and central part of the work of all congresses.... Doubtless, we have become Europeanized

in this sphere as well, because all the mentioned traits indisputably indicate the demolition of the patriarchal system of relationships between the agricultural elements, and [a move] toward external influences.[29]

Although far from being complete on the eve of the war, the process of updating the initial project of public modernization had taken a clear turn toward further professionalization and away from its populist ideological premises. The dogma of preferential treatment of the "toiling peasant family farm" (*semeino-trudovoe krest'ianskoe khoziaistvo*) was renounced by these adherents to the Organization-Production School, who worked as field practitioners.[30] Even one of the oldest and most "conscious" zemstvo agronomist organizations, the Samara province precinct agronomy, could not resist the pressure of reality, and abandoned its former ideological procommunal preferences.[31] Occasional sluggish attempts to galvanize the discussion of a desirable agrarian reform resulted in a general admission that there was no political alternative to the project of technocratic social engineering.[32] Hence, what appeared to be palliative measures in the wake of the defeat of the First Russian Revolution became the mainstream on the eve of the World War I. Educated countryside modernizers gradually accepted the idea that only the routine service of specialists would change the Russian village in the long run.

A stake on technology

The most apparent way to sustain and enhance this modernizing routine was found in putting technology in the service of rural professionals. In fact, Ekaterina Sakharova as early as October 1911 came to this conclusion in her early attempt to resolve an internal conflict between the legacy of intelligentsia activism and sober professional analysis:

> We need a different technique, that is, *cattedra ambulante*. It is a wasteful aspiration to populate our 300 districts with agronomists. This is unrealistic in our conditions. It is dangerous to speak about it, but it is true. Moving museums, train car-museums … — that is the fast, feasible, and only way toward awakening cultural aspirations in the masses.[33]

She was referring to another element of the universal culture of Progressivist reformism, this time of American rather than European progeny. A decisive event took place on February 13, 1905, some halfway around the globe from central Russia:

> [T]he first "corn train" pulled out of the Rock Island Station in Des Moines, Iowa. The train consisted of three commodious passenger coaches for audience rooms fitted with charts and diagrams, a dining car,

and a sleeping car. The train ran on schedule, switching from the main line at predetermined stations. At each stop farmers, assembled by local farm implement dealers, were shown how it was easily possible for them to … cultivate the growing plants to produce a good harvest.[34]

This idea proved to be a success. In 1906 America, such trains were running in 21 states. The peak of the agricultural trains' popularity occurred in 1911, when 71 trains operated in 28 states servicing some 995,220 people. In 1914 there were only 34 trains in 17 states, servicing 474,906 people. In a few years this service was discontinued, as it was recognized that the "moving schools" had fulfilled their mission in raising farmers' interest in improved agricultural practices.[35] While in America the fashion for agricultural trains was fading, Russian agricultural specialists found the idea of a "railway school" an exciting solution to the structural deficit in modernizers' manpower. In 1913, the senior specialist of the Department of Agriculture, A. G. Garshin (who before joining the state service was the provincial agronomist of Chernigov zemstvo in 1901–10), submitted a project for an "agricultural institute on wheels," styled after the American "corn trains," to the director of the Southwestern railroad. At the same time, the board of the Moscow–Kazan railroad decided to launch an agronomist train of its own in early 1914, at the expense of the company.[36]

As a result, in 1914 at least three agronomist trains in Russia serviced some 100,000 curious farmers in Central Russia, the Caucasus, and Siberia.[37] The Moscow train consisted of six cars (two car-museums, two cars with samples of agricultural machinery, a car with animals, and a car-auditorium), it was lighted by gas and cost about 100,000 rubles.[38] The Siberian train was less fancy. As a matter of fact, it started on June 12, 1913, merely as a third-class train car attached to train no. 26 servicing the resettling peasants on their way from inner Russia. Almost half of the passengers on train no. 26 attended lectures by the train's agricultural specialists.[39] The Vladikavkaz train operated at night and was open to the public from 8 a.m. to 7 p.m., and one may expect that the other trains operated in a similar way.[40]

The outbreak of the war did not curtail the new means of disseminating agricultural knowledge but rather stimulated it further. In September 1915, 11 cars of the new Vladikavkaz train provided information about all major branches of agriculture.[41] The autumn trip lasted 66 days, the train was visited by 41,586 peasants, 40 percent of whom were women.[42] During the autumn trip of 1916, the Vladikavkaz train consisted of 14 cars and serviced 38,695 peasants. Its manager, V. Benzin, noted that the population became accustomed to the train and sought answers to concrete problems while attending its exhibits.[43] In 1915, the Moscow train enhanced its program, adding arranged visits to the train by wounded soldiers and lectures by the train's agricultural specialists in hospitals along the way.[44] In addition to the special "agronomist" or "agricultural" trains, during the war, specialized train cars and even floating agricultural exhibitions attracted tens of thousands of agriculturists.[45] Some agricultural trains continued their activity after the Revolutions of 1917, becoming even more practice-oriented and useful to the peasantry,[46] until the Bolshevik regime transformed this idea of technocratic agricultural specialists into an important technological device of the political dictatorship.[47]

Cinema was another technical innovation that became widely used in the agricultural education of the peasantry during the World War I. The SOUKK became a pioneer in this new enterprise, as its members realized that they could not match the network of precinct agronomy in servicing the peasantry without the assistance of modern technology. On the eve of the war, the SOUKK decided to assist rural cooperatives of the province in opening village cinemas as a means of education and organization of leisure time (in 1913, there were already two village cinemas in Samara province). To compensate for a deficiency of "scientific agricultural" films, the SOUKK purchased a camera to shoot educational films at the Bezenchuk experimental station. The director of the zemstvo agricultural museum, M. R. Sinitsin, who graduated from an agricultural institute in 1912, was sent to Moscow to learn the skills of a cameraman.[48]

The virtual boom of rural cinemas began in 1914, when film projectors were purchased by local cooperatives, zemstvo agronomist departments, and agricultural trains. By the end of 1914, zemstvos owned about 40 cinemas. Not all of them targeted the village audience (in Kursk, the first projector was purchased for the zemstvo psychiatric hospital) but the countryside was a priority.[49] The Ramen district zemstvo (Poltava province) purchased a Cock film projector to be used by agronomists during lectures. Usually, four films were shown at a time: one special agricultural, two general "environmental," and a comedy. Films were rented from the Kiev branch of the Pathé Frères company at the rate of 150 rubles for a year of weekly rentals. Film projectors were bought by village cooperatives in Volyn and Samara provinces.[50] Films about various aspects of agriculture were presented on the Moscow and Vladikavkaz agronomist trains.[51]

The rapid spread of rural cinemas was facilitated by the appearance of relatively cheap and reliable film presentation hardware on the market, and by a growing network of movie retail and rental facilities.[52] In 1914, the Imperial Agricultural Museum in St. Petersburg established a special cinema department, which collected films on various agricultural subjects and shot films of its own. All films could be ordered from the museum's copy facility. By summer 1914, the museum had 33 films on 11 topics, totaling 1,400 meters on land cultivation, 1,724 meters on gardening, 835 meters on apiculture, and so on.[53] Two years later, in 1916, the Russian division of the Pathé company alone offered 80 educational films on agriculture, covering all possible branches and operations, especially emphasizing technology and depicting individual stages of production.[54] In early 1915, 16 percent of the district zemstvos already had film projectors, while the Kineshma zemstvo had two, and the Bogorodsk zemstvo even had five.[55] In late 1916, Ekaterinoslav provincial zemstvo inaugurated a special financial scheme encouraging villages to establish cinemas of their own.[56] By the end of that year, three other zemstvos had built film collections of their own,[57] and agricultural societies followed suit.[58] As it turned out, cinemas were also very profitable, covering their initial investments in several months.[59] Curiously enough, this rapid growth of the rural cinema network was not accompanied by complaints of agricultural specialists about the purely entertaining role of film presentations in the countryside. Apparently, the preceding decade had changed the attitudes of peasants and agricultural specialists: the former grew accustomed to taking the activity of agronomists seriously, while the latter recognized the importance of peasant leisure (especially in the wake of the alcohol prohibition campaign).

It was only one step from the appreciation of the role of technology in education to thinking of it as a solution to the problems of peasant production. On October 8–9, 1916, the cameraman of the St. Petersburg Agricultural Museum, V. A. Vorotilov, filmed the demonstration tests of five tractors at the Bezenchuk agricultural experimental station.[60] This was probably the culmination of the process of technological enhancement of the pre-1917 campaign for modernization of the countryside; copies of that film would be presented in many dozens of village cinemas across the country, promoting a most complicated type of agricultural machinery. The tractor, another symbol of Americanized modernization, had captured the attention of agricultural specialists long before the war,[61] and the first experiments with tractors in Russia dated back to 1911.[62] After the war began, the peasants' demand for agricultural machinery sharply increased.[63] In the situation of the severe shortage of labor and draught animals in the countryside, the idea of introducing the tractor to large private estates and peasant associations became very popular. Indeed, when the price of a light tractor became equal to the price of three pairs of oxen, tractors appeared an attractive alternative to the traditional use of draught animals.[64] In 1916, the Association of Western Zemstvos decided to organize wholesale purchases of tractors from the United States for private landowners and cooperatives.[65] The Department of Agriculture also had an extensive program of tractor imports, while the strategic goal was to begin the production of Russian tractors.

In fact, production began on the eve of the February Revolution by the Gelferikh-Sade company with an initial production plan of 200 tractors per year, subsequently growing to 500 tractors per year.[66] In 1916, 263 tractors were ordered from America. Although only 63 of them had arrived by 1917, about 70 percent of the rest were already paid for by customers, months in advance of the tractors' arrival. The Department of Agriculture introduced a two-year payment plan, with a down payment of one-third of the entire sum (1,500 rubles for a light tractor, the price of a pair of oxen).[67] In 1917, it was expected that about 500 tractors would be received by Russian agriculturists; some 60 percent of all tractors were ordered by zemstvos and cooperatives.[68] This influx of machinery into the countryside would have been unprecedented and would not have been matched until the early 1930s with the Machine and Tractor Stations (MTS) campaign. Apparently, some form of the future MTS was organized already in 1916–17 by local zemstvos and procurement committees.

What is important for our study, however, is not this promise of a technical revolution in Russian agriculture on the eve of the Revolutions of 1917, but the change of mind that this prospective revolution had actually brought about in 1916 and early 1917. At this point, those agricultural specialists who believed that "thousands of tractors are needed right now, and the need for them will increase tenfold in the future"[69] were crossing the line between the modernization of peasant economic thinking by the community of fairly independent agricultural specialists and the nationwide program of technical "rearmament" of the village. Previously, the peasant household was seen as the core element and the yardstick of modernization. Now, with the advent of technology in the plans of rural modernizers, the peasantry would have to adjust to the technical requirements of land cultivation by tractor and to the economic requirements of national procurement agencies.[70]

There was yet another consequence of the prospective "tractorization" of the Russian countryside. This measure required thousands of qualified mechanics to service the expensive and complicated equipment right away, and they were actually trained in special courses.[71] In a few years, the growth of this new type of rural professionals could significantly change the social composition of rural modernizers, as tractor operators and mechanics were of peasant or proletarian background but were paid 1,100–1,500 rubles a year, as much as many precinct agronomists.[72] The job of a village mechanic seemed very attractive even to those members of the "village intelligentsia," whose children were studying in gymnasia, and hence could aspire to higher education and a good traditional career path. According to the priest Father Kamenskii, peasants had so many agricultural machines and tools in his relatively poor and small village of Slavianka, and there was so serious a need for a qualified man to service them, that he would like his son to be trained as a mechanic after graduation from a gymnasium.[73] As of 1913, the Saratov Courses on Agricultural Machinery alone prepared annually 30–35 mechanics, who easily found jobs with a starting salary of between 40 and 100 rubles a month. Peasants and other village natives enjoyed favorable conditions of admittance to the courses.[74]

While the advance of technology promised to the village youth an efficient upward mobility without leaving the countryside, another wartime phenomenon brought many hundreds of urban youth to the villages. In 1915 the first brigades of urban high-school students volunteered to help the families of drafted soldiers during the harvest. The rules of the "student labor brigades" stated that boys who are 15 years of age and older could work up to ten hours a day in the fields, while both girls and boys who are 12 years of age and older could be employed in horticulture.[75] The goals of the students' agricultural labor were twofold: to help the peasants and to rest in the countryside.[76] Accordingly, school students were initially prohibited from accepting payment for their work from the peasants.[77] Peasants could not grasp the sense of free work and regarded the urban students as "state reapers," who were sent to the countryside for their misdeeds. Eventually, peasants realized the usefulness of the school brigades, waited in line for the next available group, and paid them, despite the organizers' regulations.[78] For example, a brigade of 26 pupils from the two Saratov gymnasia and one vocational school went to the countryside on July 11, 1916. They began work on July 14, but it was not until July 17 when the demand for their service became overwhelming among the wives of drafted heads of peasant households. Zemstvo organizers spent an average of 69 kopecks a day per teenager, while he or she earned 3–4 rubles a day with a free meal from the peasant hosts.[79] In 1916, the St. Petersburg society Labor Aid sent 288 students to the countryside in 34 brigades: 20 agricultural, 13 horticultural, and one brigade of day-nursery service.[80] At the same time, the Don-Kuban-Terek Agricultural Society organized work for 229 students from local high schools. They assisted 230 families to harvest 3,996 *desiatinas* (about 4,400 hectares) of crops.[81] Two school brigades of 25 and 11 students worked in the Moscow district in 1916.[82] These and many other school brigades (complete statistics are unavailable so far)[83] were organized through the joint efforts of local agricultural specialists and zemstvos, town agricultural societies, and school parent committees, with financial assistance from the Department of Agriculture. In late May of 1917, the initiative of school brigades was supported at the highest government level when the Provisional Government addressed the "student youth" with an appeal to participate in summer agricultural work.[84]

This new campaign of urban youth "going to the people" had enormous potential for rooting the public modernization movement even deeper into Russian society, shaping a new public discourse on agriculture. Unlike the populists of the early 1870s and the first precinct agronomists of the early post-1905 years, these boys and girls from the school brigades learned the reality of peasant life not from revolutionary brochures and textbooks on agriculture, but from their own experience of working side by side with real peasants, ten hours a day. In a few years, they could have formed a new cohort of Russian intelligentsia that would have been characterized by a more sober and uninhibited attitude toward social reality than the previous generations. The revolution and the long civil war dramatically interfered with this trend, although without changing its ultimate outcome: the demystification of the peasantry in the imagination of urban educated society, and the acquiring of firsthand experience in the rural economy.

This educational effect of the new intelligentsia "going to the people" (not as teachers this time, but as wage laborers) compensated for the "mechanization" of the agricultural specialists' assistance to peasants. As we have seen, agronomist trains, educational films, and advanced machinery greatly enhanced the influence of rural professionals over the peasants and made this influence more sustainable and far-reaching. The gain came at the cost of losing the intimate interpersonal relationships between the modernizer and the peasant "pupil" or "partner." Economic rationality overwhelmed the sociopolitical agenda. But this formalization and institutionalization of professional counseling occurred in a changing sociopolitical context. Technology (including new techniques of urban labor mobilization to meet village needs) also created new venues for social interaction, drawing new categories of educated Russians into direct contact with peasants. Thus, as the spiral of socioeconomic self-organization unwound with ever greater magnitude, the peculiar politics of modernity evolved. It found elements of its institutionalization in such modern phenomena as technology and the rationalization of labor resources management, while also acquiring a more systematic approach. Public modernizers had succeeded in mobilizing the peasantry—the next logical step was to move from rationalizing the culture of production to changing the structures of production.

Part III

Patterns of "Nationalization"

8

Nation as Motherland

The ambivalence of nation in the imperial situation

Not unlike their counterparts in the West, Russian Progressivist modernizers reached a point when the further advancement of social reform could be secured only by institutionalizing a broad public movement at the national level. They went through the stages of sacrificial amateur enthusiasm, expert counseling organized through professional corporations, and technologically enhanced mass-scale social engineering. In view of the gradual rapprochement between the economically concerned branches of the "government" and the "public," traditionally anti-statist rural professionals faced the challenge of stabilizing their relationships with the state and, probably for the first time, clarifying to themselves and their counterparts what sort of greater community they were building. The logic of the public reformist movement implied a shift in the politics of modernity: the "apolitical politics" of indirect pressure through manipulating the public discourse on agriculture and winning over the support of the zemstvos and some government agencies gave way to direct claims for political representation. The answer to the crisis of the initial modernization project was found in the transformation "of the entire army of agronomists, and each of its members" into "citizens enjoying full rights, who in addition to their general responsibility have a special [responsibility] in the realm of [their] public and professional activity."[1]

It is tempting to describe the desired sphere of citizenship as a national community, if only the notion of "nation" had any conventional meaning at the time we are talking about (or even today, for that matter). One hundred years ago, in the multiethnic and multiconfessional Russian Empire, before major theories of nationality became a topic of college textbooks, there was a much wider consensus on what "socialism" or "cooperativism" meant than on what "a nation" was. The pan-national phenomenon of *obshchestvennost'* operated predominantly using the Russian language and within the political borders of the Russian Empire, yet it was not "Russian," in the sense of a then rising Russian nationalist movement, or "patriotic," in the sense in which the government would like to see it. The more or less persistent demands for the advancement of various nationalist causes were perceived as a part of the progressive modernizing package, and to a certain extent were hardly separable from the political and economic claims of this package.

So far, the story of Russian public modernizers could be interpreted as a peculiar variation of the broad culture of Progressivist modernity: the one based on a specific local intellectual tradition and staged in a semi-autocratic country, but otherwise quite imaginable in any European or American country. The attempt at institutionalization of the reformist movement in some version of a national community added an additional dimension to this story, the one that made it a truly unique Russian experience. The initial project of "public agronomy" and the social movement that grew out of it were staged in the imperial situation. The formal historical status of Russian "empire" is less important for the topic of our study than the analytical model of the "imperial situation" as formed within the new imperial history of Russia.[2] The imperial situation is characterized by the principal heterogeneity of the social space and the absence of a single and universal frame of reference (be it a universal juridical sphere, a cultural canon, or an integrated "national" economy). Epistemologically, empire serves as a context-setting category that not only provides for the isolation of some historical experiences from others within the same polity, but also drastically changes the scope and mode of their narration depending on a chosen imperial locus or social group.[3] Even individual biographies are expressed absolutely differently depending on the master narrative chosen: the story of a leader of a nationalist movement would have only occasionally intersected with the story of the same person as an outstanding imperial scholar or a successful government employee.

This situation resulted not only in a multiplicity of loyalties available to any member of the Russian (imperial) cohort of public modernizers, but also in the possibility of adherence to more than one loyalty, giving priority to one or another under given circumstances of empire as a context-setting category (what was called in a different context "the Habsburg dilemma"[4]). The representatives of the Organization-Production School, who informally called themselves "the Artel,"[5] are a good case in point. Boris (Ber) Brutskus, Alexander Chaianov, Alexander Chelintsev, Nikolai Makarov, Konstantin (Kost) Matseevich, and Alexander Rybnikov played key roles in formulating the agenda for public modernizers in the 1910s. Together, they participated in that common sphere of Russian educated society engaged in the politics of self-organization and self-modernization. At the same time, individual members of the "Artel" were actively engaged in different specific projects, which had been "invisible" or irrelevant to other colleagues up to a certain moment. After 1917, Kost Matseevich emerged as an important Ukrainian politician: deputy secretary of agriculture in the government of the Central Rada (1917–18),[6] and later the envoy of Petlura's Ukrainian government to Budapest (1919).[7] Ber Brutskus had been involved in a number of Jewish philanthropic initiatives, was a leader of the Jewish People's Party (Volkspartei) founded by Simon Dubnow, and after his exile from Soviet Russia in 1922 eventually found himself in Palestine, where in 1936 he became a professor at the Hebrew University in Jerusalem.[8] The existence of parallel interests and loyalties by no means hampered the intensive cooperation of Matseevich and Brutskus with Chaianov, Makarov, and Rybnikov, who did not reveal any specifically "Russian" interests besides collecting antique Russian icons and books.[9] While Matseevich and Brutskus dealt with a national community defined in nationalist terms, Chaianov and Makarov tended to perceive nation as more or less coinciding with the borders of the Russian empire. This complexity of individual contexts made the seemingly universal striving for a new conscious "citizen" of a future "national community" fraught with misunderstandings and new internal conflicts. We may safely assume that the pluralism of interpretations of "national" was typical of rural modernizers, and that peasants, who were abandoning their local or regional identifications, were offered more than one version of a nation by their tutors. We can identify at least three alternative, if not competing, understandings of nationhood that affected the self-organized Russian society: nation as belonging to the state (Motherland), nation as an ethnocultural entity (the people), and nation as a community of conscious citizens ("revolutionary nation"). Below we will discuss these distinctive visions of the ideal community of self-organized individuals, focusing in this chapter on pan-imperial Motherland.

In peace

The most immediate sense of belonging to a community and territory is always very local and concrete, determined by the individual's life experience. As a villager from Samara Province put it in 1913 speaking about neighboring districts, "other localities are alien to us, the people of Semenovka."[10] The immediate feeling of one's "motherland" can be extended to embrace a territory of hundreds of square miles, but no quantitative manipulations can automatically convert local patriotism into a sense of national belonging.[11] In Late Imperial Russia, such powerful actors as the Russian Orthodox Church and the Russian state (through the local branches of its institutes and through schooling) attempted to instill an all-Russian self-identification into peasants, but this subject lies outside the focus of our study.[12] Specifically in the sphere of modernization of the countryside, these uncoordinated efforts were rather futile.[13] In this chapter, we will see how the modernizing activity of agricultural specialists made peasants aware of their belonging to a particular country (nation), and how the modernizers themselves responded to the "challenge of patriotism." By the latter, I mean not just the reaction to state-sanctioned loyalties and symbolism, but rather the more complex process of situating the individual and group political agenda and cultural values vis-à-vis the existing polity. This problem became particularly acute when the politics of self-organization of educated society acquired a truly all-imperial scale by the mid-1910s, actually appropriating some of the functions of the state through zemstvos and cooperatives, and thus inheriting some of the state's dilemmas.

To be sure, the public discourse on agriculture came into contact with the sphere of public discourse on Russian (imperial) national interests a few years before the war. As a result of the rapprochement with "official Russia" and its institutions, the community of rural professionals had gradually expanded their sphere of expertise on agricultural issues to the level of Russian economic relationships with foreign countries. Two poles of a "virtual globe," which revealed itself in the economic and political discourse of the new generation of Russian intelligentsia in the interrevolutionary period, were represented by Germany and the United States.

On November 12, 1911, the Russian Export Chamber contacted zemstvos and certain public organizations with an invitation to participate in drafting a new trade agreement with Germany. This participation would include, first of all, collecting information in the regions about what conditions would be desirable in devising a new treaty. Many agencies responded promptly. For instance, by December 14, 1911, the forty-seventh annual session of the Kazan provincial zemstvo decided to allocate 200 rubles to the Commission for the Revision of the Trade Treaty, and to authorize the zemstvo Economic Council to discuss practical matters concerning a new treaty project.[14] At this point, the initiative in forming the discourse of the zemstvo *agrarians* on Russian foreign economic policy was actually transferred to the third element, which dominated the zemstvo economic councils. They began discussing the desirable principles of a new treaty within the provincial economic council, with district zemstvo boards, and with other provincial zemstvos, and prepared a special regional zemstvo congress.[15] However, it took two years for the agriculture-concerned *obshchestvennost'* to bring the question of a new trade treaty with Germany (due to be signed in 1917) to the center of public debate.[16] In 1913, the Commission for the Revision of the Trade Treaty included representatives of 31 provincial and 88 district zemstvos, 43 agricultural societies, 25 associations of trade and industry, and so on.[17] The discussion of the future trade treaty provisions and the methods of their elaboration by local agricultural specialists moved to the pages of professional agricultural periodicals.[18] On the eve of the war, the community of agricultural specialists was aware of the international implications of their program of rural modernization: to stimulate peasant production, Russia needed lower import tariffs from Germany; in exchange, rural professionals were willing to press for lower Russian duties on imported agricultural machinery.[19] In fact, public awareness of national economic interests was much more intensive and effective than that of the government, which lagged behind the public initiative: the first meeting of the special committee for revision of trade agreements at the GUZiZ (the Imperial Ministry of Agriculture) took place only on May 14, 1913.[20]

The question of a new trade agreement with Germany stirred hostile feelings among many agricultural specialists toward German protectionist foreign economic policy. Still, the possibility of war over economic conflicts never even crossed their minds, least of all war with Germany: there is not a single hint at this in the professional periodicals or the proceedings of various congresses and conferences. In fact, the very theory of agricultural economy and biochemistry that Russian students learned at colleges and institutes was fundamentally rooted in the German scientific tradition of Johann Heinrich von Thunen, Max Sering, Justus Liebig, and others.[21] Even studies of Russian manure, which was allegedly very different in composition from its Western counterpart, were based on German research and methods.[22] This, together with a still very uneasy attitude toward the Russian state, was making the community of rural professionals potential opponents of any anti-German political moves.

While German influence over Russian agricultural specialists was achieved primarily through education and research based on German methods and discoveries, the image of the United States as an important reference point had been formed during the post-1907 decade of active modernization of the countryside. Some rural scholars saw the American farmer as a model for the future Russian peasant.[23] The stratum of veterinarians, who were semi-discriminated against, dreamed about the introduction in Russia of an analogue of the American Bureau of Animal Industry.[24]

Finally, energetic agricultural specialists of the southern provinces of Bessarabia, Samara, Poltava, Kharkov, and Ekaterinoslav were interested in American technologies of land cultivation that had proved successful in similar geographic conditions. This interest was institutionalized in the form of a special multitask service. In January 1908, an extraordinary zemstvo session of Ekaterinoslav Province supported the initiative of Ekaterinoslav Provincial Agronomist Viktor Talanov, who saw in the American experience a solution to the problems of the Russian peasant economy and decided to establish a zemstvo agricultural agency in the United States with an annual budget of 10,000 rubles (about $5,000).[25] By June, the Ekaterinoslav Zemstvo Agency in the United States hired its senior specialist Joseph (Iosif) Rosen, a graduate of the Michigan College of Agriculture,[26] while the junior specialist, agronomist A. M. Vengerovskii, was traveling around the province soliciting orders and requests for the agency.[27] In 1909, the agency compiled four informational surveys on various aspects of American agriculture, purchased agricultural machinery ordered by Ekaterinoslav farmers, and obtained orders for Ekaterinoslav seeds from American firms.[28] In the first nine months of its operations, the agency processed 70 requests for American seeds from 20 provinces.[29] For some reason, in 1911 the Ekaterinoslav zemstvo decided to liquidate its agricultural agency in the United States. Yet, the "perfect organization of the agency, its productivity and popularity among the farmers" encouraged the Kharkov Agricultural Society to take over the agency from the Ekaterinoslav zemstvo as of January 1912.[30] Perhaps, the former Kharkov zemstvo provincial agronomist Emel'ianov, now serving as senior agent of the Zemstvo Agency in the United States, was instrumental in making this decision.

One year after the establishment of the Ekaterinoslav Zemstvo Agency in the United States, the GUZiZ launched an agricultural agency of its own in the United States, in the spirit of emulating public modernizers' initiatives.[31] The original plan was to found the government Agricultural Agency in New York, but finally it was established in Saint Louis, Missouri, an important U.S. agricultural center.[32] When the war began, the government agency became a department of the Ministry of Agriculture—the Russian Procurement Committee in the United States.[33] The organization and program of the government agency very much resembled those of the Ekaterinoslav zemstvo agency: the agency was also staffed by two specialists, and its major tasks included collecting information about American agriculture, distributing information about Russian producers, and mediating bilateral commercial contacts.[34] At the same time, the government agency differed from the zemstvo enterprise: it enjoyed a larger budget, while the two "government agricultural commissioners" were incomparably worse trained than the staff of the zemstvo agency.[35] More important, the government agency did not emerge as a result of an economic initiative from below, but rather as a belated answer to the challenge of nongovernment modernizers. It was yet another modernization effort from above, although reportedly much enjoyed by the zemstvos and agriculturists of Central Russia. In the first two years of its activity, the government agency received some 1,200 letters containing different questions and requests from Russia.[36]

After the Ekaterinoslav zemstvo and the government established agricultural agencies in the United States, the Bessarabian provincial zemstvo came out with its own project for acquiring the latest American agricultural technologies. Instead of launching a costly agency overseas, the Bessarabian zemstvo decided to invite a specialist in intensive corn production from Iowa. An Iowa agricultural specialist, Louis Guy Michael, who arrived in Bessarabia in February 1910,[37] spent six years there promoting the program "More Corn for Bessarabia." Judging from his memoirs, Michael found more enthusiastic support from local agricultural specialists than from the Bessarabian second element, and succeeded much more in educating the peasant youth than in educating the local large-scale agriculturists.[38] Thus, for the rural professionals and the third element in general, American business experience appeared to be as valid as the German scientific tradition for their modernization program. Public interest in American agricultural experience, and concern about the place of rural Russia in the international arena sprang from all directions and grew constantly in intensity.[39]

However, neither German agricultural science nor American technology and organization of production as such directly influenced preferences in foreign policy. The significance of these "national" markers for the new generation of Russian intelligentsia consisted rather in associating such abstract entities as "Germany" and "America" with the very practical issues of Russian agriculture. Rural professionals learned to look beyond the boundaries of their precincts and districts, to connect in their minds the "abstract" problem of high German import duties with prospects for the spread of intensive systems of land cultivation among the local peasantry, and so on. In the 1910s, the worldview of rural professionals became more "nation-conscious" and less parochial due to an increased awareness of the international implications of their "small deeds."

Parallel to the rise of rural professionals' interest in the United States, Russian peasants "discovered America" for themselves. The emigration of peasants, usually sectarians, to the United States and Canada had been a well-known phenomenon for a few decades. The 1910s witnessed a rise in peasant economic migration. By 1921, Canada alone hosted about 35,000 Ukrainian and Byelorussian peasants from over 1,000 villages, most of whom had arrived after 1910.[40] A new trend was the short-term migration of peasants to the United States for higher paid jobs, whereas formerly they went to St. Petersburg and other Russian industrial centers. According to a contemporary observer, up to 10 percent of the peasant households in Chernigov province sent their members to earn extra cash in the United States.[41] Typically, peasants borrowed 200 rubles to sail to America (the trip cost some 140 rubles), and they managed to save up to 70 rubles a month, which made this business much more profitable than work in Russian industry.[42]

As mentioned earlier, the government-sponsored Russian Grain was founded initially to assist peasants seeking employment overseas. When the society adopted a more ideological, nationalist-modernizing agenda, its leaders became concerned with documenting the process of peasants' internal transformation during the months spent abroad on the advanced farms of Central and Northern Europe. Every peasant sent for agricultural training under the auspices of the Russian Grain (about 250 in total) was requested to write letters with accounts of the journey. Out of hundreds of those letters, over 150 were published in three volumes in 1911–15.[43] Although these letters cannot be regarded as a truly mass source suitable for broad generalizations about peasant discourse or statistical regularities, they nevertheless provide a unique insight into the process of "nationalization" of individual, quite ordinary peasants. Judging from the editorial marks on the letters selected for publication and public statements, the editors did not tamper with the original texts much beyond purely technical proofreading and style editing.[44] Those letters tell the typical story of a peasant who transcends his local identification and discovers his Russian national (imperial) identity under the stress of traveling abroad and encountering developed forms of nationalism and patriotism. This is particularly true of peasants who traveled to Moravia and the Czech lands:[45] they were duly impressed by the advanced development of local peasant farms, relatively quickly learned a new language, and soon internalized the narrative of Czech nationalist modernization:

> [F]orty years ago Moravia was as uncultured, as Russia...then the Germans began pressuring the Czechs in an attempt to Germanize them.... The Czechs...realized the threat and used the weapon of the Germans...that is, while they took on the German agricultural knowledge, they did not stop there but went farther, and currently have even outdone the Germans.[46]

> [T]hanks to harmonious unity in all kinds of unions and cooperatives, they gained power, built schools, and, thanks to the Enlightenment, the [Czech] people have developed in a complex way, and simultaneously their economic prosperity grew.[47]

While dozens of letters repeat the same story of peasants' cultural shock upon their first encounter with the highly cultivated landscape and prosperity of ordinary Czech peasants, many peasants were also surprised by the relative homogeneity and strong solidarity of the Czechs and other non-Russian nationals:

> Moravia and Czechia were saved by a national, Slav feeling.[48]

> Legality and truth are everywhere, and everybody recognizes the equality of the entire Czech people, whatever their educational background, there is no pride or arrogance, everything is decided by general consent.[49]

> The whole of Denmark represents something amalgamated into a single simmering, satisfied, cultural, and enlightened in all respects [entity].[50]

> [among the Latvians]...there is no class differentiation as we have in Russia, a gymnasium pupil, a student, a laborer, and also a woman walk and dance together.... That is why (I think) Latvians have left us behind in agriculture, because the educated class does not elevate itself to the "seventh heaven" from the people.[51]

The radical change of the sociocultural environment, the first encounter with a broad popular nationalism problematized peasants' own hitherto quite mechanical identification with their motherland "at large." As we see, some did not recognize "Russia" in Latvia, and some were ready to consider

Latvia an intricate part of Russia,[52] but, most important, they became nation-conscious and even nation-preoccupied abroad, as a rule interpreting Russia as a polity.

> Send me a lot of Russian newspapers. We must be either homesick or somewhat scared not knowing what is going on in Rus' now.[53]

> [Czechs]... probably know more about Russia than some Russians do.[54]

> In Moravia everyone reads newspapers and knows what is going on in his home country and abroad, and, in particular, they know everything about Russia better than our peasants do. And we have just darkness and ignorance.[55]

> Over a friendly conversation with these people [Czechs], one feels bitterness for the inert and overly inattentive attitude of the Russian himself toward his motherland.[56]

> When one compares what they have here [in Denmark] and what we have, one feels shame for our motherland. And an urge to get home, to work and work without rest in order to make everything flourish as it does here, in a foreign land.[57]

> Gentlemen! When I came back from abroad to native fields, I was overwhelmed by two feelings. The first one was the feeling of joy, and the second was the feeling of pity for the native village... so unattractive, unwashed, sloppy, and desolate, as it looks to me now after returning from abroad. And it seems that I had never loved my motherland so much and felt so sorry for her before, as I loved and felt sorry for her that minute.[58]

This patriotism was quite genuine, as some of the peasants were offered the real option to stay with their host families (at least in Moravia), even to marry a local girl,[59] but none had opted for this very real choice.[60] In a few instances, Ukrainian and Belorussian peasants referred to themselves as "Russians" and called their homeland "Russia," which can be explained by their awareness of their "ideal readers" in St. Petersburg, but even more so by their discovery of nationalism in the all-imperial sense.[61] While a couple of hundred of the Russian Grain "agricultural interns" were but a drop in the peasant sea of Russia, their experience suggests a direction of worldview transformation of a "new generation of Russian peasants" in general: people in their twenties, with a few grades of schooling, but most important, people who saw in agriculture not a predestined way of life, "not a yoke, not a bondage, but the best profession."[62] Many thousands of their peers were influenced by foreign experience directly or indirectly: as readers of the agropress (including publications of the Russian Grain that were distributed in tens of thousands of copies),[63] as part-time laborers abroad, or as soldiers during the World War I. These peasants and many agricultural specialists recognized themselves as members of the "imperial nation" of the Motherland after discovering a truly "foreign" experience. Vis-à-vis the interests of German modernizing society, Russian agricultural specialists identified their own interests as "Russian" (imperial); comparing the Czech society to their own sense of solidarity, peasant interns defined it in terms of empirewide Russianness.

In war

The nationalizing effect of the war over the villagers of the Russian Empire (primarily, in the sense of forging a political nation) has recently become an important topic in historiography.[64] In the context of this study, we shall focus on the major aspects of this "nationalizing" war factor over the political preferences of rural professionals.

On Saturday, July 27, 1913, one year before Russia entered the World War I, an editorial appearing in the weekly *New Economist* (formerly *Economist of Russia*) addressed the question of possible Russian involvement in a major international conflict. This general economic periodical, published by professor of Economics at Kharkov University and member of the GUZiZ council, P. P. Migulin, always sympathized with the interests of agriculture and rural professionals (as opposed to the *Trade and Industry Gazette*, which lobbied the interests of industry). Probably because of the *New Economist's* "down to earth" orientation, its editorial "War or Peace?" sounded very different from the bulk of writing in the Russian anti-German and pro-Slav press:

> It seems that the past has taught us plenty of lessons. But we continue to believe naively in the good disposition toward us of the Balkan peoples, and the communality of our interests.... We have no economic ties with the Balkan peoples: they are our competitors in the supply of agricultural goods to Western Europe.... [W]ars in defense of the Balkan Slavs have led only to the ruin of Russia. War for alien political interests is a dangerous political adventure. We see our duty as to warn the government and society against it. Let it never happen![65]

In general, foreign politics did not much interest countryside modernizers before the outbreak of war,[66] but judging from what we know about their intellectual preferences and political priorities, we can expect rural professionals rather to lean toward this point of view and not toward the militaristic ideas of the press oriented toward industry. After the beginning of the war, the emergence of a particular foreign policy agenda among rural professionals became explicit. The sense of national (imperial) unity was increasingly intensified by the emergence of the real and symbolic foreign enemy. The previously international, and in a sense cosmopolitan, project of public modernization of the Russian countryside was acquiring patriotic traits.[67]

Initially, the war was hailed by Russian society, and rural professionals were no exception. Some agricultural specialists saw in the war an opportunity to broaden their scope of "creative activity" in the sphere of the national economy and even in the international arena,[68] while cooperative activists were excited about the prospect of finally bridging the gap between the "fourth" and the "first" elements in a unifying spirit of patriotism.[69] The strengthening feeling of a pan-imperial national unity so excited rural professionals that some of them were completely overwhelmed with ideological zeal. That was the case during the campaign for the prohibition of alcohol,[70] and certainly in the initial stages of building a public discourse on the enemy and its military and economic outlook. The image of the enemy varied in the professional periodicals at the beginning of the war, curiously indicating different stages of professionalization that had been achieved by various groups of rural specialists. For example, the influential *Messenger of Public Veterinary Medicine* sounded very similar to the general popular and cooperative press that mocked the Germans and at the same time depicted them as bloodthirsty monsters.[71] Agronomists seemed to be more capable of distinguishing between professional expertise and patriotic sentiments; all major agronomist periodicals repeatedly denounced all kinds of "patriotic hysteria" and speculation about the prewar "German preponderance" in Russia's economy and culture.[72] Pedagogues working in the agricultural schools were in a more difficult situation than agronomists because the very task of schooling was to form the pupils' worldview. As one can judge from the project of a new program in geography for the agricultural schools, pedagogues found a median between patriotic propaganda and pure professional knowledge:

> About Germany, it is necessary to note the rapid growth of technical culture in conjunction with the broad spread of education and the richness of the country in coal and iron, and the emergence of German militarism, which is a result of German cultural specificity, the German character, and the geographic situation of the country. One must mention the development of agronomy in Germany and its influence on Russian agronomy, and the state of agriculture in various parts of the state.[73]

Patriotism and raging anti-German hysteria in the country might compel one to change a German-sounding name,[74] but not all patriots were eager to abandon their jobs and join the army. Quite characteristically, rural modernizers employed the arguments of professionalism and rationality to explain their desire to avoid conscription.[75] It was difficult to avoid the draft during the first period of the war, when the prospects of its ending soon had not yet faded away, and the management of human resources was absolutely thoughtless.[76] Incidentally, many representatives of the new generation of Russian intelligentsia were never called to colors. At first, when rural professionals were drafted without any concern for the importance of their jobs in the rear, the age of these already not so young people bought them time. In 1914, the minimum age of a member of the class of 1910 was about 26, while many representatives of this social generation were much older than its most typical age group born in the mid-1880s. There were still plenty of younger people, whose turn to serve came before them.

As the war dragged on, the authorities realized the importance of the agricultural specialists as cogs in the enormous and complicated food supply machine. Specialists occupying important positions were covered by the local authorities, while the Main Committee for Granting Conscription Deferments began gradually extending the right to deferment to more and more categories of rural professionals[77] and cooperative activists.[78] Now some rural professionals became protected by the important jobs that many of them had managed to secure before the war. Still, many low-ranking agricultural specialists, especially precinct agronomists, could not escape conscription.[79]

The available statistical data present a contradictory picture of the impact of conscription upon the agronomy organization of the Okhansk district (Perm province) during the first six months of the war. By spring 1915, 5 out of 11 male agricultural specialists were called up. Two of the 11 specialists had only a primary agricultural education, and both of them were drafted. The remaining nine men all had a mid-level agricultural education, including the Okhansk district agronomist Mikhail Ozharko, who had been working in the Okhansk zemstvo since 1902. We do not know the actual age of these men or their special medical conditions. We know only their service and educational backgrounds. Two of the conscripted precinct agronomists graduated from mid-level agricultural colleges in 1902 and 1904, before the Revolution of 1905, and one graduated in 1910. At the same time, young agronomists who graduated in 1912 and even in 1913 continued working in their precincts.[80] There was a similar situation in Kazan province. During the first months of the war, 4 out of 12 district agronomists were called to colors (all had graduated from mid-level colleges, two of them as early as 1900 and even 1898), while only one young precinct agronomist and three very young agricultural elders were conscripted.[81] The cases of Okhansk district and Kazan province suggest that for the military authorities no low-ranking agricultural specialist was indispensable, and the lower the position and educational background of a specialist, the more likely he was to be drafted.

Indeed, Alexander Chaianov avoided conscription by becoming a permanent faculty member at the Moscow Agricultural Institute. His classmate from the institute, Ivan Barkhatov, received a conscription deferment as the head of the government agricultural organization in Kazan province. Many of their colleagues, including Sergei Fridolin, hurried to take jobs in various procurement committees of the All-Russian Zemstvo Union, which granted them a conscription deferment.[82] Thus, one of the important impacts of the World War I on the community of countryside modernizers was a dramatic separation of the more professionally successful and socially active leaders from the less ambitious and ideologically conscious rank-and-file specialists, many of whom were called up. The war combat experience must have separated, if not completely alienated, the two groups of rural professionals, changing forever the worldview and social identity of the veterans. Even financially, those who were drafted and those who remained working were parted from each other. While the latter enjoyed salary increases and additional income from extra jobs,[83] the former, especially precinct agronomists, were actually losing money.[84] The single community of agricultural specialists–modernizers began to disintegrate under the influence of different life experiences, and so did the base of their political influence in the Russian countryside. Now they relied on institutions that protected them from the draft (zemstvos and cooperatives) as major guardians of their professional activity.

The war also seriously reconfigured the character of the activities of agricultural specialists. The imperial regime had already made steps toward accommodating rural professionals before 1914, hiring former representatives of the third element or transferring the entire government agronomy service to the zemstvos. The outbreak of the World War I greatly expedited this development, as the government realized the urgency of mobilizing the rural economy. Agricultural specialists were seen as the natural agents of government influence and wartime mobilization in the village, and hence more trustworthy than ever before.

Historians often tend to take for granted a structural predisposition of rural professionals toward greater cooperation with the state apparatus, regardless of their affiliation with the "third element" or state service. The alleged "militarization" and "mechanization" of the work of agricultural specialists during the war usually serves as a self-evident explanation of their subsequent partaking in the socialist measures in agriculture after the Revolutions of 1917.[85] The actual history of the wartime activity of the rural professionals remains largely unexplored. While some historians look at the World War I as a terminal point that halted the previous modernization campaign of agricultural specialists,[86] others underline the continuity between the war and prewar periods.[87] In his special study of the wartime involvement of agricultural specialists in government procurement agencies, Kimitaka Matsuzato discovered that this involvement was by no means universal or standardized. Out of 13 provinces representing different regions of European Russia, studied in 1916, zemstvo agronomists in three provinces and government specialists and agronomists in seven provinces did not participate in state procurement agencies at all. In other provinces, only a fraction of the available personnel was engaged in the procurement campaign.[88] And certainly the advance of technology as a major means of modernization of the countryside was not a result of the war, but, as we discussed earlier, an attempt to sustain and broaden the impact of countryside modernizers.

Leaving aside the work of some in procurement agencies, often taken reluctantly just to avoid conscription or because of a direct order, there were significant innovations in the ways that agricultural specialists exercised their expertise. First of all, the war reinforced the social component of their professional agenda. Occasional charitable actions were gradually transformed into a sustained program of social rehabilitation and professional training of wounded and handicapped soldiers who were no longer capable of performing their usual peasant work.[89] This program was initiated and implemented predominantly by rural professionals, who lectured in the hospitals and ran special courses for wounded soldiers to teach them trades that did not require hard physical labor: apiculture, cooperative management, and machinery servicing.[90] The humanitarian activity of rural professionals not only strengthened the sense of national (pan-imperial) community but also potentially paved the way for a double-standard approach toward the clients: it was possible for some people to get preferential treatment while others did not. At the same time, it was not ideological considerations that determined a special attitude toward part of the population, but rather concerns about economic rationality and optimization of the utilization of labor resources.[91]

The changing climate in the profession was already noticeable in its spiritual cradle, the Moscow Agricultural Institute. These young people who became agricultural students after the beginning of the World War I were trained for a different kind of agricultural assistance under different circumstances than were the members of the classes of 1910 and even 1914. In 1910, during the admission examination in Russian language, a prospective student of the Moscow Agricultural Institute was asked to write a composition on one of the following four topics:

- Scientific progress of our time in its utilitarian applications;
- An ideal aspect of education;
- Youth (characteristic traits of this age in respect of psychology);
- Scientific specialization and general education.[92]

In 1914, the topics of compositions sounded absolutely different:

- "The darker night, the brighter stars; The deeper sorrow, the closer God" (poet A. Maikov).
- When are our convictions the most stable and valuable?

- Can this war be regarded as accidental?
- National consciousness in its characteristic features.
- The man of duty and the man of love. Mutual characteristics.[93]

In 1910, freshmen were encouraged to think in universal terms of social engineering and the Progressivist pedagogy of John Dewey. In 1914, they were oriented toward the mysticism of *Zeitgeist* and various types of determinism, both social and psychological. Nationalism was recognized as a key theme, along with the impersonal general historical forces that determined the role and place of an individual.

The war began eroding the social composition of the new generation of Russian intelligentsia by separating the top specialists and ideologists from the rank and file. The war also had a double impact on educated society, including the stratum of public modernizers in the countryside. On the one hand, these people did not show particular enthusiasm for their military patriotic duty: we know numerous stories of attempts by rural professionals and cooperative activists to avoid conscription,[94] while no examples of their volunteering for military service circulated in the public discourse. In this sense, nationalism-as-patriotism failed within this social group: they did not support Russian empire *as the state*. On the other hand, the nongovernment wartime institutes that were created by the very same people (the All-Russian Union of Zemstvos and Towns, Military–Industrial Committees) played a crucial role in the wartime effort and were granted conscription deferment, which only increased their popularity among the agricultural specialists.[95] In fact, these institutions undermined the government by taking over the procurement campaign and public welfare programs.[96] The self-mobilized educated society took over some functions of the "state" and eventually attempted to substitute for the state. The attempt had disastrous consequences, but it succeeded in at least one respect: "acting like a state" made many rural professionals "see like a state." The head of the zemstvo "parallel government," Prince George L'vov, actually used almost identical wording in 1916 to describe the new worldview of public modernizers: "the Russian public has acquired state-like qualities," demonstrating a "supreme, state-like point of view."[97] Gradually, they learned to associate themselves with "nation's" (empire's) interests in the same way as the traditional intelligentsia identified themselves with the interests of the "people."

The project of public modernization of the Russian countryside had originated as an attempt at socioeconomic reformism through the community of rural professionals without any significant role played by the state. The "agronomist crisis" of 1913–14 softened the hostility of the New Generation of Russian intelligentsia toward such government programs in the countryside as peasant land settlement, while the growing role of technology in the activity of rural professionals reinforced the trend toward impersonal relations with the peasant masses and the necessity for the centralized coordination of that "mechanized modernization." When in 1915 it became clear that the war would not end in the near future, and not necessarily with Russia's victory,[98] many rural modernizers, including the ideologists of the movement from the Organization-Production School of rural studies, turned toward projects of direct state intervention in the rural economy. It started with calls to introduce a state monopoly on the distribution of goods of basic necessity among the population.[99] Very soon, two major themes became dominant in the specialized agricultural press: the fight against "speculation" of middlemen, and the necessity for the introduction of a set of firm prices on strategic commodities.[100] At this point, many agricultural specialists literally repeated the ideas of the populist economists of the "old generation" of Russian intelligentsia.[101]

So far, the antimarket and pro-plan proposals of agricultural specialists could be explained by the strain imposed by wartime hardships, and in part as a defensive reaction against the preferential treatment enjoyed by mobilized industry in terms of the allocation of resources and the cadres' protection from the draft. However, the very first projects for the postwar economic reconstruction of the countryside released by the young but already influential leaders of the prewar public modernization movement, Alexander Chaianov and German Tanashev, revealed their intent to make the state a key actor in the postwar rural economy. At the same time, the main thrust of assistance to the peasants was shifted from individual consultations and mass education to large-scale socioeconomic transformations. As early as March 1915, Alexander Chaianov envisioned "associations of [economic] renaissance" that would emerge after the end of the war. He argued that "hundreds of millions" of rubles in state subsidies should not be "dispersed among…individual farmers" but rather distributed among "local associations of all local agriculturists, bound by responsibility for collective property." With considerable state resources and compulsory membership, those "associations of renaissance" would carry out a number of large-scale programs, including land settlement, purchases of industrial goods and agriculture materials, and so on. Chaianov did not indicate who would direct the activities of these giant politicoeconomic conglomerates, a mixture of zemstvos and rural cooperatives. Apparently, this role was reserved for agricultural specialists, who would work under tight state control over the distribution of government funds. Already in 1915, Alexander Chaianov did not object to such a situation.[102]

German Tanashev joined the ranks of agricultural specialists a half decade earlier than Chaianov but became a prominent contributor to professional periodicals only during the war. In 1916, he criticized the legacy of Stolypinist social engineering and insisted that after the end of the war the state would have to intervene in the domain of public legal relationships with a program of land redistribution.[103] This "regression" to the rhetoric of the pre-1905 populist intelligentsia was a logical continuation of the campaign for the "mobilization of all-national forces" by the agricultural specialists.[104] They were only following the Russian "progressive public" when they demanded "the union of government authority and society by means of reforming the governance of the country on the basis of common trust and solidarity."[105] While in the sphere of politics the ultimate prize for the mobilized masses would be their larger representation in the government and especially the parliament, the agricultural specialists could offer them only radical land reform after the end of war.

Thus, the World War I brought about further "nationalization" of the politics of modernizing the Russian countryside. Before the war, Russian agricultural specialists perceived themselves as a part of the pan-national milieu of modernized "men of expertise," if not "men of letters." Russia was present on their mental map as a specific case of a general system. Russian agriculture was different from German or American agriculture but by no means "alien" to them. The war not so much alienated the imagined "Russia" from the rest of the world, as it changed the professionals' attitude toward their home polity. Not demonstrating feats of patriotism outside their sphere of expertise, they reconsidered their previous reserved position vis-à-vis the state, deeply rooted in the anti-statist tradition of Russian *obshchestvennost'* tradition and in the early Progressivist discourse.[106] Willingly or unwillingly participating in the institutions that had taken upon themselves some of the state functions, and cooperating with state agencies staffed by likeminded people, many rural professionals began integrating into the Russian nation as a polity. At the same time, millions of peasants were undergoing a similar process through more intensive integration into a larger society. The more intimate interpersonal relationships of agricultural specialists with peasants were being replaced by a more routine and mechanized mode of interaction, and public modernizers began perceiving peasants more in terms of social group categories than in terms of individual contacts. This trend dramatically changed the initial spirit of the public modernization campaign in the village, but at the same time enormously strengthened the social base of the Russian imperial nation (as the state). The real problem was with the actual state: it did not withstand the burden of wartime strains and popular expectations, and it collapsed.

9
Nation as the People

The initial program of modernizing the Russian countryside by means of collective efforts by the Russian educated public was conducted on the basis of "a-national" principles—which is not to say that they were "non-national." The object of modernizing activities was defined as the "Russian peasantry," but this did not mean that Turko-Muslim villagers of the Middle Volga region, Baltic farmers, or Ukrainian agriculturists were excluded from the scheme. The function of empire as a context-setting category makes "empire" meaningfully visible to both people living "within" and observing it from the "outside" only through the contradictions emerging from its uneven and unsystematic heterogeneity, or as a result of conscious attempts to make it more manageable and thus more rational.[1] The public modernization movement of rural professionals discovered the problem of the imperial situation rather late, when they encountered a conflict between the universal modernizing project and the diversity of its local interpretations and contexts of realization.

The "empire" did not exist for the would-be public modernizers beyond the purely administrative and legal meaning of the term, as they operated in the broad, pan-European intellectual context of agrarian studies and economic theory. Likewise, their project was not a national Russian (ethnic) endeavor. Its ultimate aim was to make the Russian (imperial) village modern, which meant, first of all, rational and self-conscious. This project was a success, but reaching a greater economic rationality had a by-product in creating the framework for rationalization of the peasants' ethnoconfessional differences and channels of expression of protonational self-consciousness. The cases of peasants in the Ukrainian and Volga-Urals region are particularly important in this respect. Both Ukrainian and Tatar-Bashkir lands belonged to the core territories of the empire,[2] which were fairly integrated (as opposed to the Baltic, Polish, or Central Asian regions) but never assimilated. While implementing their universal program, the countryside modernizers quite unexpectedly discovered for themselves the diversity of empire and its severe tensions. The politics of a universal "modernity" became engaged in complicated relationships with the modern phenomenon of nationalism.

Almost immediately after its takeoff, the public agronomy movement faced the problem of ethnoconfessional diversity of the client village population. At first, the classic imperial dilemma of a significant sociocultural gap between the benevolent educated elite and the benighted common folk concealed the national dimension of imperial generic heterogeneity. When young Samara agronomist Ryshkin envisioned his agronomist precinct populated by ethnic Russians as "a huge enemy territory," he hardly felt less estrangement from his clients than he would have had he served in an ethnically Tatar or Georgian region. As was the case with other Russian-speaking representatives of the establishment, the pioneers of precinct agronomy in ethnically non-Russian territories relied on the intermediary assistance of educated and bilingual locals. The rapid growth of the cadres of "agricultural elders," or agronomist assistants, indirectly testified to the high degree of dependency of rural professionals on the stratum of "interpreters."

As the bridge to the "new peasants" widened, actually allowing for a two-way mode of communication, the difference between the "Russian" and "alien" (*inorodcheskoe*) peasantries became more acute. Russian-speaking agricultural specialists had to struggle with the language barrier in some areas, whereas their colleagues in other regions had already found a common language with Russian-speaking villagers. Characteristically, this situation of an ethnoconfessional gap did not bother agronomists themselves too much—at least they did not mention this problem in numerous accounts and reports on their activities. Apparently, given a limited number of clients they could assist directly, they were always able to find peasants who would understand them. This does not mean that the problem did not exist. Alexander Dillon quotes from an account on a Russian agronomist's presentation in the Ukrainian village of Likhovka in the Ekaterinoslav Province:

> His talk was sincerely delivered, but lacked simplicity. Occasionally he used expressions such as "the land is of such-and-such character, or of another character," when it would have been better to say, "this land is good, that land is not so good." The agronomist spoke, as usual, in Russian, and only sometimes let out a word or two in the language of the local people—that is, in Ukrainian—when it was clear that he had to explain something again to the old folks.[3]

The Ukrainian source claims that peasants did not understand the Russian agronomist, although it is hard to tell whether this was because his language was "scientific" or Russian: it seemed that only the "old folks" needed the agronomist to speak Ukrainian, and apparently he was able to speak at least some. We witness here a situation that reveals the initial fusion of the "cultural" and "national" alienation of the representative of Russian educated society from the peasants. We suggest that, at least in this particular case, this alienation was eventually overcome: at the moment of that presentation, the Likhovka precinct agronomist was Platon Malutin, who had just graduated from a mid-level agricultural college and had been working in this capacity for hardly more than half a year.[4] This explains the "lack of simplicity" in his presentation, while the fact that he kept this position until at least 1915 suggests that he felt quite comfortable in Likhovka with its Ukrainophone population.

In the Volga-Urals region

While in the Ukrainian provinces the national (first of all, linguistic) difference between the Russian-speaking rural professionals and the Ukrainian-speaking peasants was not insurmountable, in Kazan and Ufa provinces the situation was more complicated. Tatars and Bashkirs were subject to all of the zemstvo agronomy initiatives, but the absolute majority of rural professionals serving them could not master Turkic languages with relative ease, which had been possible with Ukrainian. The ideal solution would be to train local cadres for the job, but in the short-run perspective that was not a particularly promising option: the young generation of Turko-Muslim intellectuals preferred other forms of public activism in politics or literature. A few exceptions to the rule only highlight the general trend.

Sultan-Girei Taichinov was born around 1880 to the family of a petty state official in Ufa province. In 1899, he graduated with the first class of the three-year Belebei Agricultural School of the Ufa provincial zemstvo. By 1907 he had become deputy agronomist of the Belebei district zemstvo,[5] in 1909–10, he held the position of "lecturer on agriculture in the Tatar language,"[6] and then was promoted to the post of Belebei district zemstvo agronomist.[7] However, after three years in office, Taichinov was transferred to the much lower position of precinct agronomist in neighboring Sterlitamak district.[8] Most likely, the absence of even a mid-level college diploma explained the decline of Taichinov's career: among the five precinct agronomists of the Sterlitamak district in 1915, he was the only one with just a primary professional education. His position as Belebei district zemstvo agronomist was taken by Aleksei Glushko, who had just graduated from an agricultural institute (i.e., had a higher education diploma).[9] As an experienced practitioner, Sultan-Girei Taichinov could probably have obtained an institute diploma (not to mention that of a mid-level college) with a fellowship from the Department of Agriculture, and in a shorter period than with regular schooling.[10] For some reason, he did not pursue this option and preferred to move to a less attractive job, although he had more prerequisites to succeed than other educated Bashkirs from nonprivileged families. While representatives of the Bashkir elite showed no interest in the "third element" career track,[11] ordinary

Bashkirs did not hurry to follow the suit of Sultan Taichinov and sought upward mobility in towns rather than in the village.[12]

In the absence of agricultural specialists who could speak the Tatar (Bashkir) language, the zemstvos relied on the mass circulation of special literature in Turkic languages. In 1913, the twelfth agronomist conference of the Ufa zemstvo attended by 62 representatives of the local second and third elements, concluded,

> [A]lmost 50 percent of the province's population are Muslims, and the skill of reading Tatar is widespread among them. This undoubtedly should be used in order to spread agricultural knowledge among Muslims by means of distributing special literature in the Tatar language, particularly since direct communication with them is complicated by an absence of people who have a knowledge of Tatar in the agronomist organization.[13]

The need for Tatar-speaking specialists, or at least Tatar-language publications, was felt even more urgently by the thriving credit associations. With support and encouragement provided by both the government and zemstvos, rural credit associations directly contacted more people than all the zemstvo agronomists together, and were interested in the sustained engagement of the peasant masses in their operations. The second congress of the Belebei district (Ufa province) rural credit associations in June 1910 demanded that the zemstvo create the position of Russian–Tatar interpreter, and also publish some agriculture-related articles in Tatar in the *Agricultural Leaflet of the Ufa Provincial Zemstvo*.[14] (The initial idea was to duplicate the entire periodical in Tatar, but the cost of the all-Tatar edition was found to be too high.[15]) The Belebei district zemstvo was known for its pro-Bashkir stance, and by 1910, the local zemstvo board had been reshuffled under the governor's pressure in order to eliminate "pro-Bashkir" members.[16] Still, the Ufa provincial zemstvo indeed made its major agricultural periodical bilingual, and eventually allocated additional funding to produce an all-Tatar parallel edition.[17] Thus, the politics of public modernization compelled Ufa agricultural specialists to sponsor a program of publications in the local language as a means of integrating Bashkirs into the pan-imperial dialogue of modernizers and peasants.[18] Ufa agronomists did not hesitate to work in Tatar-speaking communities, where publications in Tatar were extensively used to compensate for the existing language barrier.[19] There is evidence that Russian-speaking agronomists regarded Tatar as the language of inter-ethnic communication in that multicultural region and mastered some Tatar themselves.[20]

This "technocratic" mobilization of the nationality resource was much more powerful than direct political mobilization along nationalist lines, which was legally allowed only to Russian nationalists. For example, in that same year (1910), the Kazan chapter of the Union of the Russian People published 500 copies of the Union's propaganda brochure, while the popular manual on grass cultivation in the Tatar language was published (probably, by the zemstvo) in a print run of 2,000 copies.[21] Here, and even more so in Ufa province, "[z]emstvo leaders believed that nationality would support, not conflict with, enlightened secular values which would unite the empire's peoples."[22] The universal modernization project of rural professionals coincided with the all-inclusive nature of "zemstvo citizenship" based on the equality of taxation and representation, regardless of one's ethnoconfessional affiliation. The need to engage the Tatar-speaking part of the zemstvo constituency in various zemstvo-sponsored initiatives (including the modernization project of rural professionals) encouraged the second and third "elements" of the Ufa zemstvo to support even such a traditional institution of Bashkir society as confessional schools (*medresse* and *mektebe*).[23] The fusion of a modern curriculum with traditional ethnoconfessional education was a sure way to build mass support for the politics of modernity in Bashkir villages in the future, but it was also a precondition for the rise of modern forms of national identity in the hitherto predominantly archaic rural society. The mobilization of Bashkir villagers by rural professionals was successful: Bashkirs cooperated with agronomists, attended agricultural courses,[24] and eagerly established cooperatives.[25] By joining zemstvo-sponsored initiatives Bashkir peasants became more rational farmers and members of a larger Russian society, yet they were doing so not as abstract "new economic men" or "citizens" but as "Bashkir/Turko-Muslims," now more conscious than ever of their group distinctiveness. The spread of cooperatives in Ufa province was marked by emerging rivalry and conflicts on "religious or national grounds."[26] Absolutely unintentionally, rural modernizers contributed to creating the framework and channels of future national (ethnoconfessional) mobilization in the Bashkir countryside.

In Ukrainian lands

Whereas in the Volga-Ural region during the 1910s agricultural specialists and the "second element" were only creating the preconditions for wide-scale national mobilization, in the western parts of the empire such preconditions had already existed. Here too, as was the case on the Volga, the mobilization of the nationality resource by rural professionals was much more powerful than direct mobilization by nationalist educators, even before the takeoff of the public modernization campaign. In 1898, the archetypal text of the Ukrainian national renaissance, *Kobzar* by Taras Shevchenko, was sold in Ukrainian villages at a rate of 60 copies a month. Compare this with a Ukrainian-language brochure *Conversations About Agriculture* that was distributed in 1902 (also before liberalization of the Ukrainian-language press in 1905) in a print run of 500,000 copies.[27] Suddenly, the agropress became a stronger factor in stimulating national languages, and hence cultures, than had any direct nationalist agitation.

Furthermore, unlike Bashkirs, educated Ukrainians eagerly invested their "human capital," as agricultural specialists, teachers, or cooperative activists, in the task of modernizing and mobilizing the village.[28] While they did not need the patronizing initiative of the zemstvo and Russian-speaking rural professionals in order to begin a dialogue with the peasantry, the nature of this dialogue was problematic. The inevitable degree of populism shared by rural modernizers almost universally had a different meaning in different national contexts. In the "Russian" Russia, that populism was primarily of a socioeconomic character, expressing itself in the preferential treatment of peasants over gentry landowners, and of midsize "labor peasant farms" over capitalist farms with hired workers. In the Ukrainian context, particularly in the parlance of Ukrainian social activists, the term "people" sounded different, and its meaning changed from a social category of "common folk" to a "folkish" connotation of the "people as a nation." The tropes and slogans of the all-Russian public modernization campaign acquired a distinctive national(ist) dimension in the Ukrainian context:

> Besides material assistance, cooperatives will provide an opportunity to cultivate in the masses all the best civic instincts. And all the true national (*narodni*), conscious intelligentsia forces should help the people in these burning issues of organization and enlightenment (*prosvity*).[29]

The Ukrainian agricultural periodical *Dniprovi khvili* was a typical Russian (imperial) publication of countryside modernizers: it popularized the universal program of public modernization by means of precinct agronomy and cooperation, appealed to the stratum of "new peasants," and monitored the state of the modernizers–peasants dialogue by distributing questionnaires among its readers.[30] Yet the very fact of a "translation" of this program into the Ukrainian language and Ukrainian sociocultural context altered its character significantly. The responses of readers of *Dniprovi khvili* were equally typical, reproducing all the major tropes and clichés of the peasant letters to the editors of that period—but there was also something different in Ukrainian letters. The mere mention of the Ukrainian language issue could change the focus of an otherwise standard letter of a "new peasant" from a Ukrainian village:

> In our village there are indeed to be found people who are interested in the native language.... Unfortunately, the more educated and knowledgeable people here have not established a cooperative store, and the peasants do things the old way.... There is still a lot of darkness among us![31]

In this context, Ukrainian "darkness" differed from all-Russian "darkness." Economic rationality alone could not eliminate it; any remedy had to be in conjunction with interest "in the native language." This discourse was shared by Ukrainian "village intelligentsia" as well as by those agricultural specialists of Russian background who saw the language barriers and restrictions as serious obstacles to countryside modernization.[32] However, in general, we do not find much concern about Ukrainian-language instruction and agronomist assistance in the minutes of zemstvo sessions or in the agropress.[33] To some extent, this can be explained by the politically charged nature of the "language question," although other segments of the press widely discussed the problem with Ukrainian, while professional periodicals touched upon equally sensitive political issues. The main reason for the agricultural specialists' neutral attitude toward the nationality factor in Ukrainian villages should be seen in their primary loyalty to the original technocratic message of the politics of modernity. They went to serve the people as professionals, as representatives of international science and universal rationality. The Russian–Ukrainian language barrier was not insurmountable on either side of the dialogue, and hence there was no particular urgency to duplicate the modernizing mission in Ukrainian (in contrast to the situation in Ufa province), while government anti-Ukrainian pressure certainly was not a factor encouraging such experiments. Agronomists and veterinarians usually dealt with individuals, and their engagement in the politics of modernity up to a certain moment was not really an instance of mass politics, and surely not nationalist politics.

The booming rural cooperative movement was a different case. From a structuralist point of view, Ukraine demonstrated "the next stage" of the trend just emerging in Ufa province in the mid-1910s: a fusion of the cooperative movement with nationalist self-identification and mobilization.[34] The rising "fourth element" of cooperative activists represented no abstract technological schemes, but the people, the "Ukrainian people," and the truly massive cooperative phenomenon was so powerful that no government regulations could curtail its de facto Ukrainization. In 1908, during the takeoff of the "cooperativization" of the imperial countryside, the meeting of a regional congress of consumer cooperatives in Kiev was made possible only under the condition that its proceedings be conducted entirely in Russian. St. Petersburg cooperators expressed their indignation with the ban on Ukrainian and suggested that Ukrainian peasants-cooperative representatives had experienced linguistic problems, as

> only a few of the delegates did not know the Little Russian [Ukrainian] language. The majority of the peasant delegates only with great difficulty handled [the Great] Russian pronunciation, so difficult for them.[35]

Five years later, in 1913, the second All-Russian cooperative congress in Kiev became a virtual triumph of Ukrainian cooperation. Not only were language restrictions effectively ignored during the congress, but its Ukrainian hosts arranged for a delegation of Ukrainian cooperative unions from the *Austrian*

Galicia to participate in the *all-Russian* event. While Russian nationalists may have entertained the illusion of the Galician delegation as representatives of the big "Russian nation," the congress organizers went out of their way to stress the *Ukrainianness* of the Galician guests, speaking the *Ukrainian* language.[36] The whole cooperative congress thus resembled a pan-Ukrainian happening. One of the congress's resolutions called for the use of local languages in the day-to-day operations of cooperatives,[37] thus bringing closure to the "national question" in the cooperative movement. While the government demanded that all of the bookkeeping be done in Russian, de facto cooperative business was effectively nationalized. The leading Russian agricultural professional periodical, the *Agronomist Journal* (published in Kharkov), summarized the legacy of the Kiev cooperative congress:

> In the environment of any lively national movement, rural cooperatives get the impetus from the national movement, and at the same time themselves serve as a rich breeding ground for the sprouts of national initiative (*samodeiatel'nosti*).[38]

The Ukrainian cooperative movement, or at least its management personnel, definitely served as both a spearhead and breeding ground for the nationalist movements. In Poltava and Ekaterinoslav provinces, cooperatives were directly involved in the Ukrainian national revival movement institutionalized in the form of *Prosvita* societies.[39] Already during the 1913 Kiev congress, Ukrainian cooperative leaders argued in favor of organizational autonomy from the all-Russian central cooperative cartels, including the Moscow People's Bank—the first cooperative financial institution serving cooperative associations across the empire. The separatist cooperative agenda was discussed at a series of secret meetings and legal conferences of Ukrainian cooperative leaders in July–November 1916.[40] The nationalization of the cooperative movement found its expression in outright xenophobic statements. "Down with him! He is a *Masepinist*! I object! Go to your Austrian blue uniforms!"—shouted a Russian cooperative manager in response to a proposal from his Ukrainian colleagues that a cooperative periodical in Odessa be published "in the language of the local population," that is, in Ukrainian.[41] "Down with you! Go back to Moscow!"—booed Ukrainian cooperative activists the representatives of the Moscow People's Bank.[42] This was just an echo of naturally more reserved but quite explicit declarations that routinely appeared in the local and central cooperative press at least as of 1913.[43] After the February Revolution that triggered the process of building Ukrainian independent statehood, cooperative leaders became prominent actors in the local and central government of the Central Rada and Directory regimes (and to a lesser extent under the Hetmanate).[44] Nationalist-cooperative leaders dreamed of independent Ukraine as a "Ukrainian cooperative order."[45]

It is important to stress, however, that while the Ukrainian national movement extensively used the organizational potential of the cooperative network and the mobilization skills of its members, rural cooperatives in the Ukrainian provinces were not agents of the nationalist cause par excellence. In the words of Alexander Dillon, "Cooperation did not start out as the Ukrainian nationalists' movement; rather, it was a site in which adherents of the *various* modernization agendas worked."[46] Scholars with different research and political agendas agree that nationalist sentiment was not even widespread among the cooperative leadership in many regions of would-be Ukraine (first of all, in Chernigov, Ekaterinoslav, Kharkov, and Kherson provinces). But even in the more "nation-conscious" provinces of Kiev and Poltava and in the parts of Ekaterinoslav province where *Prosvitas* were active, peasants quite reluctantly supported the noneconomic activities of rural educators.[47] Moreover, the very "Ukrainianness" of national-minded cooperative leaders was a problematic concept. An interesting insight into this problem is provided by the statistical presentation materials arranged for the 1913 all-Russian cooperative congress in Kiev by the secretary of the Credit Subsection of the congress, a representative of the ancient Ukrainian clan, Iu. G. Gamaleia (Hamalia).

The narrative account of the congress produced a peculiar mental map of the Russian Empire as consisting of Great Russia, Ukraine (and certainly not "Little Russia"), Poland, and the Baltic Region.[48] However, the geographic grouping of statistical data related to credit associations performed by Gamaleia is more nuanced and ambivalent. Instead of "Great Russia," he distinguished ten regions, including the Caucasus and Trans-Caucasus, Central Asia, and Siberia. In his classification, there were separate Baltic, Lithuanian, and Belorussian regions (the latter consisted of Vitebsk, Minsk, and Mogilev provinces, while Grodno province was attributed to Lithuania).[49] Still more important is a "national-cooperative" geographical representation of Ukraine. On the one hand, Gamaleia constructed "the Ukrainian language region"—something he did not bother to outline for the Lithuanian, Polish, or Belorussian "cooperative nations." Four diagrams were drawn on the basis of this concept, demonstrating cooperative dynamics in nine provinces: Chernigov, Ekaterinoslav, Kiev, Kharkov, Kherson, Podoliia, Poltava, Tavriia, and Volyn.[50] The same provinces are mentioned in Table 65 "Credit Cooperatives in Ukraine,"[51] which means that Ukraine was envisioned as a cultural (language) community consisting of nine provinces. Yet at the same time, in other diagrams these "Ukrainian" territories were redistributed into three separate regions: *Little Russia* (sic! which included Chernigov, Kharkov, and Poltava);[52] *Southern* (Bessarabia, Don, Ekaterinoslav, Kherson, and Tavria);[53] and *Southwestern* (Kiev, Podoliia, and Volyn').[54] One cannot but be puzzled by this classification at first glance. The core Ukrainian territories with the most developed networks of cooperative and national movements were ascribed the technical label of "Southwestern Region" (of the Russian empire), while "Little Russia" was reduced to three provinces. Even more mysterious was the "Southern" group that declared not only the lands of New Russia (*Novorossiia*) to be a part of Ukraine but also the Cossack Don and even Bessarabia. The obvious question is, of course, how did this regional model of Ukraine correspond to the cultural-linguistic model, and why did this cooperative-national mental map include territories that clearly did not belong to the mainstream Ukrainian spatial self-perception?[55]

The story of the intracooperative confrontation between the central, pan-imperial cooperative institutions and local cooperative unions in the Ukrainian lands suggests a possible answer to the latter question. Beginning at least in 1915, when the first pan-Ukrainian cooperative trust, the "Central Agricultural Union," was founded in Kiev,[56] rural cooperators in the southern provinces of the empire showed an ever increasing gravitation toward Ukrainian institutions. Cooperators from the Don, Kuban, and southern Caucasus regions joined Ukrainian cooperative activists in the June 1916 secret meetings in Ekaterinoslav that discussed the necessity to create a Kiev-based cooperative bank. These secret meetings were followed in November by a legal cooperative congress with the same composition of participants and agenda. Characteristically, the congress took place in Rostov-on-Don, outside of culturally defined Ukraine.[57] Throughout 1917, this alliance of southern cooperative unions only grew stronger, expanding at the expense of Bessarabia cooperatives,[58] and certainly culminated with the outbreak of the civil war. With growing procurement difficulties in the empire, southern grain-producing regions were brought closer together by common economic interests and opposition to the "exploitative" grain-consuming imperial center. Apparently, these economic interests for Rostov and Stavropol cooperative activists were more important than any existing linguistic or cultural differences with the Poltava or Podoliia partners. The emerging "South-Russian" cooperative separatist movement partially overlapped with the Ukrainian nationalist movement and reinforced it, but was based on somewhat different principles of solidarity. The "cooperative nationalism" was more politicoeconomic than cultural and historical in orientation. Rural cooperatives in the Ukrainian provinces mobilized millions of peasants into a state of self-conscious economic and social actors, and integrated them into a larger society that was defined locally in national (Ukrainian) terms. Yet the national identity markers were of secondary importance in comparison with the stimuli of economic interests and interregional and inter-"national" cooperation.

"Russian" Russia and Russian empire

Last, but certainly not least in importance is the question of the Russian "Russian" (*russkii*) nationalist appropriation of the public modernization campaign in the countryside, including the cooperative movement. While to date there are no special studies of the topic comparable to case studies of the Ukrainian national-cooperative phenomenon, the available sources suggest a strikingly low degree of ethnonational Russification of the cooperative system. According to the often-cited data of Alexander Merkulov, in 1912, just 43 cooperatives in the Russian empire were controlled by the Union of Russian People and other ultranationalist ethnoconfessional Russian organizations.[59] While Ukrainian national-oriented cooperatives were greeted by Russian (imperial) cooperative ideologists and the public as a manifestation of vitality and true democracy in rural cooperation, Russian nationalist cooperatives were unanimously condemned as a reactionary deviation in the cooperative movement.[60] In general, the Russian national agenda in the form of aggressive nationalism could rarely be found in the ideology and everyday activity of countryside modernizers. Usually it expressed itself not as a set of any positive statements and actions but in the form of discrimination against minorities.[61] No other elaborated nationalist scenario was available for rural professionals that would have differed from the broad imperial nationalism or "black hundred" xenophobia advocated by ultranationalist politicians, who were treated by the Russian "progressive" educated public indiscriminately as outcasts. When Ekaterina Sakharova attempted to formulate a Russian "national" response to a probable victory of Germany in the war, she found available only anti-Semitic tropes as analytical categories, which she used in a most impersonal way.[62] Thus, when Russian cooperative ideologists spoke of the ideal society, it was envisioned within the limits of the Russian empire, while a true community of conscious people-cooperators was "national" in a civic rather than ethnic sense.

A good case in point is provided by popular publications such as cooperative calendars that targeted an extremely wide audience. The *Cooperative Desk Calendar for 1917* published by the influential Moscow Union of Consumer Associations cost only 4 kopecks while providing readers with an orthodox cooperative perspective on various issues of the day (it was published in 1916) in 113 pages. The calendar articles were pro-zemstvo and anticapitalist, and even modestly pacifist. Its most peculiar element was the most traditional and "neutral" part of any calendar: the Church calendar. The usual Orthodox Christian timetable was followed by the Armenian Gregorian, Roman Catholic, Lutheran, Muslim, and Judaic calendars.[63] Thus, a normative "Russian" (pan-imperial) cooperative worldview by definition incorporated imperial minorities as part of the general picture.

While agronomist assistance to the non-Russian population only indirectly affected the universalist, transnational worldview of rural modernizers, the nationalization of the cooperative movement presented a challenge to this universalism. Rural cooperatives became the primary locus of social self-organization and self-mobilization of the "new peasants," the ultimate goal of the initial project of "public agronomy," and the active partner of countryside modernizers. The patterns of "nationalization" as revealed by cooperatives characterized the general atmosphere in the village, and agricultural specialists had to adjust their everyday practices and general programmatic visions to these realities. In some regions, the activity of rural professionals and the zemstvo "second element" served as a foundation for the subsequent mobilization of villagers and provided modern venues for nationalist sociability by establishing cooperative associations, village schools, and libraries in the national language. That was the case in Ufa province and, to a lesser extent, in Kazan province. In other parts of the empire the public modernization movement encountered the already well-developed institutes of nationalist mobilization. While in the Polish and Baltic provinces rural professionals had to rely on the preexisting structures and cultures of "nationalization," in the Ukrainian lands the technocratic and nationalist modernization projects coexisted in a state of complementary balance. The underdevelopment of the Russian ethnoconfessional nationalist project resulted in the predominance of the pan-imperial civic version of Russian nationalism among the modernizers and "new peasants" who identified the emerging community of conscious economic actors and social activists as "Russian."

The emerging empire-wide phenomena of the "cooperative Russian nation," or "the nation of public agronomy," did not exactly coincide with the borders of the Russian Empire. Judging by the directories of agricultural specialists in the government and zemstvo service, the pan-Russian movement of rural professionals-modernizers practically ignored the Baltic region and the Polish lands. An agronomist or agricultural specialist could be expected to move from the inner Russian provinces to Ukrainian lands, to the Caucasus, Bessarabia, Siberia, and even to Central Asia, in search of a better climate or paycheck. Western borderlands were virtually off limits, and most of the personnel employed there were clearly of local origin. While this issue requires a special study, we may suggest at this point that the Baltic and Polish educated public did not see in their imperial colleagues harbingers of any appealing version of modernity, and certainly were unwilling to participate in any pan-imperial social movement. Local nationalisms served as dominant attractors of social self-organization,[64] and the "empire" of Russian modernizers did not feature Baltic and Polish territories on its mental maps. Thus, while rural professionals-activists, speaking on behalf of the universal "modernity," had experienced trouble in localizing a particular "Russian" nation as a community of rational like-minded citizens, their imagined "imperial" civic nation was confined to the territory resembling the early Soviet Union. Apparently, even a technocratic understanding of modernity as a process of rationalization and the formation of a new economic man implied a radical transfiguration of the Russian Empire.

10
Revolutionary Nation

The revival of the intelligentsia legacy

The New Year 1917 issues of Russian economic periodicals did not cheer their readers up with bright perspectives on the coming year. The *New Economist* predicted the end of the World War in 1917 due to the mutual economic and military exhaustion of the warring countries.[1] The *Kostroma Cooperative Activist* went even further in its dark predictions:

> What hopes can we talk about now, when…the existing [political] regime…has given up all of the country's hopes and dreams?

> …There is no hope—our intellect tells us. But the soul still cherishes a faint hope that "this will not happen." [Our] country faces great tasks, and it must also escape great difficulties. It is so obvious that only efforts of the whole people can cope with those difficulties. To this end, nothing should interfere with the people in their effort to create a future for Great Russia.[2]

The author of this passage was apparently echoing the earlier pessimistic economic and political prognoses of other agricultural periodicals; he blamed the government for the economic chaos and inefficiency and explicitly placed all of his hope on the mobilization of the nation—not in any ethnoconfessional or institutional sense, but rather sociopolitically (people as citizens).[3] This was the reflection of a general process of dramatic alienation of the major social actors from the regime, which allegedly demonstrated "stupidity or treason,"[4] to quote the catchy formula of Paul Miliukov. While even the most leftist agricultural specialists did not dream about any political changes in the nearest future,[5] the general atmosphere of anxiety and mistrust intensified every day.[6] Among the agricultural specialists, renewed mistrust in the state resulted in a sort of "ideological regression," as the newly formed professional and technocratic discourses gave way to the old intelligentsia "grand narratives" of "conscious *obshchestvennost'*" and "democratic zemstvos," thus indicating a retreat at least ten years back to the early post-1905 epoch.[7] The seemingly apolitical public sphere of the *obshchestvennost'* became increasingly seen as the only legitimate representative of the national interests of Russian citizens.

This trend toward the alienation of state authority on the part of both the people and the intelligentsia was noted by Peter Struve as early as January 1914. The first years of the World War I temporally bridged the gap between the tsar and the people, but in January 1917, Struve's three-year-old diagnosis sounded more up to date than ever:

> There could be only two solutions: either a gradual increase of the state crisis [*smuta*] in which the middle classes ... will be forced into the background by the elemental pressure of the popular masses inspired by the radicals, or *the normalization of authority*.[8]

While the prospects for the normalization of authority gradually faded away, many intellectuals hoped that the pessimistic part of Struve's scenario would not evolve to its full extent. This explains the unanimously positive initial reaction of agrojournalism to the February Revolution in Petrograd. The Moscow weekly *New Economist* responded to the news of the revolution on March 4, 1917. Its columnist (apparently, Professor Migulin) claimed that he was surprised not by the "coup" itself but by the easiness of it. He applauded the composition of the Provisional Government, and expressed a hope for radical economic reforms.[9] On March 15, the editorial board of the Kievan *Machine in Agriculture* greeted the revolution but added that instead of revolutionary rhetoric, "strict, sober practicality [*delovitost'*] should become the major tone of the special periodicals":

> The best guarantee of the strength of the Democratic Republic is the implementation of a certain economic program.... This is possible only when democracy is organized on professional grounds. Therefore, we will call for creative economic work.... We will call for the fastest organization [of the masses] in agriculture on the grounds of professionalism.[10]

On March 18, one of the key ideologists of the public modernization campaign in the countryside, Konstantin Matseevich, formulated new tasks for agrojournalism in the Petrograd *Agricultural Gazette*. He argued that the revolution had erased the old opposition of "us" and "them" (i.e., "society" and "the rulers"). Matseevich called upon the agricultural press to reeducate the agriculturists, teaching them to trust and support the new "popular authority."[11] This reorientation of agriculturists was needed because less than three years earlier, it was Matseevich himself who had insisted on the omnipotence of the *obshchestvennost'* and argued that the economic sphere should be completely separate from the state and even the "institutions of self-government" (i.e., zemstvos).[12] Since the entire reformist movement of the post-1905 period was structured by the self-organization and self-mobilization of *obshchestvennost'*, occasional public denouncements in the press could not weaken its grip over the political imagination of educated Russians in 1917. It was this concept of *obshchestvennost'* as the community of a self-conscious progressive public that dominated the idea of revolutionary nation in 1917 and explained much of the instability and vulnerability of the new political regime.

In February 1917, the "apolitical politics" of Progressivist reformism revealed the tremendous political potential of the organized *obshchestvennost'*. Bolsheviks were very aware of this political potential of the *obshchestvennost'*, and the deportation of intellectuals and professionals in 1922 could be seen as a direct assault on the remnants of educated elite that preserved the ethos of the independent collective initiatives.[13] Moreover, police interrogators routinely requested that those arrested as candidates for exile explain their attitude toward the role of the intelligentsia and *obshchestvennost'*. The interrogation protocol of the Moscow lawyer, Nikolai Muraviev, who provided legal counseling to several cooperative associations in Moscow, reveals exactly why Bolsheviks were afraid of the old Russian *obshchestvennost'*:

> In regard to *obshchestvennost'*. So-called public life is indispensable to the country. Without it, [state] authority cannot function properly. Under the conditions of the current cultural level (and for a long time!), *obshchestvennost'* replacing statehood is an unfulfilled dream. The coercive realization of this idea only hampers the development of statehood.[14]

Thus, even social activists without any party affiliation were perceived as a political threat because of their *obshchestvennost'* spirit. The point here is of course not Bolshevik security paranoia but an acknowledgment of the existence of a "dream" of the *obshchestvennost'* eventually undermining the state—something that Alexander Chaianov called "the organized public mind" in 1917, literally borrowing from the sociology of important Progressivist sociologist, Lester F. Ward.[15] The political potential of *obshchestvennost'* was already evident during the takeoff of the public modernization campaign. Reporting from an annual conference of the Mutual Aid Agronomist Society in 1911, a police informer noted,

> Jokingly, these November meetings of agronomists are called "the agronomist parliament." Indeed, their mutual aid society and annual November meetings accomplish much in terms of uniting this

corporation. Gradually, these meetings turn from discussing narrow and special scientific issues toward discussing general issues of peasant life and economy, and socioeconomic [issues].[16]

At that time, 750 members of the society[17] were the only constituency of the virtual "agronomist parliament." The World War I tremendously empowered the organized *obshchestvennost'* while making it virtually immune to government control: the Union of Zemstvos and the Union of Towns proved to be indispensable in organizing the war effort, and their leaders were increasingly seen as the parallel, and better, "de facto governments within the official government."[18] The reforms of the food procurement mechanism in 1916 resulted in the virtual surrender of this strategic function to the "organized mind" of provincial zemstvos and cooperative associations. By 1917, the government had demonstrated beyond a reasonable doubt its inability to coordinate political and economic life in the country, and in effect transferred its functions to the self-organized *obshchestvennost'*. Regardless of the reality of the existence of a free-mason or liberal politicians' conspiracy in February 1917, the Russian public was in charge of the country, and the revolution simply made the collapse of the state and a sudden triumph of civil society a legally registered fact.

The *obshchestvennost'* nature of the new regime predetermined its major characteristics: reliance and dependency on a general consensus as the ultimate source of its legitimacy; mistrust of the state apparatus and inability to enforce political decisions administratively; and an excessive belief in the self-sufficiency of the ideal world of principles and theories (well beyond "usual" revolutionary dogmatism and idealism).[19] The poetics of the "revolutionary nation" as it emerges from the public discourse and rich symbolic manifestations of the epoch[20] reveals its *obshchestennost'* genealogy as a fusion of revolutionary ideals and intelligentsia intellectualism. In the projections of public politicians, the ideal citizen of the new Russia was equally removed from both populist-informed "nativist" peasant socialism and Marxist internationalist proletarian revolutionarism. This citizen was expected to be so disciplined and enlightened as to wait indefinitely for the Constituent Assembly to solve the land problem; to fight German troops in the ranks of an army stripped of any remnants of military discipline; and to maintain public order in the absence of effective institutional mechanisms of social control. Most important, this citizen was not an ideal type to be matched by means of social engineering, and not a product of utopian wishful thinking, but rather the normative portrait of an *obshchestvennost'* member. The politics of modernity conducted by the Russian educated public in the early twentieth century entered its most extraordinary stage: with a strong perception of "mission accomplished," many of the former modernizers lost the sense of a distance between the social reality and the normative future. The major problem here was not the "unrealistic" ideal goals of public modernizers, but their losing sense of the actual "underdevelopment" of the objects and actors of that modernization activity. They stopped acting as mediators in the dialog between underintegrated social and ethnic groups and the larger nation (in its different meanings), and instead indulged themselves with the illusion that general consensus had already been achieved. If, as Orlando Figes suggests, "they [had] understood their task as nothing less than the creation of a *new political nation*,"[21] they would have continued the policy that had paid off during the previous decade. The actual practice suggests another interpretation. After February, public modernizers believed that the revolutionary nation *was* reality, and treated *obshchestvennost'* as the incarnation of this nation while virtually ignoring the validity of interests of other social groups.[22] The growing gap between the people and the February regime reveals the rupture or inefficiency of the intercultural dialog that had proved to be so fruitful and intensive before 1917. For the first time since 1907, the politics of modernity became radically detached from its social base and object—"the people."

The paradoxical situation of this revolutionary elitism is vividly illustrated in the following case. On April 6–8, 1917, the Odessa Soviet of Workers' Deputies, dominated by Social Democrats and Socialist Revolutionaries, organized the Congress of "Citizen-Farmers" (*grazhdan-khleborobov*). A comparison of two accounts of this event reveals a gap between the normative revolutionary discourse that politicians tried to impose on the peasantry and the confused mindset of the peasant public. An official account of the congress mentioned all the clichés characteristic of any public event of that period: "They dandled the organizers of the congress and speakers. In commemoration of the fighters for freedom, everybody stands up to sing 'Eternal Memory.' The participants in the congress slowly leave the theater…and walk along the city streets with red flags."[23] The editor in chief of the magazine the *Southern Cooperative Activist*, A. Nizhitskii, participated in the congress and was terrified of "direct democracy" in action: first of all, the organizers compelled the peasant audience to vote for the presidium of the congress composed of socialist functionaries, who were absolutely unfamiliar to the peasants. The presidium called upon the peasants to sing the revolutionary anthem "You've Fallen Victim" (*Vy zhertvoiu pali*). The peasants did not know this song, but instead sang the Orthodox Easter anthem "Christ Is Risen" (*Khristos voskrese*) with much élan. The subsequent work of the congress was orchestrated by the presidium: it encouraged those speakers who called for total land redistribution and confiscation of landowners' property, while ignoring such practical issues as the status of rented or unsown land. When Professor Alexander Lebedev attempted to explain with facts the futility of land redistribution, he was silenced by the self-proclaimed chairman of the congress.[24] Caught in the middle, the cooperative leader Nizhitskii and the agricultural specialist Lebedev (a graduate of the Novoalexandria Agricultural Institute in 1907) tried to expose the demagogy and dangerous instigation of the political agitators, but to no avail. They were ignored by the politicians and met without enthusiasm by the peasants.

In the spring of 1917, field specialists, caught between the euphoria of their leaders and the growing radicalization of peasants left to their own devices, attempted to control the spontaneous peasant unrest. On March 28, the congress of the third element of the Kherson district (Kherson province) acknowledged the necessity for rural professionals to work toward "bringing the deteriorating agrarian relations within legal limits."[25] A letter of the young agronomist from the Balashov district (Saratov province), Afanasii Levshin, written at about the same time, shows what "legal limits" meant in practical terms. In April 1917, he testified that no one around him had thought that revolution would come so soon: "it was expected that revolution would happen after the end of the war, and there would be massive disorder and bloodshed."[26] Haunted by those images, the landowners of Perevesin county did not attend the first postrevolutionary gathering of local agriculturists that elected eleven people to the County Executive Committee, which replaced the old county administration.[27] The elected peasants responded with an act of equal mistrust: obsessed with the images of machine guns allegedly hidden by the local gentry, the committee ordered all landowners to surrender their firearms.[28] The agronomist Levshin was invited to the next meeting of the Executive Committee, and was unanimously elected as its twelfth member. So far, the Perevesin peasants acted rather calmly and were against the idea of looting gentry estates, but Levshin expected a deterioration of the situation and many troubles for the local landowners in summer.[29]

The constantly intensifying agitation of urban political activists found fruitful soil in the peasants' dissatisfaction with the price they had paid for integration into national society and the economy: military conscription, procurement campaigns, and state and zemstvo taxation. In late spring, the first reports of peasant unrest began appearing in the central periodicals. In May, disturbing news about the situation of agricultural specialists came from Kharkov province. Kharkov province was known for its Agricultural Society, which published the influential *Agronomist Journal* and *South Russian Agricultural Gazette* and ran the American Agricultural Agency; for its early and successful development of the precinct agronomy network; and for its booming cooperative movement. It was in this stronghold of rural modernizers and symbol of their success that peasants most directly questioned the results of the preceding decade of modernization. With the exception of the Lebedin district, where agronomists did not participate in requisitions, several peasant gatherings throughout the province decided to expel agronomists, physicians, and teachers to avoid paying zemstvo duties. Peasants also refused to finance husbandry mating centers and other zemstvo agricultural facilities. Apparently, even the additional channels of social integration of the villagers provided by the Ukrainian nationalist movement failed to offset the trend toward village self-isolation. In some places, such as the Volchansk district, the county committees vetoed these anarchic decisions and left the local third element intact.[30] At the same time, this was not some primitive "communal reaction," a restoration of the pre-1907 status quo: in the course of revolutionary 1917, the number of rural cooperatives in Kharkov province had grown by a factor of 4.3, while the average membership increased from 354 to 411 per cooperative.[31] This growth was greater than in neighboring Ukrainian provinces. Obviously, a decade of public modernization was successful enough to assist peasants in becoming the active subjects of the rural society. Now agricultural specialists were trapped between the peasants striving for self-determination to the point of expelling "alien" social elements and institutes, and the urban society imposing over the peasants what they perceived to be an unfair social contract. In the absence of any new attractive, yet realistic incentives to sustain the costs of national integration (comparable with the prewar promise of greater productivity and profits), under the pressure of wartime hardships and radical propaganda, the peasantry "downshifted" its external social ties and obligations in the same way that peasant households simplified the structure of production in a bad year or weak economy.

The history of the "new peasantry," whose ambitious plans were toppled by army drafts, the rapidly naturalizing economy, and the alleged conservative reaction of the villagers, is yet to be written. Soviet-era studies of the fate of "Stolypin peasants" or the "peasant revolution" of 1917 tell only part of their story, ignoring all those who changed their traditional ways without changing the system of land tenure.[32] Occasional sources mention their attempts to resist the entropic influence of traditional peasant social instincts and radical propaganda and act as members of the "revolutionary nation" of citizens.[33] In the late spring of 1917, during a village gathering in Kazan province, one peasant argued that land should be divided indiscriminately, "not excluding officials, merchants, workers, Poles, Jews, nor even, perhaps, women, as they have also received equality." The opposing voices, reportedly, challenged only the rights of women.[34] On May 7, 1917, a gathering in the village of Sukhaia Reka (Kazan district) passed a resolution supporting the national war efforts and ordering all deserters hiding in the village to return to their army units.[35] In some village youngsters, the revolution inspired idealistic gestures quite comparable with the agricultural school brigade movement of urban youth. Peasant boys 9 to 14 years old from the village of Starye Patrushi (Urzhum district, Viatka province) organized a union with the objective "to eliminate in their village the stealing of apples, raspberries, and other fruits, and also of vegetables and sunflowers; … they swore by the conscience of free citizens not to play cards, smoke, or curse; to eliminate quarrels and fights among themselves." Boys shared all the small change they had, and used the resulting six rubles to subscribe to a newspaper and to buy "a flag of revolutionary freedom."[36]

These incidents, reported by a local periodical, demonstrated that an opportunity still remained to sustain the professionals–peasants dialogue along the lines of the "revolutionary nation" project. The perspectives of this dialogue were compromised both by the peasants' widespread "lack of revolutionary consciousness" and by the unrealistic turn in the intellectuals' politics of modernity in 1917. The revolution revived all of the archetypal myths of the Russian radical intelligentsia that seemed to have been substituted by the new professional ethos of the post-1905 new generation of Russian intelligentsia. Neglecting over a decade of firsthand experience of cooperation with actual peasants, ideologists of the modernizing movement referred to the peasant as some ideal (and passive) object of revolutionary politics and did little if anything to stabilize the situation in the villages. Of course, thousands of rural professionals were unable to control millions of peasants. However, while field specialists living in the village were compelled to reckon with the rising radicalization of the countryside, many of their urban colleagues actually contributed to this radicalization. The degree and pace of transformation of one's position under the influence of the revolution is illustrated in the case of economist and professor Petr Migulin, one of a few Russian economists who lobbied for the interests of a peasant market-oriented economy, as opposed to the interests of industrialist or trade lobbies. Throughout the entire interrevolutionary era, in a dozen monographs on the problems of trade and finances and in numerous articles, he developed a very sober and pro-agriculture type of economic discourse. On November 19, 1916, Migulin published an article warning Russian politicians against following the German example of a militarized "socialist" economy and curtailing free market economic mechanisms.[37] On December 3, 1916, he argued against the system of fixed prices on grain and the policy of compulsory requisitions that in his opinion inaugurated the socialist economic policy in Russia.[38] Yet, on March 4, 1917, just after the February Revolution proved to be a success, Migulin, the advocate of free-trade capitalism, made a stunning revelation:

> Bourgeois capitalism of the old type has outlived its age. Our age is the age of state socialism, probably a harbinger of the fundamental change of our entire state order.... Conservatism is a dangerous policy in economics as well as in politics. *Radical* reforms are needed.[39]

Migulin's transformation culminated on May 13, 1917, when he denounced the entire project of public modernization of the Russian economy that he had actively supported during the interrevolutionary period. He also condemned the policy of the Department of Agriculture, which after 1906 "listened to the advice of agronomist-theoreticians" aimed at the intensification of the peasant economy, instead of the compulsory confiscation and redistribution of land needed to sustain the peasantry's traditional economic practices.[40] Migulin's case proves that under the impact of revolution, even a professional economist with a set of firm liberal principles and values could make a dramatic turnabout in the course of a few days. The ideological evolution of American Progressivism during and after the war toward a more statist and manipulative version of social engineering demonstrated the same vector. It is no surprise then that many intelligentsia-turned-professionals of the post-1905 period, with no particular commitment to liberal economic values, could not withstand the temptation to regress to traditional intelligentsia attitudes and reactions.

Instead of explaining to the peasants the inevitable economic limits of any radical reform, many agricultural specialists referred to their special expertise only to enhance the credibility of their political statements. A former government agricultural specialist, who had participated in the implementation of the Stolypin land reforms, German Tanashev, presented the most striking example of placing professional expertise in the service of ideology. In the spring of 1917, in a number of articles, he condemned the Stolypin land settlement policy, speculated on the peasants' current mindset, and reconstructed peasant "agrarian programs,"—all exclusively on the grounds of publications in seven central newspapers, none of which was even a professional agricultural periodical.[41] This was a clear case of inventing a "peasant discourse" for the peasantry, and selling it to a general audience as a genuine expression of real peasants' feelings. The author's professional background would have supported the credibility of his speculations.

Tanashev was by no means alone in the submission of his professional identity (including the actual repudiation of his former activities) to the archetypal ideological discourse of the intelligentsia. It is easier to single out those agricultural specialists who, during the revolutionary euphoria of the spring of 1917, did not sacrifice their professionalism to the dominant atmosphere of triumph of the traditional radical intelligentsia's values. On March 18, the agronomist Alexander Fabrikant (born in 1880) tactfully suggested that despite all of their desires, the agricultural specialists should stay away from politics, which is beyond the sphere of their expertise, and concentrate instead on serving local agricultural needs.[42] By three weeks later, on April 8, it took Konstantin Ashin more than just the truism that the agricultural profession was incompatible with politics to support his point. The man who was first to analyze the prewar "agronomist crisis" now claimed that after the revolution too much significance was being attributed to the social conditions of coping with economic problems. Appealing to economic rationality, Ashin called for the protection of private landowners.[43] Alexander Kaufman, an agricultural economist of an older age group (born in

1864) and a member of the liberal Constitutional Democratic Party, felt that it was no longer possible to present one's expert opinion without positioning it vis-à-vis the current political spectrum. That is why, before arguing against a primitive and highly ideological popular understanding of the problem of land shortage, he made the following statement:

> I speak not as a party activist who is confined by a certain party program, ... or as an ideologist of a certain "rightist" or "leftist" sociopolitical current, but exclusively as a scholar and researcher.[44]

By late spring, as Viktor Chernov, leader of the most radical major political party, the Socialist Revolutionaries, became minister of agriculture, to speak "exclusively as a scholar and researcher" already meant taking a certain political stance. From this moment onward, the populist/socialist ideology became not just the dominant current in the public discourse on agriculture, but the official state policy. "Apolitical" professional expertise became an act of opposition to government policy. The government itself, in turn, politicized its routine operations in order to be seen as a legitimate leader of the revolutionary nation. On May 12, the newly appointed minister of agriculture invited all employees of the ministry to visit him at 1 p.m. Only 400 people could find places in the Great Hall of the Ministry of Agriculture. Many could not see Chernov, so they shouted "On the table, on the table." Minister Chernov climbed up on a table and delivered a short speech about how the Ministry of Agriculture must lead the popular movement. He then retreated to his office, while the meeting continued, featuring the speeches of peasants who had come to see Chernov.[45]

When the GUZiZ embarked on the Stolypin land reforms, the agricultural specialists opposing those measures on the grounds of their professional expertise found in the project of public agronomy the vehicle for an alternative reformist activity. In May 1917, spreading peasant unrest paralyzed any possibility of professional activity that was not approved by a local village committee, while the Ministry of Agriculture under Viktor Chernov tried to keep up with peasant revolutionarism. One step necessitated the next. Those who in late May defended the nonpartisan character of the agricultural assistance and cooperative movement were also compelled to criticize the project of a socialist economy.[46] The agricultural professionals were literally squeezed into the sphere of public politics.

Rural professionals into public politicians

In 1917, the agricultural specialists explored three major strategies of participation in national politics: acting beyond the government, for the government, and in the government. The perception of the revolution as the victory of the *obshchestvennost'* initially led to an attempt to institutionalize the political influence of public modernizers in the form of a broad public association. In the spring of 1917, intellectuals and professionals revived the most ambitious projects that had been suppressed by the old regime. Leading agricultural specialists of various political orientations were united in the League for Agrarian Reforms. The league was founded on March 25 during its constituent conference that elected an Organization Committee of eight leading agricultural specialists: A. V. Chaianov, N. P. Makarov, P. P. Maslov, S. L. Maslov, K. A. Matseevich, N. P. Oganovskii, A. V. Peshekhonov, and M. I. Tugan-Baranovskii.[47] Apparently, the initiative for the league came from the "the Artel" of the most active leaders of the public modernization campaign in the countryside (Chaianov, A. N. Chelintsev, Makarov, Maslov, Matseevich, and A. A. Rybnikov). With B. D. Brutskus (and excluding Maslov and Matseevich), this group is known as the Organization-Production School in Russian rural studies. There is evidence that all organizational work was done by this core.[48]

The first congress of the League for Agrarian Reform in Moscow on April 16–18 was attended by 136 delegates from 20 provinces.[49] According to the statute of the league, its membership was open to all agricultural nongovernment organizations as well as to individuals known for their research or practical experience in the realm of agriculture. Speaking on behalf of the league's organization committee, Alexander Chaianov formulated the aim of this public association as the elaboration of a common agrarian program that would transcend all party differences.[50] Following the congress, local branches of the league were established throughout the country.[51] In a true spirit of *obshchestvennost'*, members of the league deliberately distanced themselves from any direct involvement with the government. In the first critical months of the revolution, when the Provisional Government had at least some authority in the village, Russian public modernizers chose to play the role of independent experts relying exclusively on their moral and professional reputations and speaking directly to the people and the rulers.[52]

As it turned out, the niche was already occupied by party agitators and freelance ideologues, who appealed to the mass audience with simple arguments and lavish offers. Economists, agronomists, and cooperative managers could not compete with their propaganda and soon found themselves alienated—not unlike Lebedev and cooperative manager Nizhitskii at the Odessa Congress of "Citizen-Farmers" in April. Therefore, the second strategy of rural professionals' participation in national politics came to the fore in late spring of 1917 when agricultural specialists joined the Chief Land Committee. The committee, established by the Provisional Government on April 21 to elaborate a project of land reform, included all of the key figures of the League for Agrarian Reforms.[53] While the league was a free association of agricultural experts shaping and correcting public discourse on the agrarian question, the Chief Land Committee had the potential power to influence state policy. For the members of the Chief Land Committee, the price of their greater influence over politics was their greater dependence on the current political mainstream. The constituent meeting of the committee (May 19) adopted the major principle of the future reform, which from the very beginning predetermined its character and objectives: "all arable land should be transferred to the laboring agricultural population," that is, to the peasantry.[54] Thus, the role of the agricultural specialists was merely to clarify the technical details of the transfer of land.

The entire "Artel" of the ideologists of the past decade's modernization movement participated in the Chief Committee. They controlled the most important Commission for Redistribution of the Land Fund, one of five commissions that studied different aspects of the future reform. Semen Maslov was chair of the commission, Alexander Chelintsev was deputy chair.[55] Ber Brutskus, Nikolai Makarov, and Alexander Rybnikov were also among the key figures on the commission, elaborating the principles for determining the individual quotas of land possession.[56] The activity of the commission from July through September revealed the insignificance of the existing theoretical disagreements among its members in the face of the predetermined political agenda. In his presentation on July 28, Brutskus was first to recognize the futility of all attempts to constrain peasant land redistribution activity by any "scientific" norms and rules.[57] On August 1, he was confronted by Alexander Chelintsev: "there is a general impression from the presentation that B. D. [Brutskus] is hypnotized by the facts."[58] Indeed, heated discussions concerning what a "semicapitalist farm" is or whether rent actually exists could be sustained only in an atmosphere of isolation from the country's political and economic reality. While acquiring the semiofficial status of government advisers, members of the Chief Land Committee scarcely had any real influence over actual politics. In the rapidly radicalizing atmosphere of the summer of 1917, the government itself gradually lost control over the countryside.

Meanwhile, many agricultural specialists turned from the revolutionary euphoria of the spring period to a more sober attitude toward reality in the summer. In early June, the dominant tone of agrojournalism changed from support for radical reforms in agriculture to sharp criticism of any signs of anarchy in the economic and juridical relationships in the countryside. On May 27, the *New Economist* published an article by the Kiev economist, 40-year-old old professor K. G. Voblyi, who demonstrated the economic impossibility of universal land redistribution, and actually rehabilitated the initial program of public modernization.[59] On June 10, Petr Migulin condemned the radical approach toward agriculture that he himself had demonstrated only one month earlier:

> While dreaming about turning upside down firmly established social relationships, projecting grandiose reforms, building castles in the air, we forget living reality. While we will "nationalize," "socialize," and

"municipalize" all the land in Russia,…universal hunger will strike, followed by hunger riots, civil war, a general Jacquerie, and similar excesses. Demagogy is a dangerous thing. Ignorance is much worse; mass madness is even more so. The greatest caution is needed in setting and solving the agrarian problem.[60]

The same day, the prominent cooperative activist Andrei Evdokimov, a former member of the Russian Social Democratic Party, attacked the infantile economic consciousness of the intelligentsia, which allowed them to believe "in the magic of socialization, municipalization, and so on."[61] This turn against extremism in dealing with the agrarian question gained momentum during the second Congress of the League for Agrarian Reforms. Closing the three-day congress on June 26, Alexander Chaianov suggested that this day be recognized as a very significant date:

> Today, economists belonging to socialist parties had to descend from the skies to earth. As soon as they were compelled to account for the means of realization [of reform] before the representatives of economic science, they immediately found it necessary to correct their program dogmas by the data of actual economic reality.[62]

Throughout the summer, virtually all of the flamboyant springtime agitators for radical action in the countryside from the ranks of rural professionals retreated to a much more sober position in the face of the growing economic and political crisis. Even the former eloquent defender of a "black repartition" of all land among the peasantry, German Tanashev, felt compelled to oppose the realism of "populist economists" to the "dogmatic fanaticism, doctrinairism, and dangerous optimism" of the Russian radical socialist parties.[63] In a situation where the League for Agrarian Reforms lacked direct influence over politics, and the agricultural specialists of the Chief Land Committee were proving their inability to correct the preset political agenda of that institution, the third strategy for participation in political life emerged: agricultural specialists turning into politicians.

This trend also originated in the late spring of 1917. On May 2, A. N. Chelintsev, the founding father of the Organization-Production school and a member of the "Artel," was appointed director of the Department of Agricultural Economy and Statistics in the Ministry of Agriculture. The populist economist, zemstvo agronomist, and statistician Panteleimon Vikhliaev (born in 1869) was appointed deputy minister of agriculture.[64] However, any possibility of their influence on ministerial affairs was overshadowed by the radical position taken by Minister Chernov. By mid-May, frustration about the situation in the countryside and the absence of any political leverage over it reached its peak.[65] The *obshchestvennost'* did not include the entire population of Russia, the people did not behave as "conscious citizens," and the government could not sustain "revolutionary order." In the opinion of high-ranking cooperative activist Evdokimov, the main problem was that the village was undermined by the "dictatorship of towns" both economically and ideologically:

> In fact, only townspeople speak on behalf of the village.... Cooperative activists could construct the ideological apparatus of the New Russia's village. To this end, they must be firm and courageous. Because elemental forces are against them. The dictatorship of towns is against them. The demagogy of "throats of cast iron" is against them. The main currents of intelligentsia psychology, aimed at *redistribution* instead of *production*, are against them.[66]

The idea of turning the extensive cooperative network into a political force representing the "productive" economic interests of the countryside was crystallizing throughout the summer. Evdokimov, editor and publisher of the journal *Messenger of Cooperative Unions*, was among the most active propagators of this new turn.[67] In the provinces, cooperative activists also recognized a new role for the cooperative movement in tutoring the peasantry for elections to the Constituent Assembly.[68] In early October 1917, the All-Russian Cooperative Congress in Moscow decided that cooperatives should directly participate in the elections as a separate political movement.[69] At the same time, members of the "Artel" were taking key positions in the Ministry of Agriculture.[70] There were hopes for a peaceful transfer of state authority to the "organized mind" in the form of a magnificent pyramid of cooperative associations and unions. After the deep political crisis of the summer of 1917 (the failed July Bolshevik and the August Kornilov putsches) and the final split within the politically active *obshchestvennost'*, the extensive cooperative network appeared as a legitimate candidate for the embodiment of a "revolutionary nation." Only the persistence of Russian radical tradition and populist-socialist revolutionarism can explain why this shift had not occurred much earlier,[71] as there were ready visions and scenarios of a "cooperativist regime," and the Russian cooperative movement entertained the utopian ideal of changing the political and economic system in the country. "Cooperativism" became publicly visible only when the dominant political scenarios and actors compromised themselves during the first six months of the revolution.

Yet, it was too late for agricultural specialists to use such a direct political leverage as parliamentary politics. The vast population did not know the names of cooperative activists-turned-politicians, and the election campaign itself was hastily and superficially organized.[72] Almost all cooperative candidates were also members of local procurement agencies, which did not contribute to their popularity among the peasants.[73] According to the *Nizhegorod Zemstvo Gazette*, during the elections to the Constituent Assembly, many peasants reacted with much irritation to the party agitation: "those agronomists should all be killed."[74] Ukrainian cooperative unions called upon the population to vote for Ukrainian socialists rather than Russian cooperative activists.[75] Eventually, the cooperative politicians failed in their election campaign,[76] while the team of Semen Maslov in the Ministry of Agriculture did not take any decisive actions before the Bolshevik October Revolution.

Thus, cooperative leaders and rural professionals performed no better than other social groups in the Russian political arena in 1917. None of the three strategies of political activity proved successful, given the sociopolitical conjuncture and timing of their implementation. Contrary to the dominant historiographical view, "agricultural experts" never actually enjoyed a position of authority after the February Revolution.[77] Their attempts to institutionalize the politics of modernity as conducted by educated members of the *obshchestvennost'* in the course of 1917 evolved from establishing a representative public association, through membership in semiofficial advisory bodies, to direct participation in the government and the parliament. However, they were always one step behind the actual political conjuncture, offering solutions to yesterday's problems.

The revolution had further splintered the socio-ideological unity of the group of public modernizers both as a social generation cohort and as a semipolitical movement. The embattled local agricultural specialists had increasingly been separating from their theorizing urban peers, populist-minded economists from liberal economists, and socialist revolutionaries from constitutional democrats. The October Revolution threatened to split even the "Artel" of their recognized leaders. Immediately after the October Revolution, the deputy minister of agriculture, Alexander Chaianov, left Moscow for Kharkov, instead of returning to his ministry in Petrograd, and resigned his post.[78] Nikolai Makarov, on the contrary, publicly called on all ministers and deputy ministers to keep their positions and resist the Bolshevik "impostors" by all means.[79]

After eight months of revolution, the major achievement of many agricultural specialists was their disillusionment with populist and socialist ideas. Four days before the October Revolution in Petrograd, the agronomist Petr Rybalov, who joined the ranks of rural professionals around 1910, published an article elaborating a new program of action for agricultural specialists. He concluded that agronomists should use their expertise to explain to peasants the limited effect of any land redistribution, but immediately recognized the unreality of this measure: "But who now has the courage to deliver such speeches (or who can bring himself to break those illusions)?"[80]

The third congress of the League for Agrarian Reforms held on November 21–23 revealed this deep turn away from all projects of land redistribution and confiscation as the means of agrarian reform. In the words of the Menshevik economist P. P. Maslov, the debates following the opening presentations by A. A. Rybnikov and N. P. Makarov had the character "of a trial of the revolution and the land apparatus that it created."[81] All speakers more or less vigorously condemned all socialist experiments in the countryside, from the policy of Chernov's Ministry of Agriculture to the Bolshevik land decrees. Closing the congress, Nikolai Makarov suggested that its role was to replace populism in agrarian discourse with more up-to-date sociopolitical theories.[82] The third (and last) congress of the League for Agrarian Reforms reacted to the recent October Revolution with the following resolution:

> The uprising of Bolsheviks...detained and obstructed solution of the agrarian question by breaching the planned character of the preparatory work by land committees, and by pushing the rural population to unauthorized resolution of the agrarian question without any plan for the entire country. Despite these difficulties, which have increased with the disintegration of the state because of the Bolshevik uprising, the congress of the League finds necessary the most energetic continuation of the preparatory works by the Chief Land Committee and local land committees.[83]

In fact, the post-1905 crisis situation repeated itself, but this time not to the advantage of agricultural specialists: they lost an opportunity to counterbalance political extremism with their professional expertise, and their belated return to the program of limited social engineering coincided with the outbreak of civil war in Russia. At the same time, the situation of rural professionals in 1917 differed significantly from their situation in 1907. The preceding decade elevated a loose, small group of agricultural specialists to the position of a distinctive and influential social cohort with many thousands of members, consolidated by common professional and ideological values. The fiasco of this social group and its politics was the result, first of all, of its inability to institutionalize its influence under new circumstances. An *obshchestvennost'* phenomenon par excellence, the movement of rural modernizers could not compete with revolutionary party politics. Yet it also failed to mobilize its own unique resources (professional expertise and an extensive cooperative network) during the first months of the revolution, when the anti-system activity had not yet become the only popularly recognized form of politics. In this respect, public modernizers followed the path of Russian *obshchestvennost'*, which was powerful enough to topple the old regime but was unable to accommodate the contradictory social, economic, and national interests in the empire under the duress of total war. This was also a crisis of the general Progressivist scenario of piecemeal reformism. An ideologist of the substitution of the state by the self-organized *obshchestvennost'*, Alexander Chaianov testified one year after the February Revolution:

> We…made a stake on the popular independent initiative [*samodeia-tel'nost'*]…. But it turned out, to our deep regret, that this stake failed. Russian society, or perhaps European society in general, is still incapable of an enduring independent initiative, an enduring effort.[84]

The main lesson of the 1917 Revolution for most rural professionals was a deep disillusionment with leftist ideology and, most important, with the "revolutionary nation." "The people" (this time as a social category, including peasants) did not behave "right," as ideologists of the public modernization campaign would envision in the spring of 1917. As we have seen, field specialists did not have many illusions regarding the revolutionary consciousness of the peasants from the very beginning, but they were not allowed to influence public discourse until late in the summer. The powerful socioeconomic mobilization in the village caught both precinct agronomists and city theoreticians in the city off guard. The ideal of the dialogue with the "new peasant" was compromised by what was seen as a "communal reaction" that threatened both agricultural specialists and peasants engaged in rational farming. The old model of "economic man" did not fit the new reality, the program of gradual rationalization and intensification of the peasant economy seemed obsolete in the situation of collapse of the market and total land redistribution. The very presence of rural professionals in the village had become problematic. The army of agricultural specialists and cooperative managers found itself with no clear sense of the mission, and detached from its clients and partners: "the people." The potential for society's self-organization and integration into a "nation" (in its various meanings) was exhausted, at least temporarily. The country was sliding into civil war, with no prospects for a coherent politics of public modernization by means of the joint efforts of the autonomous *obshchestvennost'*. As they put it during the third Congress of the League for Agrarian Reform, "the belief in the might of the 'organized public mind' was shattered."[85]

11
The Dissolution of the "Imagined Community": Nationalization as Expropriation

The role of the public modernizer and the project of piecemeal social reformism became redundant in the situation of "permanent revolution." With the constant radicalization of the regime and the masses such attempts inevitably come to be seen as "conservative" vis-à-vis the revolutionary agenda. The corporation of agricultural specialists structured by the ethos of piecemeal reformism of the peasant economy and by high social demand for their services as expressed through the booming job market found themselves in a social and ideological void. The editorial opening in the very first issue of the leading *Messenger of Agriculture* in 1918 expressed the utter frustration of countryside modernizers, acknowledging the return to "ground zero" of the post-1905 *Bezvremenie* (literally: an epoch "out of time") period, only without any positive program of action or hope:

> We came into the new year along the path of such a *bezvremenie* that one cannot even think about any logically substantiated program whatsoever.[1]

The cohort of rural professionals united by a common program and life experience, with the core consisting of the post-1905 new generation of Russian intelligentsia, was the major actor in public mobilization and modernization after the failed Revolution of 1905. By 1918, socially disintegrated and ideologically confused, these people were going through a painful demobilization, gradually ceasing to be a distinctive group and emerging as individuals acting largely on their own. In a most sober and astute way, Alexander Stebut, director of the Agricultural Experimental Station in Saratov (and son of the patriarch of Russian agrarian studies Ivan Stebut), analyzed the situation of rural professionals vis-à-vis the nation one year after the February Revolution:

> We think in vain, though with understandable irritation, that Bolshevism is nothing more than a repulsive party slogan. No. Bolshevism is an upsurge of life itself; it stirred up and raised from the bottom all that evil, despair, hatred, and bitterness of which the people's soul is capable in moments of disastrous misfortune.... We are afraid to acknowledge that not individuals but the whole nation [*narod*] is guilty of the current events.... Yet if one can run away from Bolsheviks but not from life, is it worth it even to think about escape? When life collapses, it crushes much and many. You may perish. But there is a creative element in life that one should not forget about. With this belief in the capacity of life to renovate itself it is possible to endure even the darkest time, to search for a way out, and to fight. Do not flee, comrades![2]

Long before Nikolai Ustrialov and Nikolai Berdiaev acknowledged the deep "national" roots of the Bolshevik Revolution and accepted it as a historically inevitable event, Russian agricultural experts, particularly field specialists, realized the fundamental nature of the revolution. But unlike city intellectuals, they were not ready to accept the destruction of the fruits of a decade of modernizing activity and embrace "the people" as a historically given reality. Witnessing peasants' indiscriminate rampaging and burning of landlords' mansions and agronomist facilities,[3] agronomists and cooperative activists knew that the very same people could be different: rational farmers and law-abiding villagers. There was no reason to idealize peasants in their present mindset, "the dark people, confused and unable to somehow rationally distinguish their friends from their foes and altogether rejecting their main life force [*zhivuiu silu*], their intelligentsia."[4] Any populist illusions were compromised in the course of 1917, and a key pillar of the initial public modernization program (a service to the people)[5] lost ground in the situation of the revolutionary village. Besides personal survival, the only public mission now accepted by rural professionals was to keep their knowledge and experience for the future.[6]

Left without any positive program of their own, deprived of institutionalization with the abolition of zemstvos in 1918, and disillusioned with the revolutionary nation, rural modernizers-turned-Soviet-citizens became objects of manipulation by major social actors of the time—first of all, the political regime. In the situation of social anomie and revolutionary excesses, a regular state authority was perceived by many educated people on the postimperial territory as the only positive and highly desirable outcome of the revolution. These people were instrumental in transforming a regime into a state,[7] and reintegrating the "population" into a "nation," thus their modernizing potential was utilized in one way or another. This new nation would coincide with the state and be defined by the state-sponsored ideology, and the process of national integration would take the form of an expropriation of classes and nationalities by the only major political actor left on the scene, the political regime. However, its ideological outlook and actual performance were influenced to a certain degree by the "old regime" personnel and by its relations with different groups of the population. Therefore, we shall examine more closely the interaction of the former public modernizers with the new political and socioeconomic realities.

Demobilization and hopes for revival

The initial reaction of many agricultural specialists who opposed the October Revolution was to concentrate on their professional activities and to ignore the Bolshevik political regime, which was not expected to last long and was not actually present on much of the former empire's territory. The apparatus of rural cooperatives promised a particularly suitable platform for independent socioeconomic activity. At the beginning of 1918, a member of the "Artel" and a prominent leader of the public modernization campaign of the past decade, Alexander Minin, explicitly stated, Russia is ruined. "Cooperatives should take over the task of state building and allocate from their milieu people capable of state work."[8] Cooperative leaders called themselves "active agents arranging for a new Russia."[9] The cooperative superstructure, then still largely intact within the broadening economic crisis, looked especially attractive to high-ranking agricultural specialists who could aspire to positions in the offices of all-national cooperative associations.[10] The first All-Russian Cooperative Congress in Moscow in February 1918 expressed the intention of the agricultural specialists involved in cooperative business to strive for economic domination while ignoring the ongoing political struggle. In its final political resolution, the congress condemned the Soviet government and the spreading civil war.[11] However, the congress's major theme was the economic activity of the cooperative movement. Although highly critical of any nationalization projects, cooperative activists were anxious to monopolize all Russian agricultural exports themselves at the expense of the weakened private trade.[12] Closing the congress, the cooperative activist Alexander Merkulov attempted to offer a new agenda for the community of agricultural specialists that found itself *nationless*:

> The great Russian people died, what remains is simply Russian people.... We, the bridge builders, should pave the path to a better future.... Recognizing our guilt for the destruction of the motherland, we will work in closed ranks for the reconstruction of great Russia.[13]

Merkulov almost repeated the words of Alexander Stebut that had been published in the *Messenger of Agriculture* a few days earlier. While Stebut saw the role of agricultural specialists as rather passive, as "keepers," Merkulov envisioned a more active position for cooperative activists still relying on a certain social base. For a while, the central cooperative apparatus was successful in its activities. Despite difficult political conditions, in 1918–19 the central cooperative agencies established effective trade contacts abroad, toppling the economic blockade imposed by Western countries and thus contributing to the stability of the Soviet regime.[14] The old dream of cooperative ideologists to produce highly qualified cadres for the cooperative network was becoming a reality.[15] In 1917, the first Cooperative Institute was founded in Petrograd, and in 1918 the Cooperative Institute in Moscow followed suit,[16] despite protests from critics who saw in cooperative institutes a purely symbolic gesture underlying the high status of the cooperative movement.[17] In 1918, there was even hope that the strong cooperative associations would take over the former zemstvo agronomist structures, which were now all but paralyzed with the abolition of the zemstvo. Indeed, the first experiments in this direction looked quite optimistic. According to local cooperative periodicals, special agronomist departments were founded by the Samara, Astrakhan, Viatka, Kaluga, and Ekaterinburg provincial unions of cooperatives. The Moscow District Association of Cooperatives in Mediation (*kooperativov po posrednichestvy*) went even further and decided to finance the entire former zemstvo district agronomy network with a budget of 120,000 rubles.[18]

While cooperative activists on the Bolshevik-controlled territories entertained projects to build a system parallel to the Soviet regime and the collapsed "revolutionary nation," their Ukrainian colleagues successfully worked for the "national-revolutionary" cause. The network of cooperative associations and unions provided the popular basis for the Central Rada regime, whose administration included cooperative leaders at various levels.[19] Former "Artel" member Kost Matseevich, along with cooperative ideologist Mikhail Tugan-Baranovskii and former Poltava zemstvo instructor of cooperation, now Ukrainian prime minister, Borys Martos, founded the Ukrainian Cooperative Institute in Kiev (which opened on January 1, 1920).[20] Borys Martos characterized the regime of independent Ukraine in 1917–18 as a "Ukrainian cooperative order."[21]

Yet, the success of the cooperative superstructure did not last. In Ukraine, the authoritarian regime of Hetman Skoropadsky forcefully replaced the government of the Central Rada on April 29, 1918. The new regime responded with great suspicion to the political activity of cooperative leaders, who very often were members of the Socialist Revolutionary Party or just socialist sympathizers. The occupation of Ukraine by German troops undermined the sense of national independence and irritated cooperative activists. In the RSFSR, by mid-1920, with the nationalization of the financial backbone of the cooperative movement—the Moscow People's Bank—and of the consumer cooperatives, and with the closing of the Cooperative Institutes, all hopes for the nationwide influence of the cooperative apparatus faded. Cooperative publications suffered from economic crisis as well as severe censorship.[22] At the same time, the local cooperative associations grew extremely unhappy about the costly and authoritarian cooperative superstructure in the form of all-national unions.[23] Unlike their leaders, local cooperative activists could not afford simply to ignore the peasants who formed rural cooperatives. The situation of rural professionals in the countryside deteriorated every day. The Viatka cooperative instructor, former St. Petersburg librarian Alexander Gusakov, whose story was related earlier, was shocked by the changes that took place in the Viatka countryside between August 1917 and March 1918: radicalized soldiers controlled life in the village, cooperative property was often seized and redistributed by peasants, and normal economic relations were all but paralyzed. His assessment of the situation literally repeated Merkulov's conclusion:

> Cooperative activists lose all hope and energy, and feel bitterness from the thoughtlessness [*nesoznatel'nost'*] of the people.[24]

> All creativity comes only from visiting intelligentsia-cooperative activists. It has always been this way, and it is still so.[25]

The Board of Instructors of the Viatka Credit Union tried to calm the rural population, which began withdrawing deposits from cooperative banks: the Bolsheviks would not dare to requisition cooperative funds—no government would; soldiers could only threaten to destroy cooperatives—that was just rubbish, spread by confused people.[26] The Chistopol' district (Kazan province) cooperative instructor Konstantin Pyzhenkov, who graduated from a mid-level agricultural college in 1911, urged consumer cooperatives to use the prohibition of private trade to their profit, quite in line with the statements of the Moscow Cooperative Congress.[27] Boleslav Okushenko, a Chistopol' precinct agronomist, who also received a mid-level agricultural education in 1911, offered the initial program of public modernization, almost unmodified, as guidance for local agricultural specialists.[28] The Slobodskoi district agronomist (Viatka province), Arkadii Zonov (who had attempted to embody the ideal of agricultural expert as jack of all trades in 1913), desperately appealed to peasants, arguing that unless they started to yearn for a better life, nothing and nobody would help them.[29] These plans of local agricultural specialists to continue their activity, independently from the regime, proved unrealistic during the summer of 1918, as the Russian countryside sank into civil war and the state established its policy of compulsory food procurement. The first cases of suicide committed by cooperative functionaries on political grounds were registered at about the same time.[30] Neither agronomist precincts, nor experimental stations, nor rural cooperatives were safe in the revolutionary countryside. During the first nine months of 1918 alone, cooperative periodicals registered 328 cases of repression against local cooperatives, from unauthorized audits to requisitions, arrests, and nationalization.[31]

On the territories of the former Russian Empire not under Soviet control, agricultural specialists experienced the same disillusionment in the people (as the social category and the embodiment of the nation), and the same dashing of their hopes for independent activity, as that experienced by their colleagues living under Bolshevik rule. The first All-Ukrainian Agronomist-Economic Congress in autumn 1917 already observed that agricultural specialists had become an alien element in the Ukrainian village (but not yet in politics).[32] The peasants' hostile attitudes toward the village intelligentsia noticed in Kharkov province by May 1917 now spread throughout Ukraine. Rural professionals responded to peasant hostility with deep disillusionment concerning their former populist biases. In January 1918, specialist of the Don-Kuban-Terek Agricultural Society, Peresypkin, published an article in which he explicitly stated that peasant needs no longer corresponded to national economic interests:

> I do not... want to say that peasants are inferior to us—no, they are not inferior to us, but no better either.... Russia cannot afford the luxury of a primitive culture, nonintensive industry, and noneconomical allocation of human labor. The small laboring peasant economic unit does not meet... the demands of life at the present time.... [I] do not defend the large estates. As an agricultural specialist, I defend the interests of agriculture... from the vantage point of economic necessity and state rationality. At the same time, I cannot ignore a strange and curious phenomenon: 50 percent of the cultured population of Russia, who can understand our state-building, if they want to, takes the point of view of the peasant, who sees the key to his economic welfare exclusively in the increase of land use.[33]

In spring 1918, in a situation where peasants actually sabotaged zemstvos and the third element, a group of the most enterprising Kharkov agricultural specialists organized the South Russian Artel of Agronomists. This artel offered the expertise of its members to any public or government agency that wished to organize or improve the performance of an agricultural enterprise of any specialization.[34] From a formal point of view, it was only natural that rural professionals sold their services to any interested customers. The problem was, whether there was any demand for their professional expertise with the demise of the zemstvo and state programs of agricultural innovations, under severe political and economic crisis and structural de-intensification of peasant economy. We do not have much information about the performance of this artel. However, we know that a few months later, a different type of artel emerged in Novocherkassk (the Don region), which indirectly indicates that the Kharkov initiative could not become a big success:

The Proclamation of the Don Labor Artel:

Regional agriculture is experiencing a huge shortage of farm laborers.

The Artel aims: to organize the unemployed masses of the intelligentsia, student youth, and individuals of other professions, and to send them to do agricultural work.... Agriculturists—Cossacks, peasants, small and large estate owners, horticulturists, vine-growers, agricultural societies, experimental fields... et al., who need agricultural laborers, should forward their requests to the following address...[35]

And so, the ten-year-long cycle of rural professional activities reached its final point: voluntary teaching of the peasantry; professional and well-paid coaching of the peasantry; the selling of professional expertise to anyone who could pay for it; and the selling of the intelligentsia's labor as wage workers. By 1919, when an egalitarian land reform was introduced even in the Cossack regions, all hopes of the agricultural specialists for independent professional activity disappeared in both Red and White Russia.[36]

As soon as all prospects for the independent coexistence of the rural professionals with the political regime and major social actors had evaporated, other strategies of social interaction were tested in rapid succession. During the winter of 1918–19, some local agricultural specialists attempted a cautious limited cooperation with the Soviet regime. In January 1919, the well-known Moscow agronomist Dmitrii Shorygin[37] observed the results of the previous year almost with satisfaction: soldiers returned to their villages, the agricultural specialists began cooperation with Narkomzem (the People's Commissariat of Agriculture, the successor to the Ministry of Agriculture).[38] Though less optimistic, the Tambov district agronomist conference, which took place on January 16 and 17, also expressed belief in the possibility of cooperation with the existing authorities.[39]

With the steady disintegration and demobilization of the community of public modernizers and the growing self-isolation of the peasant population, a benevolent political regime similar to the Provisional Government of 1917 was the only hope of rural professionals. Their optimism was a political rather than an emotional phenomenon, a friendly gesture toward the authorities and an invitation—or rather a plea—to endorse the activity of public modernizers, and thus save its fruits from complete destruction. This optimism did not last long.

Expropriation and attempts to avoid it

Any dreams of collaborative partnership with the dominant political regime were brought to an end by the decree of the Soviet government on January 25, 1919, "On the Registration and Mobilization of Agricultural Specialists." This decree demanded that all citizens of the RSFSR with mid-level and higher agricultural education, all people whose work required any specific agricultural skills, everybody who had held managerial positions in agriculture for two years or more, and also all agricultural students in the last two years of their education must register within a ten-day period under the threat of arrest. All those registered could be "drafted into service in agriculture" and assigned to any position or task. In return, the drafted agricultural specialists received a guaranteed salary and (most important) food ration, and could be arrested, searched, and expelled from their homes only with the consent of the state agency that employed them (usually, the People's Commissariat of Agriculture).[40] This decree eliminated any possibility for an agricultural specialist to claim equal "partnership" or even "cooperation" with the regime. As expressed by the commissar of agriculture, S. P. Sereda, the old distinction between the state and public agronomy ceased to exist.[41] The bulk of RSFSR agricultural specialists officially became servants of the state. In autumn 1919, at least 2,977 field specialists were compelled to carry out the Soviet land settlement program in 32 provinces of Soviet Russia, not to mention managerial personnel at the district and province levels.[42] The political self-mobilization of public modernizers as a community of activists was completely replaced by the state's military-type mobilization of individual agricultural specialists. They were not allowed to exercise any initiative, but only to execute the orders of the authorities. The mobilized experts were not required to demonstrate their political loyalty, to have any specific citizenship, or prove their "nonbourgeois" family background.[43] Not invited to integrate into any political or national community, they were just expropriated as a form of human capital for the needs of the regime.

In fact, the history of the Russian civil war does not suggest that Bolsheviks operated with categories of "nation" as a political or cultural entity. On the contrary, even if some of their measures could be identified as a very crude social engineering and elemental social policy (such as experiments with establishing Committees of the Poor, *kombedy*), growing social polarization was the ultimate goal and the immediate outcome. Regardless of the dominant ideological rhetoric, "class" or "political party" hardly served as decisive principles of social unity either. It seems that Bolshevik mobilization was based on the identification of enemy groups rather than on the offering of a clear image of the homogenous community of solidarity action. There was a regularly updated hierarchy of sociopolitical enemies, and a social group could perceive its position as "safe" insofar as there were other groups considered more dangerous by the authorities. Some categories of peasants could be recognized as a friendly class when needed, but even workers could fall out of favor if they expressed any interests contradictory to those of the regime. While the problem of the nature of Bolshevik political mobilization lies far beyond the scope of the present study, we may tentatively suggest that its mechanism was different from the models advanced by both revisionist and totalitarian traditions.[44] The Bolshevik-led regime certainly was not a one-party enterprise, but it did not have a social base in the usual understanding of this notion: as a certain social group with a set of interests endorsing a political force. Even if they wished to, Bolsheviks could not serve as an authority representative of anybody in the situation of the post-1917 disintegration of Russian society, economic collapse, and growing social anomie, which the Bolsheviks wholly sponsored by dismantling the remnants of social institutions and by direct terror. Both the rhetoric and practices of the regime aimed not at the social integration, but at the annihilation of its most active opponents. It mobilized selected social groups and categories not "into," but "from": from the disintegrating social milieu, de facto outlawed by the regime. Those mobilized were granted recognition as members of society and officially entitled "to all rights granted by the decree of the Soviet of People's Commissars,"[45] while others were left to their own devices. The former received a modest ration and were entitled to formal prosecution; the latter could starve or be hunted down indiscriminately.[46]

As of February 1919, agricultural specialists had no other way of legal coexistence with the regime. At the same time, this ruthless subordination of agricultural specialists to the regime had a psychological effect. Working directly for the regime made many people associate themselves with the regime's goals, or at least understand its rationale.[47] This gradual transformation of a former vehement anti-Bolshevik into an understanding supporter of the regime can be traced in the diary of the prominent historian Iurii Got'e,[48] or in the notebooks of the "ordinary" Petrograd *intelligent* G. Kniazev.[49] Throughout 1919, Kniazev oscillated between condemning Bolshevism and recognizing its supreme truth. The main reason for the acceptance of Bolshevism was its "national" character and its successful struggle with foreign occupants and internal separatism.[50] By 1920, after years of war and revolutionary violence, Kniazev was ready to accept any regime that would guarantee peace and economic stability.[51]

The publications in the magazine *Southeastern Farmer* of the Don-Kuban-Terek Agricultural Society suggest that the same transformation happened to many rural professionals living in the non-Soviet territories. The magazine's editor was Nikolai Sokolov, director of the Rostov-Nakhichevan' agricultural experimental station, who had graduated from Kiev Polytech in 1907 along with Ivan Emel'ianov. In January 1918, the general tone of the *Southeastern Farmer* was very anti-Bolshevik.[52] This mood only intensified in spring 1918, after a brief occupation of the region by Soviet troops, who confiscated

11 horses and fodder from the Rostov-Nakhichevan agricultural experimental station. The retreat of the Reds was followed by a pogrom on the station by the local villagers, who were looking for denatured alcohol.[53] The battle for Rostov severely damaged the structures of the agricultural station. The subsequent German occupation brought no relief to the agricultural specialists working at the station. German soldiers were posted on its premises, which was not the case under the Soviets. The cattle were now kept on the street, while Germans confiscated all transport and the remaining fodder from the station. Instead of protecting the property of the station, German soldiers along with local villagers stole its crops.[54] On March 25, 1918, the precinct agronomist A. G. Storozhev (Valki district, Kharkov province) had a similar experience with the German occupation forces. Germans requisitioned everything useful from the precinct and the village cattle-breeding association (probably also a result of agronomist activities), promising to pay in the future. They never did, as in numerous other cases.[55] It is unknown how this incident changed the worldview of agronomist Storozhev, who began his professional activity after receiving a mid-level agricultural education in about 1910, but in 1923 we see him working in the same capacity in the same village under the Soviet regime.[56] Humiliated by the foreign occupation, and alienated from the people who "betrayed" them, many field specialists longed for a strong "national" (this time, in the sense of state sovereignty) regime that would recognize their importance and use their expertise.

As a matter of fact, honest service to the Soviet regime could be rewarding, thus tying a person to the regime economically. Alexander Chaianov, who for a long time kept a distance with the Bolsheviks,[57] fully made use of the economic opportunities provided by loyal and enthusiastic service to the new government. On October 20, 1920, the official *Izvestia VTsIK* published a new instruction of the Moscow Housing Department, granting privileges to important scholars who cooperated with the Bolshevik government. Just two days later, Chaianov obtained an official letter from Sverdlov Communist University, where he taught part-time, asking the Moscow authorities not to confiscate any part of his spacious apartment.[58] On October 24, Chaianov succeeded in securing support from the Moscow Land Department for his plans to purchase a house in the countryside—something quite unusual during the epoch of War Communism.[59] In 1921, when Chaianov occupied a high-ranking position within the Commissariat of Agriculture, he even received a Nagan revolver—a symbol of authority under War Communism in Russia.[60] On the other hand, close relationships with the authorities could be useful in helping less fortunate colleagues. Chaianov himself used his Soviet government connections to ease the situation of many people.[61]

In the atmosphere of political terror, there were still economic experts who took the risk of participating in clandestine organizations, and even drafting an economic program for post-Bolshevik Russia. In 1919, two coevals and colleagues of Alexander Chaianov, Lev Kafengauz and Iakov Bukshpan, joined forces to put into writing the program of Russian economic revival for the underground political organization the National Center.[62] Allegedly, they did not know about the military anti-Bolshevik activity of the center coordinated by Anton Denikin's army from Ekaterinodar, but they clearly understood that they were trespassing against the draconian Soviet legal system by their very engagement in the intellectual opposition.[63] The point of departure of the economic program of Kafengauz and Bukshpan was the priority of the "national" (in this case meaning the country's) economy over the interests of any particular group or class. Instead of creating greenhouse conditions for certain economic actors, the maximization of economic performance was proclaimed as the main goal of future economic policy.[64] In the sphere of agriculture, they actually subscribed to the main ideological principles of the Stolypin agrarian reforms:

> There is no doubt that for anybody who loves the peasant economy more than the populist books about the peasant economy, the real program of the peasant agrarian policy [includes]: private peasant property in land, land settlement, transition from communal land tenure to enclosures, [and] transition toward more intensive [field] cultures.[65]

Before the October Revolution, on May 11, 1917, Bukshpan made a presentation to the Free Economic Society, greeting the grain monopoly as a step toward state socialism.[66] Only two years later, Bukshpan and Kafengauz declared that the socialist experiment had brought nothing but misery, and the state should play a limited role in the economy.[67] The only socialist measure they were willing to include in their program was comprehensive health and accident insurance for workers.[68] Bukshpan and Kafengauz called their employer, the VSNKh (the Supreme Economic Council), the state agency responsible for the organization and regulation of the nationalized economy, a harmful and absurd utopia, "the most bureaucratic agency" in history, which should be condemned together with the idea of monopolies.[69] Thus, formal service to the regime did not exclude dissident, if not subversive, application of one's professional expertise.

The turmoil of civil war did not provide many opportunities to completely ignore the sociopolitical environment. Still, some people managed to avoid any sustained interaction with the authorities. During the years of civil war Sergei Fridolin, whom we have already mentioned on several occasions, moved to Moscow from Kiev, where he held an important position on the provincial procurement board. He worked on the Moscow procurement committee for a while, but then, unlike the majority of his high-ranking colleagues, left the starving capital for a farm in the depths of Tver province. Fridolin was not mobilized because of his poor health (or, perhaps, because the central authorities could not reach him), and he survived the civil war on his farm as an agriculturist, earning extra income by providing veterinarian assistance to his neighbors.[70] Sergei Fridolin returned to Moscow only after 1921, "with the change in economic policy" and under "new conditions."[71]

Fridolin had to pay a high price for his elevation from a semi-peasant existence to professional activity after years of self-isolation. Besides loyal work as an expert, he was required to take certain political steps in support of ideological campaigns of the Soviet authorities. This activity culminated in 1925, when the Commissariat of Agriculture published his highly biased memoirs *An Agronomist's Confession.*[72] It was possible for some agronomists, as Fridolin did, to avoid mobilization during the period of War Communism, only to be mobilized by the regime later on.

Emigration was the last resort of those who could not adapt to any of the strategies of social interaction mentioned above. One might expect rural professionals to be more closely "tied to the soil" of their homeland than other categories of intellectuals and professionals. In fact, we do not have accurate statistics on the emigration rates of agricultural specialists after the 1917 Revolutions. The available data indicate, however, that a significant proportion of leaders of the public modernization campaign in the countryside emigrated from Russia, or were at least considering emigration. During the civil war, the founder of the Organization-Production school, Alexander Chelintsev, found his way to Constantinople. Another key figure of the "Artel," Nikolai Makarov, made a difficult trip abroad through Siberia. Later, they reunited in Prague, where in the early 1920s V. E. Brunst, F. I. Kolesov, N. P. Makarov, S. S. Maslov, and P. A. Sorokin founded the Russian Institute of Agricultural Cooperation.[73] When in spring 1921 Alexander Chaianov was sent by the Commissariat of Agriculture to Western Europe, one of his tasks was to persuade Chelintsev and Makarov to return to Russia.[74] Yet, during this trip, Chaianov himself considered the possibility of not returning to Moscow.[75] B. D. Brutskus, another important economist of the Organization-Production school, was expelled from Russia in 1922, while K. A. Matseevich, editor of the influential *Agronomist Journal,* and later a prominent politician in Ukraine, left the country in 1919 as the envoy of Petlura's Ukrainian government to Budapest.[76]

The postrevolutionary exodus was not confined only to prominent economists and cooperative activists. A certain number of "ordinary" agronomists and other agricultural specialists left Russia as well. In 1938, at least 14 of them joined the Agricultural Association of North America, which also included former Russian landowners and cooperative activists. Quite surprisingly, the majority of these emigrants, even those with mid-level agricultural education, found jobs according to their specialization in the United States and Canada, in agricultural colleges and firms.[77] Thus, emigration presented a serious option even for agricultural specialists, especially for theoreticians and prominent leaders with sound international academic reputations.

The different strategies of interaction with the political regime in the country led in different directions. Therefore, it is all the more surprising to find a striking similarity in the worldview of people who chose different paths after 1917. Those who were mobilized by the Bolshevik-led regime and those who found themselves abroad agreed in their reevaluation of the state's role and the cult of "national interests" (where "nation" is understood as a geopolitical entity). Alexander Chaianov, a prominent collaborator with the Soviet regime, who considered not returning to Russia from his business trip abroad in 1922–23, believed that it was necessary "to firmly and definitely distinguish Russia from the USSR,"[78] separating the "regime" from the "nation." His former colleagues and friends who chose to stay abroad and even joined the ranks of anti-Soviet political organizations were thinking in the same categories. Some of the scholars and activists associated with the Prague Russian Institute of Agricultural Cooperation formed a semi-virtual political organization "Peasant Russia" headed by Sergei Maslov, who was one year older than Chaianov and formerly worked with him in various cooperative structures. Maslov and his collaborators believed that Russia's defeat in the World War I and the Reds' victory in the civil war were explained by the lack of national consciousness among the peasants.[79] In the Program of their organization, which was later declared a political party, Prague émigrés stated,

> The field of our work is Russia. It is our "home," … its interests are superior and constant for us.
>
> … We are nationally oriented [*natsional'ny*], but not nationalist. We consider the all-Russian culture and state integrity to be supreme values, but … regard as unquestionable and indisputable the [c]ultural-national rights of the peoples living in Russia.[80]

These words seem very typical of Russian intellectuals of the 1920s, regardless of their political affiliation and residence. Yet they were all but unthinkable for Russian educated society before the civil war, with its supreme ideal of *obshchestvennost'* as an anational community, the *république des lettrés*. Among the most fateful results of the revolutionary epoch of 1917–21 for the community of rural modernizers was not only their mobilization into state service but also their expropriation of the ideal of *obshchestvennost'* as a political community based on the civil society. VChK interrogators worried in vain about a hypothetical plot of *obshchestvennost'*. While the professional corporation and public associations were seen by the remnants of the already not-so-new post-1905 generation of Russian intelligentsia as useful tools of social engineering,[81] now their goal was a unified Russian nation-state. Something that did not yet exist.

Normalization and demodernization

The years of the World War I, the turmoil of revolutionary 1917, and the chaotic period of civil war disintegrated the social movement united by the program of peaceful professional modernization of a politically stable society with a free economy. New loyalties and value orientations were formed, and a distinctive social generation of new Russian intelligentsia ceased to exist, as the initial formative impact of the post-1905 worldview crisis became all but ephemeral. The trauma of the revolution and civil war served as a new formative experience, forging a new sense of community among the survivors. But this drastic change was not apparent to agricultural specialists who believed that the end of War Communism and forceful mobilization of rural professionals would restore the status quo ante bellum.

Indeed, the first postwar years generated a feeling of déjà vu among older members of the Russian educated public, who cherished recollections of the high social status and relative independence of the former *obshchestvennost'*. The new epoch began with the famine relief efforts of the *obshchestvennost'* united in the *Pomgol*[82] and brought about a certain improvement in rural professionals' status—exactly as had happened during the period 1891–1914. In fact, there appeared the hope that history had given the former public modernizers a second chance. Many important positions in the People's Commissariat of Agriculture were now occupied by representatives of the professional elite.[83] In March 1922, for the first time in years, an All-Russian Agricultural Congress was convoked in Moscow. This representative forum reflected the new situation of the rural professionals vis-à-vis the authorities. On the one hand, the presidium of the congress featured the most prominent figures of public agronomy left in Russia: A. V. Chaianov, A. G. Doiarenko, A. N. Minin, A. I. Sigirskii, D. M. Shorygin, I. P. Stepanov, and A. V. Teitel. On the other hand, high-ranking officials of the Commissariat of Agriculture, P. A. Mesiatsev and M. E. Shefler, were included in the presidium to supervise and control the work of the congress.[84] Under the old regime, in the audience of every large public gathering there was a police officer with the authority to stop any politically "subversive" activity, but no government official could steer a congress or influence its decisions. In 1922, this was already a common practice. In his speech, the Soviet functionary Mesiatsev explicitly stated that the agricultural specialists were expected to execute the government's plans.[85]

The assembled rural professionals were themselves less clear about their own program. The lecturer at the Moscow Agricultural Institute and the editor of the *Messenger of Agriculture*, Aleksei Doiarenko, attempted to make a distinction between professional assistance to the rural population and the carrying out of government orders (in line with the old opposition between the government and public agronomy).[86] The agronomist Ivan Stepanov, the graduate of an institute in 1914, who served as a Tsivilsk district agronomist (Kazan province) during the war, went even further, actually proposing the restoration of the original program of public modernization of the countryside.[87] The former agronomist of the government land settlement commission of the Orel district, Lev Nemirovskii, supported the old project of precinct agronomy,[88] while the former zemstvo agronomist and now professor at the Voronezh Agricultural Institute, Alexander Minin, claimed that the system of precinct agronomy, centered on the figure of a versatile agronomist, was no longer viable.[89] Amid these theoretical discussions, the speech of the former government senior agricultural specialist in Saratov province, Pavel Gratsianov, sounded like a cry for help. Rather than remodeling old *obshchestvennost'* projects, he asked for some basic conditions of service to be secured for agricultural specialists: to release them from nonagronomist functions; to legalize regular agronomist conferences and congresses; to allow agricultural specialists to choose jobs and specializations voluntarily; and simply to improve their material situation.[90] Hence, the congress revealed the absence of a clear program of social engineering and professional service under the new circumstances, the kind of program that had made the community of prerevolutionary agricultural specialists a distinctive social movement after 1907. It was obvious that old professional strategies and social practices could not be automatically restored in the new historical context.

Seven months later, the Agricultural Congress in Moscow was followed by the first Conference of Representatives of the All-Russian Union of Agricultural Cooperatives (*Sel'skosoiuz*). Similarly to the agricultural congress, representatives of the old cohort constituted the driving force of the cooperative conference.[91] And like the congress of agronomists, the cooperative conference was tightly controlled by the Soviet authorities. In fact, all delegates belonged to one of two factions: the faction of the RCP(b) members or the nonpartisan faction. Each faction drafted a set of resolutions for the conference, and proposed candidates to the top *Sel'skosoiiuz* offices.[92] Thus, the restoration of the dual foundation of public modernization reform activity (agronomy plus rural cooperatives) was accompanied by a new phenomenon: the former domination of *obshchestvennost'* in the sphere of agricultural discourse and activity were challenged by the new Bolshevik educated elite.

Initially, this interference of the authorities in the sphere of discourse on agriculture was overshadowed by government's direct administrative control, which left rural professionals with the impression that their ranks were still united against the alien, though partially liberalized, state. The recognized speakers for the community of rural professionals attempted to revive the model of public activism that was autonomous from the regime.

The All-Russian Agricultural Congress in Moscow, dominated by old cadres, adopted resolutions echoing the post-1917 revisions of the initial program of the politics of modernity in the countryside. The congress demanded the legalization of all forms of property, development of a market economy, liquidation of the state monopoly on foreign trade, restoration of an effective local self-government system, and the free development of rural cooperatives and industry.[93] In 1922 such a program presented a direct challenge to Bolshevik policy.

The reaction of the Soviet authorities to the challenge of the old public activists as an alternative modernization force differed from that of the tsarist officials. They did not create a parallel structure of government agronomy or a new system of agricultural assistance to the population. Instead, after a period of political persecution in 1921–22, the Bolshevik leaders literally appropriated the key elements of the *obshchestvennost'* program while presenting the "old agronomists" as conservative administrators-Stolypinists. At the beginning of 1923, the prominent Bolshevik leader, Peter Stuchka, who back in 1918, as the commissar of justice, authorized the policy of "Red Terror," published a remarkable article in the official party magazine, the *Communist Revolution*. He admitted that the project of large collective farms had failed, and suggested that the peasantry be acknowledged as the backbone of national agriculture. He sketched out a plan of intensification and modernization of the peasant economy, a plan very similar to the program of "public agronomy." Stuchka further asserted that the majority of "old" Russian agricultural specialists would oppose this approach, because they had allegedly always resisted the idea of intensive peasant farming: "The Russian agronomists en masse are brought up by Prussian theory, and then by Russian gentry practice."[94] The regime had deprived agricultural specialists of their professional autonomy and now attempted to expropriate their past, smearing everything that could not be usefully incorporated into the state policy (Fridolin's memoirs were a part of this campaign). Depriving the former rural modernizers of their past, the regime sealed their future.

While the updated program of former rural professionals acknowledged the necessity of specialization, and the futility of the wager on a universal agronomist, Soviet agricultural leaders emphasized the general modernization role of the agricultural specialists à la Fortunatov:

> The Red agronomist...should secure the alliance with the peasantry.... Thus, the task of the agronomist is not confined to dealing only with narrow, special questions, but is broader, and more complicated...it is the task of involving...the broad masses of the peasantry in public activity.[95]

> The Soviet agronomist is first of all a public activist and organizer of the masses. He organizes not so much individual farms, but rather the

whole peasant economy, he is a direct conductor of Soviet agricultural politics in the village.... [S]oviet agronomy differs dramatically from the former zemstvo one, which proclaimed its apolitical nature and cultural mission, and served predominantly the interests of individual farms.[96]

The real problem was not the rhetoric itself, but its level of circulation. In 1923, it was articulated by the high-ranked party functionary, P. Stuchka. In 1924, senior activist of the trade-union of agricultural specialists, a certain Bragin, was formulating the task of the "Red agronomist" in the first of the two preceding quotations. Finally, in 1926, it was the Samara provincial agronomist, V. Kopeikin, who speculated on the nature of the "Soviet agronomist." It was only a matter of time before the ordinary field specialists would follow suit on the eve of collectivization and cultural revolution, accusing the "old specialists" of neglecting a broader social agenda. The layer of well-positioned keepers of the *obshchestvennost'* modernization legacy now found themselves isolated not just from the regime and the peasants but also from the rank-and-file agricultural specialists, including those who had actively participated in the movement before 1917–18 and belonged to the same "new" post-1905 generation of Russian intelligentsia. Only when the long-awaited demobilization of agricultural specialists in the summer of 1922 inaugurated a new epoch of "normalization" did the impact of the preceding five years on the community of rural professionals become obvious. While the bulk of agricultural specialists remained in place, and physically they were the same people as in 1913 or 1915, the stratum of village intellectual elite disappeared as a social phenomenon. Those who survived in the revolutionary village became incorporated into the peasant world in terms of living standards, labor practices, and socialization. While occupying privileged positions in the village hierarchy, particularly with their improved material status throughout the 1920s, field specialists were no longer the messengers of urban wisdom in the backward countryside. They could still teach peasants and even execute the unpopular decisions of the authorities, but not as representatives of any informal pan-Russian public movement.

A closer look at the situation of agricultural specialists in Ukraine substantiates this thesis. The Ukrainian case is paradigmatic in many aspects. Before the revolution, Ukrainian provinces had more agricultural specialists than any other region of the empire. Just three provinces, Kiev, Poltava, and Kharkov together employed about 1,000 field specialists.[97] On the other hand, Ukraine experienced a succession of political regimes after 1917, and suffered from civil war more than other parts of the former Russian empire. Consequently, agricultural specialists on this territory were more likely to be killed or expelled from the village, to retreat with the German or White armies and emigrate, or to be promoted by the new authorities looking for cadres who were not compromised by service to the previous regime. The available data reveal a different picture.

The demobilization of agricultural specialists in 1922 de facto restored the zemstvo-epoch system of precinct agronomy. As it turned out, by January 1, 1923, there were 727 precinct agronomists in the Ukrainian SSR, and 546 senior agronomists and different specialists (altogether, 1,273 people). Paradoxically, these figures did not differ much from the prerevolutionary period, when on January 1, 1915, 1,370 agronomists and specialists were employed on the same territory (1,139 in zemstvo service and 231 in government). Thus, after eight years of war and political turmoil, the cadres of agricultural specialists in Ukrainian lands had decreased by just 7 percent.[98] The Commissariat of Agriculture planned a reform of this system that would drastically cut the number of employed specialists, who became a serious burden for the budget once the government decided to pay them regular salaries, and to restructure their positions. The system of "precinct agronomy" was to be replaced by "district agronomy," with one agronomist district replacing two to three former precincts. Altogether, the Ukrainian government was ready to employ 900 agricultural specialists out of the 1,273 currently in state service: 300 would become "district agronomists," and the remaining 600 were to work as specialists and instructors on the tasks formulated by agronomists.[99] These remarkably round figures betray the purely bureaucratic nature of this reform, hardly taking into consideration the actual administrative, demographic, and topographic realities of the Ukrainian countryside, and most likely motivated by budget considerations.

According to the local agricultural administration, the majority of the most experienced agricultural specialists in Volyn province had left low-level positions in the field after the demobilization in July 1922.[100] Still, on average in 1923, over 60 percent of rural professionals throughout the Ukrainian Republic had prerevolutionary work experience (seven years or more); that is, they were trained in the tradition of public agronomy and had begun their professional activity in the system dominated by the ideology of the public modernization campaign. Figures varied by province, time (cf. dramatic changes occurred in Ekaterinoslav province in the course of a few months), and, most important, by the educational background of the personnel as can be seen in Table 11.1, which includes data from four out of ten provinces of Soviet Ukraine.

With the exception of Kharkov province, where the capital of the republic was located and political control was particularly tight, old cadres dominated the ranks of agricultural specialists. A person with mid-level or higher professional education was most likely to have pre-1917 work experience. What is important here is not the hypothetical political influence that those people might have exercised over peasants or the system in general, but the very fact of successive service of the majority of agricultural specialists in

Table 11.1 The percentage of agricultural specialists in four provinces of Ukrainian SSR with prerevolutionary training and work experience, among major education groups, 1923[101]

Province	Educational background (%)		
	Higher	Mid-level	Primary
Poltava, April 1, 1923	61	55	70
Ekaterinoslav, [April ?] 1923	66	59	n/a
Ekaterinoslav, October 1923	40	45	25
Kharkov, fall 1923	28	34	22
Podoliia, fall 1923	53	48	30

the same capacity. This means that for many years these people lived in a situation of very limited horizontal and upward mobility and socialized within the territory of their precinct, despite all of the historical calamities they had managed to survive. A close analysis of one province reveals these patterns of socialization and mobility.

As in 1915, in 1923 Poltava province had more agricultural specialists than other provinces did (209 in January, 120 after the reform). Two-thirds (67 percent) of these specialists began their service before 1917. I was able to find additional background information for about half of them. In studying biographies of Poltava agricultural specialists, two parallel trends are revealed: former precinct agronomists became 1923-type "district agronomists" (not to be confused with the old *uezd* agronomists), while former agricultural elders and village cooperative managers (local peasant folks) were promoted to positions as "instructors" and "specialists." For peasants with only primary education this was certainly upward mobility, their practical experience was converted into the status of rural professionals. Former precinct agronomists with mid-level and higher education got stuck in a position that before 1918 was but a first stage in the hierarchy of rural professionals. The absolute majority of specialists representing both categories began their careers between 1909 and 1913, during the triumphant takeoff of the public modernization campaign (although four agronomists had gotten their degrees before 1890). And all of them served in 1923 at exactly the same location as before the war, or nearby.[102] This means that people spent at least 10 to 15 years in the same place, in the same capacity.[103]

Similar dynamics were present in Ekaterinoslav province, for which I was able to reconstruct the biographies of 80 percent of old cadres.[104] Almost everyone remained in the same localities, moving at most 40 miles from their old place of work.[105] Almost all former precinct agronomists became new "district" agronomists, but in one instance the former agricultural elder with primary education, peasant Petr Namliev, became "district" agronomist in his native village of Chernigovka. This case shows that upward mobility for the former peasant assistants of zemstvo agronomists was not limited to instructor positions.[106] Another interesting pattern can be seen in the Aleksandriiskii district (*uezd*) of Ekaterinoslav province: there four out of five agronomists kept their positions for over ten years. Pavel Pertsev had graduated from the Moscow Agricultural Institute, and in 1906 became Aleksandriiskii district zemstvo agronomist.[107] In 1913, he was joined by three precinct agronomists: Trukhin, Millers, and Slavov, who in 1923 also became "district agronomists." It seems reasonable to suggest that despite their age and educational differences, these men composed the core of local society, a stable network uniting people on the basis of friendship, status, and common interests.

I found another cluster of old cadres that clearly held together through wars and revolutions in Kharkov province. In 1911, Mikhail Pushkarev, who received mid-level agricultural training in 1902, became a precinct agronomist in Vakovskii district. He was joined by a young fellow, Petr Shilo, who finished a few classes of primary school in 1908 and got a job as assistant to the gardening instructor in the same village, also in 1911.[108] Twelve years later, they still worked together, as an agronomist and an agricultural technician, in the same village. Their former boss, the Valki district agronomist Aleksei Ivitskii, who graduated from an agricultural institute in 1903 and was probably only a few years older than Pushkarev,[109] in 1923 occupied almost the same position.[110] Thus a "broken link" of the old community of countryside modernizers could preserve both "horizontal" and "vertical" ties.

On the other side of the European part of the USSR, in Arkhangelsk province, there was a similar situation, although with local specificity. The available data suggest that agricultural specialists with prerevolutionary experience did not receive any significant promotions between 1914 and 1924. Indeed, Ivan Benevolenskii, who graduated from St. Petersburg University in 1910, was senior government specialist in marshes, serving in Vologda province in 1914.[111] In 1924, he was director of the "Arkhangelsk experimental swamp field,"[112] which could be regarded as a position lower than the one he formerly occupied. Mikhail Neliubin graduated from a mid-level agricultural college in 1913 and became deputy agronomist of the Penza district land settlement commission.[113] Ten years later, he was secretary of the agricultural department of the Arkhangelsk provincial land administration, at approximately the same level of importance.[114] Vasilii Bogomolov received a mid-level agricultural education in 1905, and in 1914 was working as a Samara precinct agronomist.[115] In 1924, he held the more prestigious job of Arkhangelsk district (*uezdnyi*) agronomist.[116] In his case, we see a moderate promotion. Characteristically, all of these people were recent migrants to the province, which was not typical in Ukraine. This can be explained by the absence of a zemstvo network of public agronomy in Arkhangelsk before the revolution, and a limited number of government experts employed in the province: in 1915, there were 309 agricultural specialists in Poltava province and only 44 in Arkhangelsk.[117] The latter were all government officers, which made their chances of surviving in the countryside during the revolution and civil war negligible.

The vacuum of rural professionals in the province was filled by the unprecedented upward mobility of people from the very bottom of the pre-1917 professional hierarchy. In 1914, Dmitrii Khristoliubov was an agricultural elder with a primary education in Viatka province.[118] Ten years later, he became a precinct agronomist in Arkhangelsk province.[119] Petr Tufanov had received only a primary education in 1905, and worked as an agricultural technician before the World War I.[120] After the civil war, he was appointed Shenkursk district agronomist in Arkhangelsk province.[121] Even more spectacular was the career of Petr Grunbaum. A former agricultural elder with a primary education, he became the Arkhangelsk provincial agronomist.[122] The peculiar social composition of the Arkhangelsk agronomist community accounted for its ideological and professional ethos, which was very different from that of the "old zemstvo" provinces. As *vydvizhenets* (worker promoted to an administrative post) P. Tufanov declared, the intelligentsia in general, and agronomists in particular, had accepted the Soviet regime, and hence they could carry out the task of implementing the Bolshevik political agenda in the countryside.[123] Grunbaum went even further, claiming that agricultural specialists must work in contact with the Communist Party and Komsomol.[124] He also described the ideal Soviet agricultural specialist:

> The first and basic requirement for a Soviet agronomist is that he must be from the laboring peasant milieu, and not formally, but factually.... The modern agronomist should be a publicly minded person.[125]

The roots of the cultural revolution, driven by intraprofessional conflicts and the striving of low-grade technicians for upward mobility, were already in place in the early 1920s.[126] But well before the open conflict erupted, it was clear that even at the level of field specialists there was no nationwide political solidarity and similarity of ideals, such as those characteristic of the prerevolutionary public modernization movement. Deprived for many years of prospects for upward mobility and professional socialization at congresses (in the "agronomist parliament") and through the agropress, agricultural specialists in different regions could only formally be regarded as members of a single professional community. The ties of "vertical" communication were disrupted no less seriously than the "horizontal." Field specialists not only worked but also ate and dressed differently from their high-positioned leaders. When agronomist leaders in Moscow lectured to students, advised the government, and purchased old icons and books (some of Alexander Chaianov's hobbies), precinct agronomists, according to an official statement, documented the propriety of nationalized estates, supervised peasant land cultivation on those estates "under a second edition of the corvée system," collected statistics, and participated in propaganda campaigns. "Not getting any certain pay from the agricultural administration [*zemorganov*], this personnel obtained means of subsistence in various ways: by reselling clothes, from trade, and at best from agriculture, personally tilling plots of land."[127] Economically, they hardly differed from neighboring peasants.

This ruralization and pauperization of agricultural specialists eventually brought them "closer to the people," but not in the way ideologists of the public modernization campaign had dreamed about. The majority of former modernizers ceased to embody any normative modernity. Without the zemstvo-, cooperative-, and government-sponsored infrastructure, and in a hostile political atmosphere, former public agronomists were no longer on a mission to modernize. As an occupation group, they underwent complex *demodernization*, in terms of both their individual and public outlooks. Perhaps, with the restoration of the old system of agricultural assistance to the population they would have been reintegrated into the profession of agricultural experts as a distinctive corporation. But their relations with peasants would never have been the same, as they had become an integral part of the "people," if not of the "nation," and could become actors of modernization again only as insider "revolutionaries," not as modernizers from outside.

Sovietization and deprofessionalization

Inauguration of the NEP (New Economic Policy) changed the miserable conditions of rural professionals, although they still could only dream of the prerevolutionary period. At the meeting of senior agronomist personnel of the newly formed Donetsk province in Ukraine on October 26, 1923, it was decided that the current salaries of agricultural specialists were well beyond the subsistence wage, not to exceed 15 "commodity rubles." Agronomists decided to ask for a raise: for the senior personnel (the level of former zemstvo district agronomists) to 45–50 "commodity rubles," for precinct (in Ukraine renamed "district") agronomists and specialists to 32 rubles, and for agricultural elders-turned-"instructors" of agriculture to 25 rubles.[128] These groups of agricultural specialists were assigned, respectively, the fifteenth, fourteenth, and thirteenth categories on the 17-category tariff template,[129] which made an instructor in agriculture with primary education equal to an engineer (the thirteenth category was the highest for "mid-level administrative or technical personnel"), and an agronomist in the town of Sumy or Izium was only two steps below the people's commissar of agriculture.[130] Yet even a high tariff category did not secure high living standards. In the situation of early Soviet hyperinflation, "commodity rubles" were just a mathematically produced figure, equal to one-tenth of the once a week locally calculated cost of a "consumer basket," which included 24 basic goods. Money became a meaningful indicator with the introduction of the golden currency standard in 1924. The 50 "commodity rubles" that the Donetsk agronomists dreamed about became 65 new rubles, which in 1924 had approximately the same purchasing power as 36 rubles in 1913.[131] Thus, agricultural specialists at the rank of former district agronomists, whose minimal salary was 200 rubles a month in 1913, in 1923 asked for a salary of 36 "old" rubles (it should be recalled that 36–40 rubles a month, or 460 per annum, was the starting salary of technical personnel with primary education before the World War I). Former precinct agronomists were expected to be given a raise to a minimum of 42 chervonets rubles, or 24 "old" prewar rubles—while no precinct agronomists were paid less than 100 rubles a month before 1914. Even though before 1924 field specialists had been paid two to three times less, and the actual salaries might be higher than the indicated minimal rates, their income during the period of early Soviet "normalization" did not exceed 20–25 percent of their prewar salaries.

In fact, actual living standards could be even lower than the indexed figures of salaries suggest, for instance, due to poorer housing and a more primitive diet. In 1926, the Samara provincial agronomist acknowledged that field specialists were not compensated for their housing expenses and transportation—something that was taken almost for granted before the war.[132] Hence, one may assume that agricultural specialists paid for these expenses out of their own small salaries. In addition, a considerable portion of their income was withheld by multiple semipublic associations in compulsory membership fees.

The case of agricultural specialist V. I. Popov (probably, eligible for no less than the fifteenth tariff category, living in the vicinity of Moscow) vividly illustrates the general statistical data. He graduated from an institution of higher learning in 1913, and in the spring of 1914 entered state service at a starting salary of 125 rubles per month. Three months later (apparently, after the outbreak of war), his salary was increased to 200 rubles per month,[133] which was the basic salary of a district agronomist, and sometimes even of a precinct agronomist. Ten years later, in the early spring of 1924, he received a salary of approximately 84.5 prewar rubles per month (according to the current official index), or 42 percent of his 1914 salary.[134] In addition, 15–20 percent of his total income came from author's fees for publications in professional periodicals.[135] Although Popov worked in town (evidently, in Moscow), his family could only afford to rent a peasant house four miles from his place of work, which took 18 percent of his annual income.[136] Clothes and shoes for the entire family (three adults and two children) took only 5 percent of his income, which can be explained by the deficit of consumer goods in Soviet Russia. Indeed, as Popov himself admitted, his family was still wearing clothes that had been purchased before the war, or that were obtained during his military service (e.g., a trench coat).[137] Food accounted for 35 percent of total expenses.[138] The bulk of the ration for this relatively well-to-do family consisted of various cereals.[139] Popov spent some 8 percent of his annual income on books and periodicals.[140] Before the war, these types of expenses were often covered by zemstvos and other employers of agricultural specialists. Every year, a sum equal to about two-thirds of one month's salary was taken by various public funds, including the trade union, the Voluntary Society of Chemical Defense (Dobrokhim), and the International Society of Aid to Workers (MOPR).[141] Thus, Popov's actual living standard in 1924 was no more than 20–25 percent of its prewar level.

In the mid-1920s, the salary of agricultural specialists increased, and in many provinces a system of periodical raises was introduced.[142] In fact, in the second half of the 1920s, agronomists were the highest paid category of rural professionals, as can be seen in table 11.2.

Agronomists earned one and a half times more than rural policemen and even judges.[144] Their salaries increased every year at the highest rate among rural professionals. This did not prevent the old specialists from perceiving their economic situation as the ultimate deterioration in comparison with the prewar period, while it gave a totally different perspective to younger Soviet agricultural specialists recruited after 1917, who even in Ukraine constituted 35–40 percent of agricultural personnel. Young agricultural specialists who joined the profession after the revolution, and who did not know any other social reality, perceived the NEP period as one of continuous progress in their social and economic status. It was "their" time and "their" system, while the "old specialists" of the post-1905 generation had had their hour of triumph before the revolution but still continued to hold senior positions.

The restoration of the system of agronomy assistance, even in a modified fashion, and the steady increase in agricultural specialists' salaries in the mid-1920s produced the structural preconditions for the revival of the corporation of rural professionals. According to the 1926 population census, there were a total of 8,795 agronomists, 11,603 land surveyors and topographers,

Table 11.2   The average monthly salary of rural specialists in the USSR, 1925–1928 (in chervonets rubles)[143]

	1925/26	1926/27	1927/28	Increase since 1925/26, in %	
				1926/27	1927/28
Physicians	33.0	38.7	43.5	17.3	31.5
Veterinarians	42.2	50.4	54.7	19.4	29.6
Agronomists	52.5	67.7	74.2	29.0	41.3

and 4,656 veterinarians in the USSR,[145] almost 10,000 of whom held college or university degrees.[146] These figures were close to prerevolutionary statistics, but this similarity was only superficial: any attempts to discuss the need for public self-organization were closely monitored and prosecuted, while the loose community of the NEP-era agricultural specialists was undergoing a process of disintegration rather than professionalization. Sociologically, we may even speak about a *deprofessionalization* of this occupation group: people were still experts in their field, but the patterns of professional socialization and integration were becoming increasingly corrupted.

The standard practice of the new community of agricultural specialists seemed to be a parody of the prerevolutionary agricultural *obshchestvennost'*. The core event in the life of a provincial agronomist organization, the annual agronomist conference, was restored in many regions, but it had changed its former meaning and composition. The case of Samara province provides an example of this pseudocontinuity. The fourth provincial agronomist conference took place on December 16–20, 1914, and was the last one before the World War I and the Revolutions of 1917. The fifth conference was convoked only on September 13–19, 1926. The very timing of this conference indicated a departure from the prerevolutionary rationale of such events. Previously, agronomists saw in local professional meetings an opportunity to discuss their professional issues, and hence they were the main driving force behind their organization. Naturally, they could meet only during the winter recess in peasant agricultural activity, which consumed all of their attention from early spring to mid-fall. In 1926, 206 agricultural specialists attended the agronomist conference on September 13, when harvesting work was in full progress. This means that the organizers of the conference were more concerned with the procedure itself than with the interests of agricultural specialists and their clients: peasants. The conference agenda also violated the traditions of the third element. It began with an endless report by the provincial agronomist V. I. Kopeikin (47 pages of small type), whose prerevolutionary service in the state or zemstvo agricultural structures I was unable to track.[147] Kopeikin was followed by other agricultural specialists who accounted for their activities. No discussion of these reports or any preliminary agenda was documented in the voluminous proceedings of the conference. Needless to say, the opposite was the norm in the old profession, when all agronomist conferences were designed to discuss urgent issues, rather than to account for activities that had already been undertaken.

The published proceedings of the fifth Samara agronomist conference did not mention any procedural details. One can assume that they did not differ much from the well-documented 1924 Arkhangelsk conference. The Arkhangelsk meeting began with the speech of a high-level Moscow official and the election of the presidium.[148] The latter action had a purely decorative meaning; presidium members did not rotate during the week of the conference's activity. In the prerevolutionary third element tradition, the chairman and his assistants were normally elected for just one day or even one session. Then, the telegram of greetings, signed by the presidium, was sent to the commissar of agriculture and the Central Committee of the Union of Agricultural Specialists (*Vserabotzemles*).[149] The commissar's answer was laconic: "Please forward my gratitude to the agroconference. Period."[150] Thus, even semiotically, the revived agronomist conferences "quoted" the forms of the pre-1905 conventions of civil servants, rather than the public culture of a professional association of the interrevolutionary period.

The cooperative network, another important vehicle of the prerevolutionary public politics of modernity, was also revived in the early 1920s from virtual collapse. Despite formal continuity with the old institution, the newly emerged structures only superficially resembled the prerevolutionary originals. In 1926, the share of "old cooperative activists" with pre-1914 experience in Soviet cooperative associations did not exceed 1.2 percent (at the same time, 16.3 percent of "Red cooperative managers" had been engaged in private "capitalist" trade in 1913).[151] The most striking case of "inflation" was provided by rural cooperatives. The third Conference of Representatives of the All-Russian Union of Agricultural Cooperatives (June 1–6, 1925) revealed the illusory character of the resurrection of the agricultural cooperative network in the 1920s. The very account of this conference of a supposedly independent socioeconomic public association was printed by the OGPU (the Soviet political police) press. Similar camouflage of the new Soviet reality by classical forms characterized the cooperative movement itself. The conference elected Alexander Chaianov, Nikolai Makarov, Semen Maslov, Mikhail Vonzblein, and Alexander Shvetsov as members of the Council of the *Sel'skosoiiuz*. They had all become leading cooperative activists during the public modernization campaign at the beginning of the century.[152] These cooperative "generals" could be proud of the resurrected cooperative apparatus that was no less powerful than the prerevolutionary all-Russian unions. At the same time, that strong centralized apparatus united local rural cooperatives with a total membership four times smaller than before the Revolutions of 1917. Out of 224,000 local cooperative managers in 1925, only 1.3 percent had begun their cooperative activity in 1914 or earlier.[153] The "fourth element" of cooperative managers was not a formally organized professional group, unlike agronomists, and this explains the remarkable fluidity of cooperative cadres. But even more important were systematic attempts by the Soviet regime to achieve total control over cooperatives and tame this potentially dangerous channel of financial and social mobilization. The soviet rural cooperative hierarchy virtually did not have financial capital of its own, and operated with funds loaned by the Soviet government.[154] Thus, old cooperative leaders presided over a cooperative movement that had a social composition, financial basis, and political agenda provided by the Soviet state. The community of cooperative activists as a distinctive social group with its own ideals and interests did not exist anymore.

Graduates of the Moscow Agricultural Institute class of 1910 embodied the vanguard of public modernization movements of the early 1910s. Most of them were below 40 years in the mid-1920s, and even their senior colleagues were quite active and not considering retirement anytime soon. These people formerly constituted the core of their occupation group, and their gradual upward mobility promised to reshape the entire profession in accordance with their standards. This did not happen, as the ties of social generation did not survive the rupture of wars and revolution, and early Soviet normalization further differentiated the composition of agricultural specialists. By the late 1920s, the veterans of public agronomy were not members of any professional community, just an age cohort with a certain pattern of employment that increasingly irritated both party authorities and their junior Sovietized colleagues.

The hierarchy of agricultural specialists included three main stages: Commissariats of Agriculture of union republics (e.g., the RSFSR or Ukraine); regional administrations (approximately former provincial level); and local specialists and practitioners. By October 1929, all three levels demonstrated a high correlation between age and occupation, with two groups particularly visible: the age group that dominated the 1910 class of the Moscow Agricultural Institute (born between 1885 and 1894) that had begun their professional activity most likely before 1914, and the younger generation of people who were born after 1900 and matured after the revolution. They had the highest percentage of people with mid-level education, and Komsomol members.[155]

At the local level, the majority of administrative positions were occupied by people born between 1885 and 1894 (43 percent), while the younger generation born after 1900 dominated the ranks of "researchers and consultants with no administrative functions" in the provinces, who actually shaped the discourse of the agricultural specialists (the same proportion, 43 percent). At the middle level, the majority of administrators also belonged to the 1885–94 age group, and an even higher share of top experts (43 percent), while the only comparable concentration of coevals was demonstrated again by "researchers and consultants" born after 1900 (41 percent). Finally, in the central apparatus, the older cadres comprised three-quarters of the personnel in all categories, with the specialists born between 1885 and 1894 specifically occupying almost 50 percent of administrative positions.[156] The generational hierarchy coincided with social and professional stratification, thus itself becoming a political factor.

Thus the "new" post-1905 generation of Russian intelligentsia who had turned into "old specialists" in the new social system lost direct contact with the peasantry and leadership in the sphere of elaborating professional discourse and public agenda to the young agricultural specialists who matured under the Soviet regime and sought faster upward mobility. By 1929, the coexistence of two absolutely different age and social cohorts within the single group of agricultural specialists became explosive. The policy of the cultural revolution in the late 1920s triggered an open confrontation.[157] The physical destruction of the old public modernizers followed soon after their social disintegration. The arrests of 1930, later explained by the alleged threat of the Labor Peasant Party, isolated absolutely all leaders of the old public agronomy who had not emigrated or who had returned from emigration.

The hypothetical Peasant Labor Party was not a complete invention of the OGPU. The idea of the necessity to create a truly peasant party had already originated in spring 1917, when the journal *Messenger of Cooperative Unions* published a strange unsigned document: the manifesto of the Broad Peasant Party. The manifesto suggested that neither the party of Socialist Revolutionaries nor the peasant Soviets really represented the peasantry, and stressed three main means of realizing "true" peasant interests: the cooperative movement, enlightenment, and self-government. Hence, it claimed that "The county zemstvo is the first cell of the peasant political union—the *Broad Peasant Party*," and called for the creation of a party with the assistance of rural professionals: teachers, agronomists, physicians, and clergy.[158] The broad peasant party, obviously a product of the agricultural specialists' imagination, failed to emerge, but the idea survived the civil war. It can be found in the writings of Alexander Chaianov in 1918–19, before he entered Bolshevik service about 1921.[159] After the civil war, agricultural specialists in the USSR were too deeply disappointed in the absence of political common sense among the peasantry to cherish any illusions about the desirability of a peasant party. It was one of the founders of the Russian Agricultural Cooperative Institute in Prague, Socialist Revolutionary S. S. Maslov, who revived this old idea under the name of the Labor Peasant Party. In 1928, he declared that the Prague émigré group Peasant Russia, which in the early 1920s published a few collections of social and economic essays under the same title, was just a cover for the political party Peasant Russia–Labor Peasant Party (LPP) with underground branches in Russia. He traced the history of this imaginary organization to 1921, when the core of the LPP was formed in Moscow (apparently, simultaneously with the emergence of the Pomgol Committee). To prove the reality of the LPP's existence, Maslov published a collection of its programmatic materials.[160] While it is not likely that any political organization of the LPP existed before 1928, the propaganda activity of Maslov and his associates yielded its fruit, and a number of circles of the LPP emerged in various emigrant colonies.[161] The most active and noisy was the Far East Committee of the LPP. In April 1930, just a few months before wide-scale arrests among the leading agricultural specialists in the USSR, the magazine of the committee called upon the Soviet peasants and the intelligentsia to resist authority by means of economic sabotage and political terror.[162] Still, the noisy activity of the emigrant LPP did not become the cause of repressions against the remnants of old agricultural specialists. The first arrests took place in mid-June 1930, while reference to the LPP emerged in the interrogation materials only two months later.[163]

The significance of the "LPP" prosecution goes far beyond the trial of a dozen high-positioned specialists,[164] luminaries of public agronomy. By September 21, 1931, throughout the USSR, at least 1,296 agricultural specialists had been arrested and sentenced on charges of membership in the counterrevolutionary Labor Peasant Party.[165] The geographic distribution of the purges suggests that they precisely targeted the group of agricultural experts with a prerevolutionary record, who were overrepresented in the former "old zemstvo" provinces. They were hunted down not only in the upper echelons of the agronomist administration but also among the field specialists. According to the database of the Memorial Society, 43 percent of the village agronomists arrested in the 1930s (97 out of 227) had been born before 1895.[166]

The scale of the purges and their precise targeting can be explained by the fact that the intraprofessional conflicts among Soviet agricultural specialists that intensified during the cultural revolution in 1928–29 had done the homework for the OGPU. "Old cadres," "bourgeois specialists," "counterrevolutionaries," and simply "conservatives" had been detected by their junior colleagues during recurrent ideological campaigns. Every local agronomist organization identified "Kondratievists" and "Chaianovists" in their midst. The OGPU used this information to prosecute "old agronomists" as a social group.

The new generation of Russian intelligentsia that staffed and inspired the public modernization campaign in the Russian countryside after the 1905 Revolution finally ceased to exist as a social movement or even as a cohort of like-minded individuals. As their example shows, the new Soviet community was built by expropriating the individual from all social ties and loyalties, and mobilizing him or her into the service of the political regime, under threat of being assigned the status of an "alien element." To become a member of the emerging Soviet nation one had to renounce any claims to possessing an alternative social vision, and to quit any nonofficial public associations. Demodernization and deprofessionalization were essential for mixing with the people and joining the mainstream. All publicly visible old agronomists did exactly that, renouncing their old views and steering clear of any professional activities other than their direct duties in 1927–29. Their "nationalization" had been practically completed, and their engagement in the public politics of modernity in the countryside became an episode of their past. Unfortunately for them, a constellation of different circumstances in 1930–31 (the escalation of cultural revolution conflicts, the failure of collectivization, and the logic of the intraparty struggle) brought about their physical liquidation, several years after their ideological and intellectual annihilation.

Postscript

After the Bolshevik October Revolution, Ekaterina Sakharova, now Vavilova, wrote in her diary:

> Third day of the uprising...Three days. They provide extensive material for a sociologist. In the future, some postulates will become unquestionable elements in the sphere of administration. Why did the Bolsheviks prevail? Or, rather, who else could prevail after the deepening of revolution?...To give the multimillion people control over the state means abolition of the state itself. We did not grab the state and it fell....This is so painfully easy.[1]

As always, in a most condensed way she rationalized the feelings that preoccupied many of her colleagues. The dominant sense was one of catastrophe that undermined the results of a decade of their collective efforts and inaugurated a decade of their gradual disintegration as a social group. Bolsheviks were not granted any decisive role in the tragedy featuring the main protagonists of the intelligentsia mythology: the educated elite, the people, and the state. This worldview crisis was comparable in intensity to the post-1905 disillusionment in radical politics, although it had graver consequences. The new generation of Russian intelligentsia, and the bulk of the educated elite in general, breached the old tradition of Russian anti-statist *obshchestvesnnost'* and embraced the idea of the powerful state as an ultimate social value.

For almost a century, the Russian educated class had been searching for a way to integrate a sedimentary imperial society into a modern community of emancipated subjects, which today we call a "nation." To radical populists, this was a nation of the "people" as a social category. Only toiling people were regarded as essentially self-sufficient, and hence eligible to form a free union of truly independent citizens. This nation was expected to grow from below, through self-organization, and therefore the formalized and repressive institute of state and politics as a mechanism of distributing authority were perceived as redundant categories. The pan-imperial stratum of *obshchestvennost'* inherited this archetypal worldview and preserved its main traits after 1905, when the collapse of radical politics and the influence of Progressivist ideology formed a new ideal of the nation. The ideal nation of the post-1905 generation of Russian intelligentsia was not necessarily about common folks. The new imagined community was a community of the modernized: of creative individuals and rational economic producers. The state was seen as a technical instrument of budget policy, and not a subject of any large-scale welfare reform. As it was put in 1915 in the article introducing the concept of "social engineering" in Russia, "The state organization, as any other, may embark on the path of new social building [*sotsial'nogo stroitel'stva*]."[2] That is, the state can play a role in social engineering, but not necessarily, and not the main one.

The former clear ideal of the "new republic" of rational subjects was problematized by the very success of public modernizers. The logic of social mobilization in different regions and among various groups of population in the empire revealed at least two different visions of the imagined yet seemingly so self-evident community: the pan-imperial nation of the motherland and the ethnocultural nation of minorities and separatists. We can only speculate as to how deadly the contradiction was between the two scenarios of nation-building, and whether the principle of bottom-up self-organization as practiced by the cooperative movement could compromise the two in some federalist accommodation.

The February Revolution of 1917 was widely perceived as a victory of the efficient and well-mobilized *obshchestvennost'* over the weak and inefficient state. The timing could not be worse for experiments in stateless social reformism as preached by *obshchestvennost'* activists, while the revolution revived the old archetypal ideological biases of the intelligentsia. The new ideal of a revolutionary nation envisioned not only an "imagined," but also a utopian community. While other scenarios of nationhood expected an ordinary citizen to be loyal to the empire, or to love one's national (ethnic) culture, the revolutionary nation expected everyone to behave as revolutionaries—subjects of historical process. This stake on sustained revolutionarism of the masses failed in 1905, and it failed again in 1917, despite the genuine enthusiasm of the people and self-sacrificial heroism of the many. When public modernizers decided to resort to the state as an instrument of social engineering and control, it was too late, and the worn-out instrument did not work.

The not-so-new generation of Russian intelligentsia reemerged from the chaos of civil war with a strong fixation on the state. That is when the endless mind games of the intelligentsia originated: how to distinguish a political regime from the state, or how to assist the people without hurting the state. The state became the ultimate embodiment of the modern nation built on the ruins of Russian empire: it reconciled loyalty to the country as a geopolitical entity with allegiance to a particular national (ethnic) community; it provided political regimes with legitimacy and asserted a historical continuity with the glorious past. Even before the Soviet intelligentsia's "pact" with the Stalinist regime in the 1930s, people like Sakharova-Vavilova or Chaianov after 1917 worked hard to construct a positive image of the state as the truest of all "imagined communities." But what about the old *obshchestvennost'* cult of self-organization and persistent anti-statism? The bitterness expressed in Sakharova-Vavilova's diary anticipates the sacramental question famously documented by Marc Bloch under different circumstances: "Are we to believe that history has betrayed us?"[3] "For our entire crash eventually only the 'intelligentsia' should be blamed and not the 'people'—yes, beginning with Tolstoy, we have dug a grave for ourselves."[4] In a sense, history did play a bad joke on Russian Progressivist modernizers, as the initial projects of social reformism through public initiatives in one way or another had failed everywhere by the early 1920s, in Europe and the United States, by evolving into ominous "schemes to improve the human condition." The legacy of Progressivism and the practices of social engineering became strongly associated with the welfare state in its various incarnations. Russian public modernizers also drifted toward a greater appreciation of the state as an efficient instrument of social engineering during the World War I, but this evolution had not been completed by 1917, and their subsequent surrender to the supremacy of "state interests" was not that of ambitious reformers but that of frustrated nation-builders. In the 1920s, the remnants of *obshchestvennost'* attempted to revive the old ideals of public activism, but this time they wanted to serve the "national interests" meaning the interests of the "true" state as some superior "eternal nation." As Chaianov put it during his trip abroad in 1923:

> It is necessary to firmly and definitely distinguish Russia from the USSR. It is necessary to accept live processes in the economy and even assistance to these processes by the intelligentsia working with the Soviet regime.[5]

He did not speak of "Russian people," or "Russian peasants," or even an "economic man." The new attitude of intelligentsia toward the state was both more "mature" than their former naive anarchism and suicidal, as it discarded the model of civil society that was successfully developed and implemented by *obshchestvennost'* during the interrevolutionary period. All subsequent attempts to revive the traditions of Russian public activism have invariably been trapped by the identification of the "nation" with the state. "Historical legacy" is a misleading explanatory category that implies the determination of present events by the past, without a clear explanation of the "medium" of that legacy, or how exactly the mechanism of determination works, or why the "future" is so different from the "past" in all other respects. However, the intelligentsia's fixation on the state during the early Soviet period has had a decisive impact on Russian society up to today. Russian public discourse simply lacks any concept of a community of solidarity of civic action (i.e., nation) beyond and beside the state. The imported concept of "civil society" does not find a distinctive and in any way substantial social group to take root, and is easily compromised by the manipulative regime as an "antinational" (because anti-statist) phenomenon. The alleged "legacy of totalitarianism" can be blamed for the persistence of the omnipotence of state in Russian society, but the divergent political trajectories of other post-Soviet societies suggest another explanation. Whereas Ukrainian or Georgian intellectuals struggle with different versions of the nation as a subject of political process, be it an ethnically cleansed "national body" or a civic body politic, Russian educated society cannot overcome the dilemma of the new generation of Russian intelligentsia. Frustrated by the catastrophic consequences of their own success, public modernizers renounced any claims to independent social initiative, thus making the state the only subject of history and social development, the embodiment of national community. Was this the only logical conclusion after a decade of modernization through social self-mobilization?

As the history of the public modernization campaign in the Russian countryside shows, by the mid-1910s, a community of solidarity of civic action began to emerge as the dialogue and partnership between rural professionals and peasants grew in scale. As many public activists discovered, to their surprise, that community (or communities) did not limit its scope by socioeconomic mobilization, as had initially been envisioned, but stimulated political and nationalist mobilization as well—as one would expect of an emerging "nation." As soon as Russian villagers became rational "peasants" and self-conscious individuals, they began identifying themselves with one or several imagined communities of solidarity of civic action: "cooperativists," "Russian patriots," "Ukrainians," "Tatars," "revolutionaries," and so on. In the absence of a general political and legal framework that adequately accommodated different versions of voluntary groupness and "nationness," the remarkably successful mobilization of peasants into citizens became a serious factor in destabilizing late imperial society. Russian imperial society, self-mobilized and self-organized into the public spheres of *obshchestvennost'*, political nation, and nationalist political movements, outrivaled the inefficient state, but could not preserve the delicate balance of a general consensus of all social actors in the face of the challenge of total war, and later, of revolution. While this course of events was not accidental, it does not mean that interpretations of the past should serve as a self-fulfilling prophecy of the future.

The spontaneous, unqualified "Russianness" of *obshchestvennost'* had not been critically analyzed by the post-1905 generation of intelligentsia, but because of its discursive "neutrality" it managed to accommodate very different connotations without losing its mobilizing and integrative potential. Our analysis of social practices of *obshchestvennost'* suggests that at least as of 1905, Russian imperial educated society functioned as a community of "bilinguals." The universal sphere of common language was gradually evolving to accommodate the challenges of modernity on an empire-wide level, while at the same time some members of the *obshchestvennost'* also acted within narrow circles of emerging nationalist movements on a local scale and in local languages. This situation resulted not only in a multiplicity of loyalties available to any member of the Russian (imperial) intelligentsia but also in the possibility of adherence to more than one loyalty, giving priority to one or another under given circumstances. The "Artel" of Organization-Production economists is an illustration of this thesis on the level of individual contacts, while the complicated relations between "Ukrainian" and "Moscow" cooperative networks is an example at the structural level.

The performance of the *obshchestvennost'* as a socially active community of civic-minded individuals resulted in the catastrophic events of 1917 and afterward, which also testifies to its efficiency and strength. With its political mythology of the "land" (*zemshchina*) opposing the "state," *obshchestvennost'* accommodated the diversity of the Russian imperial situation by including different types of "minorities" in the quite imagined community of civic action, that is, nation. The only alternative to this project has been the "imperial" ideal of the state as equally distanced from and disinterested in its various categories of citizens. The history of the Progressivist movement of public modernization during the interrevolutionary period shows that both *obshchestvennost'* and semantically neutral social institutes can be adjusted to different ideological programs and cooperate quite productively. They differed not so much in their goals as in their approaches to social mobilization: the former from "below," and the latter from "above." The seemingly "technological" difference leads to important political consequences: *obshchestvennost'* implies the building of a national community by the people and upon their initiative; a state-sponsored national project mobilizes individuals as objects of further manipulation by the state as a superior social actor.

An attempt to turn Russian imperial villagers into self-conscious peasants through the "apolitical politics" of self-organization quite unexpectedly led to the creation of a new political nation by means of society's self-mobilization. This process was complicated and controversial, but very dynamic. It was crushed by the burden of its own success, causing many contemporaries and latter-day historians to assess it as a failure. However, looking at the dynamics and outcome of the alternative projects of nation-building from above, by political regimes (from Stolypin to Putin), can we regard them as a greater success than the *obshchestvennost'* project?

Notes

Introduction

1. Marina Mogilner, *Mifologiia podpol'nogo cheloveka* (Moscow: New Literary Review, 1999).
2. E. N. Sakharova-Vavilova, "Dnevnikovye zapisi," in RGAE, f. 328 op. 1, ed. khr. 8, ll. 30, 31, 36. All translations by the author, unless otherwise indicated.
3. E. N. Sakharova-Vavilova, "Dnevnikovye zapisi," l. 158.
4. *Moskovskii oblastnoi s"ezd deiatelei agronomicheskoi pomoshch'i naseleniiu. Trudy S"ezda*, vol. 1 (Moscow, 1911), pp. 50–51.
5. E. N. Sakharova-Vavilova, "Dnevnikovye zapisi," l. 166.
6. See NART, f. 256, Kazanskoi gubernskoi zemleustroitel'noi komissii, op. 7, d. 43. "O prokhozhdenii sluzhby Gubernskogo agronoma pri Kazanskoi gubernskoi zemleustroitel'noi komissii po vol'nomu naimu uchenogo agronoma 1-go razriada I. S. Barkhatova," ll. 11, 28 ob.

1 Becoming "Progressive": Structural Settings and Mental Mapping of Reformism

1. See Laurie Manchester, *Holy Fathers, Secular Sons: Clergy, Intelligentsia, and the Modern Self in Revolutionary Russia* (DeKalb: Northern Illinois University Press, 2008).
2. The first agricultural society was established in 1723 in Scotland, the second in 1736 in Ireland, the next appeared in 1747 in Switzerland, and in 1753 the Royal Agricultural Society was founded in London. In the 1760s, France and Germany followed suit. V. V. Morachevskii, ed., *Agronomicheskaia pomoshch' v Rossii* (Petrograd: Department of Agriculture, 1914), p. 104. In the United States, the first agricultural society was the Philadelphia Society, founded in 1785. By the end of the century, agricultural societies had appeared in South Carolina, Maine, New York, and Massachusetts. Alfred Charles True, *A History of Agricultural Education in the United States, 1785–1925* (New York: Arno Press and the New York Times, 1929), pp. 7–17. There is a strange parallel between the order in which agricultural societies burgeoned throughout the continent and the stages of freemasonry's spread a few decades earlier. Perhaps these very different institutions reflected the stages of "unfolding" of the emerging public sphere, its differentiation among various group interests.
3. Thus, when the Philadelphia Society for Promoting Agriculture appeared in March 1785, its main object was to promote "a greater increase of the products of land." In less than 10 years, however, the society associated activity for the promotion of agriculture with "the education of youth in the knowledge of that most important art," thus shifting the focus of their prime concern to the dissemination of knowledge "suitable for the agricultural citizens of the State." See True, *A History of Agricultural Education in the United States*, pp. 7–8.

4. On the history of the Free Economic Society and its various activities, see Joan Klobe Pratt, "The Russian Free Economic Society, 1765–1915" (Ph.D. diss., Columbia: University of Missouri, 1983).

5. By April 1911, 3,103 agricultural socities had been registered in the Russian Empire, of which 68 percent emerged after 1906. V. V. Morachevskii, "Obshchie statisticheskie vyvody," in V. V. Morachevskii, ed., *Spravochnue svedeniia o sel'skokhoziaistvennykh obshchestvakh*, vol. 1 (St. Petersburg: GUZiZ, Department Zemledeliia, 1911), pp. i, xii. There were almost 4,000 agricultural societies in Russia on the eve of the World War I. Cf. Morachevskii, ed., *Agronomicheskaia pomoshch' v Rossii*, p. 108.

6. Ibid., pp. 223, 226, 227.

7. For a study of the intelligentsia's experiments with agricultural cooperatives in the nineteenth century, see Yanni Kotsonis, "Agricultural Cooperatives and the Agrarian Question in Russia, 1861–1914" (Ph.D. diss., Columbia University, 1994), pp. 21–63.

8. Cathy A. Frierson, *Peasant Icons: Representations of Rural People in Late Nineteenth-Century Russia* (New York: Oxford University Press, 1993).

9. M. Vit-", "Sel'skohoziaistvennaia pechat' v Rossii (k ee piatidesiatiletiiu)," *Agronomicheskii zhurnal*, no. 7–8 (1915): 75.

10. The Ministry of Agriculture had founded *Zemledel'cheskaia gazeta* in 1834 (which coincided with the reform of Count Kiselev) and *Sel'skoe khoziaistvo i lesovodstvo* in 1865 (after the reform of 1861). The Main Moscow Sheep-breeding Society founded *Zhurnal dlia ovtsevodov* in 1833. The same year *Lesnoii zhurnal* was founded by the Society for the Encouragement of Forestry.

11. See Richard G. Robbins Jr., *Famine in Russia 1891–1892: The Imperial Government Responds to a Crisis* (New York and London: Columbia University Press, 1975), pp. 176–83; W. Bruce Lincoln, *In War's Dark Shadow: The Russians Before the Great War* (New York: Dial Press, 1983), p. 26; Ben Eklof, *Russian Peasant Schools: Officialdom, Village Culture, and Popular Pedagogy, 1861–1914* (Berkeley and Los Angeles: University of California Press, 1986), p. 97; David Kerans, "Agricultural Evolution and the Peasantry in Russia, Tambov Province, 1880–1915" (Ph.D. diss., University of Pennsylvania, 1994), p. 432.

12. In 1894 *Viatskaia gazeta* and *Saratovskaia zemskaia sel'skohoziaistvennaia gazeta* were founded; in 1896 the weekly *Khutorianin* appeared; in 1897, *Kazanskaia sel'skohoziaistvennaia gazeta*. *Khutorianin* was published by the Poltava agricultural society, and all of the others, by by local zemstvos.

13. Based on data derived from M. Vit-", "Sel'skohoziaistvennaia pechat' v Rossii," 76; Morachevskii, *Agronomicheskaia pomoshch'*, p. 344.

14. V. V. Morachevskii, ed., *Spravochnik po sel'skohoziaistvennoi periodicheskoi pechati, 1916* (Petrograd: Spravochno-izdatel'skoe biuro pri Departamente Zemledeliia, 1916), pp. xxii–xxiii.

15. I. V. Vol'fson, ed., *Gazetnyi mir na 1911 god: Adresnaia i spravochnaia kniga* (St. Petersburg, n.d.), columns 329–30.

16. I. V. Vol'fson, ed., *Gazetnyi mir: Adresnaia i spravochnaia kniga*, 2nd edn (St. Petersburg, 1912), columns 525–26. Each group was represented by 10 titles, or 0.4 percent of the total.

17. For instance, in 1905–09, readers of the St. Petersburg Public Library requested 54,616 volumes on economic and legal themes. Over the next five years, 1910–14, 58,157 volumes in the same category were requested. N. A. Efimova, "Chitateli publichnoi biblioteki v Peterburge i organizatsiia ikh obsluzhivaniia v

1814–1917," in *Trudy GPB im. M. E. Saltykova-Shchedrina*, vol. vi (9) (Leningrad: GPB im. M. E. Saltykova-Shchedrina, 1958), p. 130.

18. Vol'fson, *Gazetnyi mir: Adresnaia i spravochnaia kniga*, columns 522–23, 525. Thus, the average percentage of all periodicals in Yiddish was almost three times higher than the figure for the agrarian press. And quite the opposite, Lithuanian was used twice as often in agrarian periodicals than it was in the general press. This indicates a higher degree of the Lithuanian (and also Estonian and Ukranian) national intelligentsia's involvement in the agrarian question than was common among many other nationalities.

19. Morachevskii, *Spravochnik po sel'skohoziaistvennoi periodicheskoi pechati*, p. xxxiii.

20. "Vnutrennee obozrenie," *Agronomicheskii zhurnal*, no. 8 (1913): 129–30.

21. Based on data published in Vol'fson, *Gazetnyi mir: Adresnaia i spravochnaia kniga*, columns 525–26.

22. Provincial activists bitterly complained that "the [l]iterature circles of both capitals apparently refuse to service the countryside, which is engaged in cultural work. While the province mobilizes all its forces to service the village in its needs and demands, the literature of the capitals more and more often replies with a stream of cold water." D. Mikheev, "Krest'ianskaia kul'tura ili 'khuliganstvo'," *Agronomicheskii zhurnal*, no. 4 (1913): 99. See a similar complaint by the Samara provincial agronomist, Alexander Teitel: A. Teitel', "Po povodu vozobnovleniia 'Zemledel'cheskoi gazety,'" *Zemskii agronom*, no. 8–9 (1913): 64.

23. As the editor of an influential provincial periodical, the Kharkov *Agronomy Journal*, triumphantly wrote in 1913: "The past five years…became a period of extremely rapid and intensive development of provincial activity of a general cultural character. It is as if all the social energy, which could not find an outlet in the sphere of politics, has moved here. The new forces, which were born and brought up by a new period of our social life, could express their creative work here with maximum benefit and success." See K. Matseevich, "Naibolee vazhnye cherty sovremennoi obshchestvennoi agronomii," *Agronomicheskii zhurnal*, no. 1 (1913): 114.

24. Morachevskii, *Spravochnik po sel'skohoziaistvennoi periodicheskoi pechati*, p. xviii.

25. Based on data published in ibid., p. xxvi.

26. GARF, f. 102 Departament politsii, D-4, op. 119, d. 237, ll. 19, 24, 43.

27. We can only indirectly assess the number of actual readers of the agricultural press. Not all issues of journals and magazines reached the readers; at the same time, many periodicals were subscribed to by collectives, such as zemstvo boards, cooperatives, or village libraries, hence any single issue was read by a number of people.

    The larger part of the agricultural periodicals was distributed via subscription and hence did not miss their readers. In 1906, 83.42 percent issues of the magazine *Vestnik obshchestvennoi veterinarii* were distributed (95.42 percent by subscription). The same year, all issues of the magazine *Veterinarnyi feldsher* were distributed (88.6 percent by subscription). See *Vestnik obshchestvennoi veterinari*, no. 11–12 (1907): columns 393, 396. In 1909, the annual revenue from subscriptions to the magazine *Batumskoe sel'skoe khoziaistvo* was 10 times higher than the revenue from retail sale (153.25 rubles vs. 15.75 rubles). See *Batumskoe sel'skoe khoziaistvo*, no. 1 (January 1910): 3. In 1912, out of 2,400 copies of the second issue of the Kazan magazine *Trudovoi soiuz*, only 100 were left in the office (4.2 percent). See *Trudovoi soiuz*, no. 2 (1912): 14. Thus, we may hypothesize that on average, between 90 and 95 percent of the copies reached readers.

Collective subscribers built up significant demand for agricultural periodicals. For instance, the Ekaterinoslav provincial zemstvo in 1911 spent 600 rubles for several hundred annual subscriptions to the magazine *Khutorianin*. See *Samarskii zemledelets*, no. 1 (January 15, 1911): 19. The Moscow Literacy Society subsidized 169–328 newspaper subscriptions in the villages in different periods throughout 1915, and in addition distributed 25–102 copies of used periodicals. The agricultural the *New Spike* was one of the three titles distributed by the society. See VI. Murinov, "Gazeta v derevne (po pis'mam v Komissiu po rasprostraneniiu gazet pri Moskovskom obshchestve gramotnosti)," *Vestnik vospitaniia*, no. 4 (1916): 177, 182. The Birsk Union of Credit and Savings Associations allocated 100 rubles for subscriptions to cooperative literature and periodicals in 1916. See I. St-ov, "Uchreditel'noe sobranie Birskogo soiuza kreditnykh i ssudosberegatel'nykh tovarishchestv," *Vestnik kooperativnykh soiuzov*, no. 12 (December 1915): 634. By 1915, there were 13–25,000 rural popular libraries, with an estimated 3,000,000 patrons. See A. Reitblat, *Ot Bovy k Bal'montu: ocherki po istorii chteniia v Rossii vo vtoroi polovine XIX veka* (Moscow: Izdatel'stvo MPI, 1991), pp. 173, 181.

28. A. D. Pedashenko, *Ukazatel' knig, zhurnal'nyh i gazetnyh statei po sel'skomy khoziaistvu za…god* (St. Petersburg/Petrograd). Alexander Pedashenko classified all agriculture-related publications into 22 categories, and with such a tight net it is not likely that many of those publications had escaped his attention.

29. Calculations are made on the basis of data published in Pedashenko, *Ukazatel' knig*, for 1901–16.

30. This was the case even when the agronomist, Ashot Atanasiants, applied for permission to publish a weekly agricultural periodical *Gukhatnetes* (Agriculturist) in the Armenian language, in Tiflis (Tbilisi) in 1908. The rising Armenian national movement made authorities particularly suspicious of the Armenian-language press, especially in such provincial capitals in the Caucasus as Tiflis, where the presence of large Armenian communities was a matter of grave concern. Nevertheless, the Tiflis governor granted permission to publish the new magazine within the required two-month period after the application. RGIA, f. 776, op. 21, Chast' 2, ed. khr. 215, ll. 1, 2–4.

31. One of the most notable cases involved the leading *Agronomicheskii zhurnal* (Agronomist Journal). The Kharkov Circuit Court attorney ordered the confiscation of the second issue of the journal in 1914 for publishing an article by female agronomist, Iuliia Eremeeva, who blamed the government for all the troubles of countryside, including land shortage. The following passage provoked the particular outrage of the attorney: "Any strive of peasants toward human life is being carefully suppressed…by the rivalry with church schools…. Libraries are being cut in number. Books and various lectures in the native language of the population are prohibited, folk Ukrainian songs are being expelled from schools." This was more dangerous in the eyes of the authorities than a critique of the economic policies of the government: agitation for the Ukrainian national cause. Yet, the Kharkov Circuit Court overruled the order of its attorney on June 2, 1914, and acquitted the publisher of the journal and the author of the article in question of all charges. TsDIAU, f. 1680, op. 1, ed. khr. 343, ll. 1–3, 18.

32. In June 1906, amid village riots and political unrest in towns, the Pskov governor authorized the publication of a new weekly newspaper *Krestianskoe delo* (The Peasant Cause). The unsold run of the very first issue was confiscated, for all eight of its pages were dedicated to revolutionary propaganda, and only two

small articles (8 percent of the issue) discussed rye harvest and apiculture. The paper was officially shut down only when the same story was repeated in the second issue, and the third. RGIA, f. 776, op. 9, Chast' 2, ed. khr. 451, ll. 1–4, 11, 14.

33. See Esther Kingston-Mann, *In Search of the True West: Culture, Economics, And Problems of Russian Development* (Princeton, NJ: Princeton University Press, 1999), p. 4.

34. On Struve's intellectual evolution, see Richard Pipes, *Struve: Liberal on the Right, 1905–1944* (Cambridge, MA: Harvard University Press, 1980).

35. Petr Struve, " 'Sovremennost' i 'elementarnost' russkoi revolutsii," in P. B. Struve, *Patriotica: politika, kultura, religiia, sotsialism* (Moscow: Respublika, 1997), p. 23.

36. Daniel T. Rodgers, *Atlantic Crossings: Social Politics in a Progressive Age* (Cambridge, MA: Belknap Press of Harvard University Press, 1998), p. 3.

37. Ibid., p. 5.

38. John M. Jordan, *Machine-Age Ideology: Social Engineering and American Liberalism, 1911–1939* (Chapel Hill: University of North Carolina Press, 1994), pp. 7, 13.

39. James T. Kloppenberg, *Uncertain Victory: Social Democracy and Progressivism in European and American Thought, 1870–1920* (Oxford: Oxford University Press, 1986), p. 362.

40. A key figure of French social reformism, Georges Benoît-Lévy, in 1903, founded the Association française des cités-jardins (the French Association of Garden Cities) to develop and promote the new ecology-friendly urbanization models. The first mention of "garden cities" in Russia can be found in an anonymous magazine article as early as 1904; this became a central theme of several periodicals dedicated to architecture and town management, for example, *Zodchii* (Architect) and *Gorodskoe delo* (Town Affairs). See Mark Meerovich, "Rozhdenie i smert' goroda-sada: deistvuiushchie litsa i motivy ubiistva," *Vestnik Evrazii*, no. 1 (2007).

41. Cf. V. Dodonov, *Sotsialism bez politiki: goroda-sady budushchego v nastoiashchem* (Moscow: Kushnerev and Co., 1913).

42. The leader of the Russian Party of Constitutional Democrats, Paul Miliukov, was widely regarded as the chief architect of the bloc. Miliukov was known for his political connections in the United States, which he had visited three times before the war and where he had socialized with leading American Progressivists. On the other hand, the core of the bloc was composed of former Moscow City Council members known as the "Progressive Group," who pioneered urban reforms in Moscow at the beginning of the century, quite in line with (and keeping an eye on) urban reformers in the United States and Europe. P. N. Miliukov, *Vospominaniia, 1859–1917*, vol. 2 (New York, 1955), pp. 24–27, 207–16; Robert W. Thurston, *Liberal City, Conservative State: Moscow and Russia's Urban Crisis, 1906–1914* (Oxford: Oxford University Press, 1987).

43. Populist journalist, Nikolai Shelgunov, expressed this position in a series of essays *Sketches of Russian Life*, particularly in those written in the late 1880s. See N. V. Shelgunov, *Sobranie sochinenii*, vol. 3 (St. Petersburg: O. N. Popova, 1904), pp. 651, 677, 683.

44. On the revival of popularity of the "small deeds" theory and one of its original ideologists, Ia. V. Abramov, see V. V. Zverev, "Marksizm i genezis neonarodnich-estva: Po materialam perepiski V. M. Chernova s N. F. Danielsonom v kontse

90-h godov XIX v.," in N. V. Samover, ed., *Rossiia i reformy: Sbornik statei*, Vyp. 4 (Moscow, 1997), pp. 123–24.

45. The patriarch of the Russian cooperative movement, Vakhan Totomiants, recalled in his memoirs how at the beginning of the twentieth century he used a penname to sign an article propagating "small deeds" because he was afraid to compromise himself in the eyes of the St. Petersburg radical intelligentsia. "So great was the desire among not only Marxists but also the leftist Populists to distance themselves from the 'small deeds,' which could interfere with the 'great deeds,' that is, the preparation of revolution in Russia." See Vakhan Totomiants, "Iz moikh vospominanii," manuscript in BAR, Vakhan Totomiantz Collection, p. 40.

46. See Kloppenberg, *Uncertain Victory*.

47. Petr Struve, "Intelligentsia i narodnoe khoziaistvo," in P. B. Struve, *Patriotica: politika, kultura, religiia, sotsialism* (Moscow: Respublika, 1997), pp. 203, 205.

48. Since the publication of Leopold Haimson's groundbreaking two-part article "The Problem of Social Stability in Urban Russia, 1905–1917," *Slavic Review* 23 (December 1964): 619–42; vol. 24 (March 1965): 1–22.

49. Cf. Edith W. Clowes, Samuel D. Kassow, and James L. West, eds., *Between Tsar and People. Educated Society and the Quest for Public Identity in Late Imperial Russia* (Princeton, NJ: Princeton University Press, 1991); Christine Ruane, *Gender, Class, and the Professionalization of Russian City Teachers, 1860–1914* (Pittsburgh, PA: University of Pittsburgh Press, 1994); Harley D. Balzer, ed., *Russia's Missing Middle Class: The Professions in Russian History* (Armonk, NY and London, England: M. E. Sharpe, 1996); Boris Mironov and Ben Eklof, *Social History of Imperial Russia, 1700–1917* (Boulder, CO: Westview Press, 2000); Elise Kimerling Wirtschafter, *Structure of Society: Imperial Russia's People of Various Ranks* (Dekalb: Northern Illinois University Press, 1994); idem, *Social Identity in Imperial Russia* (DeKalb: Northern Illinois University Press, 1997).

50. Alfred J. Rieber, "The Sedimentary Society," in *Between Tsar and People*, 343–66. First published as Alfred J. Rieber, "The Sedimentary Society," *Russian History* 16 (1989): 353–76.

51. A catalogue of fatally fragmented social entities in pre-1917 Russia can be found in a 1996 article by Robert McKean, which summarized the first stage in the revision of the Haimsonian tradition in historiography and triggered an important discussion in *Revolutionary Russia*: "One of the outstanding features of the Duma political system was the fragmentation of politics...the nobility was ethnically and religiously diversified; it was divided by occupational allegiances;...it was stratified by income and landownership.... Furthermore, ethnic diversity and regional economic rivalries divided merchants and entrepreneurs.... Politically divided in 1905, the professions fragmented after the failure of that revolution.... Furthermore, there existed a deep divorce and a gulf of suspicion and misunderstanding between the professionals and the intelligentsia...and the industrial middle class." Robert B. McKean, "Constitutional Russia," *Revolutionary Russia* 9 (June 1996): 34–36.

52. Cf. "...imperial Russia was slow to produce a massive political and industrial transformation comparable to that which had fundamentally altered much of west and east-central Europe by the time of the First World War. Most important in this regard were the traditional fragmentation of Russian society, including its mobility and porous boundaries; the limited development of

formal (though not necessarily informal) societal structures; and the accompanying dispersion of resources and cadres." Wirtschafter, *Social Identity in Imperial Russia*, p. 97.

53. Laura Engelstein, "Combined Underdevelopment: Discipline and the Law in Imperial and Soviet Russia," *American Historical Review* 98 (April 1993): 349.

54. Ibid., 348.

55. Peter Holquist has brought these themes of the existence of some "aggregated Europe" and its "normative historical scenarios" to a lapidary formula: "The absence of the institutions necessary to secure a true civil society... [b]ecause Russian society telescoped phases of development that unfolded sequentially in western Europe." Peter Holquist, *Making War, Forging Revolution: Russia's Continuum of Crisis, 1914–1921* (Cambridge, MA: Harvard University Press, 2002), p. 14. Note the sources of these generalizations: Leon Trotsky, *The Russian Revolution*, 3 vols. (New York, 1932), 1, pp. 4–6; Laura Engelstein, "Combined Underdevelopment."

56. Jürgen Habermas, *The Structural Transformation of the Public Sphere: An Inquiry into a Category of Bourgois Society* (Cambridge, MA: MIT Press, 1989).

57. Scott J. Seregny, "A Wager on the Peasantry: Anti-Zemstvo Riots and Adult Education in Stavropol Province, 1913–1916," *Slavonic and East European Review* 79 (January 2001): 90–126; idem, "Zemstvos, Peasants, and Citizenship: The Russian Adult Education Movement and World War I," *Slavic Review* 59 (Summer 2001): 290–315; idem, "Peasants, Nation, and Local Government in Wartime Russia," *Slavic Review* 59 (Summer 2001): 336–42.

58. David Moon, "Peasants into Russian Citizens? A Comparative Perspective," *Revolutionary Russia* 9 (June 1996): 43–81.

59. Eugen Weber, *Peasants into Frenchmen: The Modernization of Rural France, 1870–1914* (Stanford, CA: Stanford University Press, 1976).

60. Bertrand Patenaude was perhaps the first to apply the Weberian formula to the Russian case, although more as a rhetorical device than a coherent concept, and in the context of the Russian Civil War. See Bertrand M. Patenaude, "Peasants into Russians: The Utopian Essence of War Communism," *Russian Review* 54 (October 1995): 562. Jane Burbank made an indirect reference to this formula in a 1995 article by discussing an evolution "from peasant to citizen," and in a much more elaborated and nuanced form: Jane Burbank, "A Question of Dignity: Peasant Legal Culture in Late Imperial Russia," *Continuity and Change* 10 (1995): 391–404. The actual appropriation of the explanatory paradigm "Peasants into Frenchmen" by historians of Russia in the formula "Peasants into Russians" took place over the past decade, particularly following discussions in *Revolutionary Russia* (June 1996) and *Slavic Review* (Summer 2001).

61. For a comprehensive presentation of a new paradigm in the United States "Russian zemstvo studies," see Thomas Porter and William Gleason, "The Zemstvo and the Transformation of Russian Society," and "The Democratization of the Zemstvo During the First World War," in M. S. Conroy, ed., *Emerging Democracy in Late Imperial Russia* (Niwot, CO, 1998), pp. 60–87, 228–42. For a new approach to the topic in Russian and Japanese historiography, see Vitalii Abramov, Kimitaka Matsuzato, and Andrei Yartsev, *Zemskii fenomen: politologicheskii podhod* (Sapporo, 2001).

62. Seregny, "Zemstvos, Peasants, and Citizenship," 314; idem., "A Wager on the Peasantry," 125.

63. Boris Mironov should be mentioned as the champion of this argument. The Russian edition of his *Social History of Imperial Russia* has a telling subtitle: *The Genesis of Personality, the Democratic Family, Civil Society and the Rule of Law* (St. Petersburg: Dm. Bulanin, 1999).

64. Cf. Holquist, *Making War, Forging Revolution*, p. 14. See also the article by Laura Engelstein in the collection that attempted to solve the puzzle of the "true" civil society using a comparative approach: Laura Engelstein, "The Dream of Civil Society in Tsarist Russia: Law, State, and Religion," in Philip Nord and Nancy Bermeo, eds., *Civil Society before Democracy: Lessons from Nineteenth-Century Europe* (Lanham, MD: Rowman and Littlefield, 2000), pp. 23–41.

65. Josh Sanborn, "A Mobilization of 1914 and the Question of the Russian Nation: A Reexamination," *Slavic Review* 59 (Summer 2001), esp. 280, 289. He contextualized this approach in the book *Drafting the Russian Nation: Military Conscription, Total War, and Mass Politics, 1905–1925* (DeKalb: Northern Illinois University Press, 2003).

66. Joseph Bradley, "Subjects into Citizens: Societies, Civil Society, and Autocracy in Tsarist Russia," *American Historical Review* 107 (October 2002): 1094–123.

67. Harold Mah, "Phantasies of the Public Sphere: Rethinking the Habermas of Historians," *Journal of Modern History* 72 (March 2000): 155, 168.

68. *The Bibliography of Russian Obshchestvennost'* published in 1927 mentioned over 2,500 books and articles specifically dedicated to this phenomenon on 62 pages of tiny print. N. M. Somov, ed., *Bibliografiia russkoi obshchestvennosti* (Moscow: Published by Author, 1927).

69. Cf. Anastasiia Tumanova, *Obshchestvennye organizatsii goroda Tambova na rubezhe xix–xx vekov* (Tambov: Izdatel'stvo Tambovskogo gosudarstvennogo universiteta, 1999); idem, *Deiatel'nost' Ministerstva vnutrennikh del Rossiiskoi imperii po osushchestvleniu svobody soiiuzov* (Tambov: Izdatel'stvo Tambovskogo gosudarstvennogo universiteta, 2003); Joseph C. Bradley, "Voluntary Associations, Civic Culture and Obshchestvennost' in Moscow," in Edith Clowes et al., eds., *Between Tsar and People*, pp. 131–48; idem, "Russia's Parliament of Public Opinion: Association, Assembly and the Autocracy, 1906–1914," in Theodore Taranovski, ed., *Reform in Modern Russian History: Progress or Cycle?* (Cambridge: Cambridge University Press, 1995), pp. 212–36; idem, "Subjects into Citizens"; Lutz Häfner, *Gesellschaft als lokale Veranstaltung. Die Wolgastädte Kazan'und Saratov (1870–1914)* (Cologne: Böhlau Verlag, 2004). The scholar of Russian public associations, Tumanova explicitly equates the history of Russia to the history of Russian statehood. See Tumanova, *Deiatel'nost' Ministerstva*, pp. 3, et seq.

70. Words of the renowned lawyer, S. A. Kotliarevskii, quoted by Tumanova in *Deiatel'nost' Ministerstva*, p. 29.

71. Ibid., pp. 29–31.

72. See James C. McClelland, "Diversification in Russian-Soviet Education," in Konrad H. Jarausch, ed., *The Transformation of Higher Learning 1860–1930: Expansion, Diversification, Social Opening, and Professionalization in England, Germany, Russia, and the United States* (Chicago: University of Chicago Press, 1983), p. 184.

73. Cf. Yohanan Petrovsky-Shtern, *Jews in the Russian Army, 1827–1917: Drafted into Modernity* (Cambridge: Cambridge University Press, 2008); Marina Mogilner, *Homo Imperii. Istoriia fizicheskoi antropologii v Rossii* (Moscow: NLO, 2008), esp. chap. 10 "Army as Empire."

74. As was reflected in the legislation, *Polnoe Sobranie zakonov Rossiiskoi Imperii. Sobranie Vtoroe*, Vol. III, p. 331.

75. As was the case during the notorious "Chigirin affair" in Poltava province in Ukraine in the late 1870s. See Daniel Field, *Rebels in the Name of the Tsar* (Boston: Houghton Mifflin, 1976).

76. Cf. Daniel Field, *The End of Serfdom: Nobility and Bureaucracy in Russia, 1855–1861* (Cambridge, MA: Harvard University Press, 1976); William Bruce Lincoln, *The Great Reforms: Autocracy, Bureaucracy, and the Politics of Change in Imperial Russia* (De Kalb: Northern Illinois University Press, 1990); L. G. Zakharova, *Samoderzhavie i otmena krepostnogo prava. 1856–1861* (Moscow: MGU, 1984); David Macey, *Government and Peasant in Russia, 1861–1906. The Prehistory of the Stolypin Reforms* (De Kalb: Northern Illinois University Press, 1987); Francis William Wcislo, *Reforming Rural Russia: State, Local Society, and National Politics, 1855–1914* (Princeton, NJ: Princeton University Press, 1990).

77. Mikhail Dolbilov, "The Emancipation Reform of 1861 in Russia and the Nationalism of the Imperial Bureaucracy," in Tadayuki Hayashi, ed., *The Construction and Deconstruction of National Histories in Slavic Eurasia* (Sapporo: SRC, 2003), pp. 135–36.

78. David Moon, *The Russian Peasantry: The World the Peasants Made* (London: Longman, 1999), p. 16.

79. Ibid., p. 18.

80. For an astute analysis of the ethnopolitically determined class relations in right-bank Ukraine, see Kimitaka Matsuzato, "Pol'skii faktor v Pravoberezhnoi Ukraine s 19 po nachalo 20 veka," *Ab Imperio*, no. 1 (Spring 2000): 123–44.

81. Cf. Zenon Kohut, "The Ukrainian Elite in the 18th Century and Its Integration into the Russian Nobility," in Ivo Banac and Paul Bushkovitch, eds., *The Nobility in Russia and Eastern Europe* (New Haven, CT: Yale University Press, 1983), pp. 65–98.

82. Cf. Serhy Yekelchyk, "The Body and National Myth: Motifs from the Ukrainian National Revival in the Nineteenth Century," *Australian Slavonic and East European Studies* 7, no. 2 (1993): 31–59.

83. Cf. "Although the Jadids spoke of 'the 10 million Muslims of Turkestan,'...the rural population remained virtually invisible in their writings." Adeeb Khalid, *The Politics of Muslim Cultural Reform: Jadidism in Central Asia* (Berkeley: University of California Press, 1998), p. 221. On the urban locus and Islamic orientation of Turkic protonationalism in the Caucasus, see Jörg Baberowski, "Tsivilizatorskaia missiia i natsionalizm v Zakavkazie: 1828–1914," in Ilya Gerasimov et al., eds., *Novaia imperskaia istoriia postsovetskogo prostranstva* (Kazan: TsINI, 2004), pp. 307–52.

84. This problem was initially dubbed the "Russian dilemma" by Hans Rogger, who influenced Geoffrey Hosking's perception of the empire as the primary reason of Russian nationalism's fatal underdevelopment. See Hans Rogger, "Nationalism and the State: A Russian Dilemma," *Comparative Studies in Society and History* 4, no. 3 (April 1962): 253–64; Geoffrey Hosking, *Russia: People and Empire 1552–1917* (London: HarperCollins, 1997). For a recent comprehensive discussion of this problem, see the thematic issue "Searching for the Center: Russian Nationalism," *Ab Imperio*, no. 3 (2003).

85. Ethnographic studies have suggested that even at the turn of the twentieth century peasant economic activities were tightly structured by traditional rituals, and "production" was inseparable from ethnoconfessional symbolic activities. See L. A. Tul'tseva, "Obshchina i agrarnaia obriadnost' riazanskikh krest'ian na

rubezhe XIX–XX vv.," in M. M. Gromyko and T. A. Listov, eds., *Russkie: semeinyi i obshchestvennyi byt* (Moscow: Nauka, 1989), pp. 50–52.

86. During the postemancipation period, Russian Orthodox peasants had been constantly increasing the number of nonworking days under the pretext of observing the religious calendar. In his study of the labor/leisure balance in the Russian pre-1917 village, Boris Mironov concluded that the average number of working days per year had decreased from 140 in the 1850s to 107 in 1902 in the Russian Orthodox communes. He explained that by 1905 Russian peasants were working only 29–34 percent of the available time (as opposed to 38–54 percent during the preemancipation period) by the initiative of the peasants themselves, who were no longer controlled by their lords and suffered from relative overpopulation. The last argument, however, looks questionable, as the Volga Muslims living in the same socioeconomic conditions had about 75 holidays as opposed to 240 of the Russian Orthodox, while Catholics had 90–100 holidays, and Protestants only 65–75. See B. N. Mironov, " 'Vsiakaia dusha prazdniku rada': trud i otdykh v russkoi derevne vtoroi poloviny XIX–nachala XX v.," in A. N. Tsamutali, ed., *Problemy sotsial'no-ekonomicheskoi i politicheskoi istorii Rossii XIX–XX veka* (St. Petersburg: Aleteiia, 1999), pp. 200–10. In some villages, all labor was prohibited on Fridays even during the most intensive summer period of work in the fields. Tul'tseva, "Obshchina i agrarnaia obriadnost'," p. 52.

87. V. V. Ferdinandov, "Neskol'ko slov o melkikh sel'skokhoziaistvennykh obshestvakh," *Veterinarnaia khronika Voronezhskoi gubernii*, no. 3 (March 1905): 142–43. After 1917 Ferdinandov became a renowned scholar, one of the founders and professors of the Voronezh Zoological and Veterinary Institute, chair of its Zootechnicks Department in 1935–41.

88. This point has been more than exhaustively elaborated by Yanni Kotsonis in *Making Peasants Backward: Agricultural Cooperatives and the Agrarian Question in Russia, 1861–1914* (New York: St. Martin's Press, 1999).

89. A. Grigorovskii. "Na rasput'i (K sovremennomy zemskomy krizisu)," *Agronomicheskii zhurnal*, no. 3 (1913): 10.

90. Alfred Rieber, *Merchants and Entrepreneurs in Imperial Russia* (Chapel Hill: University of North Carolina Press, 1982).

91. Quoted in Achadov, *Tretii element. Ego znachenie i organizatsiia. (Sluzhashie po naimy v gorodskih i zemskikh uchrezhdeniiah)* (Moscow: Biblioteka Samoupravlenie, 1906), p. 3.

92. Cf. A. N. Naumov, *Iz utselevshikh vospominanii, 1868–1917* (New York: Russian Printing House, 1954), vol. 2, p. 14.

93. Compare, for instance, Talcott Parsons, "The Professions and Social Structure," in Talcott Parsons, *Essays in Sociological Theory*, rev. edn (New York: Free Press, 1954), pp. 34–49.

94. Cf. L-g, "Fakty i vyvody," *Veterinarnoe obozrenie*, no. 13–14 (1907): 366.

95. S. P. Fridolin, *Ispoved' agronoma* (Moscow: Novaia derevnia, 1925), p. 43.

96. GARF, f. 102 Departament politsii, D-4, op. 119, d. 406, l. 15.

97. Boris Veselovskii, *Istoriia zemstva*, vol. 3 (St. Petersburg: Popova, 1911), p. 465.

98. Achadov, *Tretii element*, p. 11.

99. See, for instance, Charles E. Timberlake, "The Zemstvo and the Development of a Russian Middle Class," in Clowes et al., eds., *Between Tsar and People*, pp. 165, 169–71.

100. N. V-ev, "O polozhenii i roli tret'ego elementa v nyneshenem zemstve," *Zemskoe Delo*, no. 9 (May 5, 1912): 611.

101. Achadov, *Tretii element*, p. 21.
102. N. V-ev, "O polozhenii i roli," 613.
103. At this point, the interests of the third element coincided with the ambitions of the elected deputies of the zemstvo, who thought that the county zemstvo would become a bridge to then inaccessible peasants. That was a recurring theme in zemstvo periodicals and the minutes of local zemstvo sessions. For a most elaborated presentation of this idea, see A. M. Shingarev, *Melkaia zemskaia edinitsa ili volostnoe zemstvo* (Moscow: Ia. Levenshtein, 1907); K. F. Golovin, *K voprosu o volostnom zemstve* (St. Petersburg, 1912).
104. Cf. K. F. Pavpertov, "Agronomiia i veterinariia," *Vestnik obshchestvennoi veterinarii*, no. 1 (1914): 21–27.
105. Grigorovskii, "Na rasput'i", 6, 9.
106. Ibid., 10–13.
107. *Sbornik dokladov Birskoi uezdnoi zemskoi upravy i postanovlenii zemskikh sobranii ocherednoi sessii 20 sentiabria 1914 goda i chrezvychainykh zemskikh sobranii XXII sessii 19–20 fevralia i XXIII sessii 2 avgusta 1914 goda* (Ufa, 1915), pp. 9–10.
108. The seemingly different mood of the Iaroslavl province agronomist conference in March 1914, two weeks after the Birsk celebration, only proves this thesis. Zemstvo deputies, agronomists, and agricultural specialists in state service debated the notoriously poor performance of zemstvo agricultural assistance in the Rybinsk district of the province. All parties involved blamed each other for the failure, but the importance of rural professionals and their high status was beyond doubt to everyone. In general, the problems in Rybinsk were attributed to miscoordination between the "elements," which implied that the more common stories of success were explained by efficient cooperation and rapprochement between the third and the second elements. See *Trudy XI-go agronomicheskogo soveshchaniia pri Iaroslavskoi gubernskoi zemskoi uprave 3–8 marta 1914 g.* (Iaroslavl, 1914), pp. 28–29.
109. "Vnutrennee obozrenie," *Agronomicheskii zhurnal*, no. 9–10 (1913): 131, 132–33.
110. On the dynamics of the cooperative movement during this period, see Kotsonis, *Agricultural Cooperatives*, esp. chap. 5; V. V. Kabanov, *Krest'ianskaia obshchina i kooperatsiia Rossii xx veka* (Moscow: Institute of Russian History, 1997).
111. A typical set of arguments in support of these claims of the cooperative movement can be found in I. Kudriashev, "Otnoshenie kooperativov k zemstvu," *Kostromskoi kooperator*, no. 20 (November 1915): 8–9. Such claims and complaints against zemstvos were common during the mid-1910s at cooperative meetings of all levels. The competition for government procurement orders during the war added a strong economic component to this political battle. See S. Pichkurov, "Kooperativnyi s'ezd v Odesse 21–25 oktiabria (vpechatleniia uchastnika)," *Iuzhnyi kooperator*, no. 2 (January 31, 1916): 44.

2   Bringing Up a New Generation of Intelligentsia

1. "Vysshee sel'skokhoziaistvennoe obrazovanie v Rossii," *Russikie vedomosti*, no. 268 (1904): 4. Between 1850 and 1903, just over 2,000 people graduated from a few Russian institutions of higher agricultural learning.
2. K. Barantsevich, "Moi pomidor," *Studencheskaia zhizn'*, no. 14–15 (April 18, 1910): 13. Vladimir Purishkevich (1870–1920) was a notorious right-wing Russian nationalist; Ivan Dumbadze (1851–1916), a scandalous reactionary public figure, army commander, and Yalta town governor.

3. I. A. Stebut, *Sel'skohoziaistvennoe znanie i sel'skohoziaistvennoe obrazovanie. Sbornik statei* (St. Petersburg, 1889), p. 99.

4. I. I. Meshcherskii, *Vysshee sel'skohoziaistvennoe obrazovanie v Rossii i za granitsei* (St. Petersburg, 1893), p. xxiii.

5. A small number of agronomists was also trained at the agricultural departments of some polytechnic institutes, for example, in Kiev, Novocherkassk, and Riga. See A. E. Ivanov, *Vysshaia shkola Rossii v kontse XIX–nachale XX veka* (Moscow: Institut istorii AN SSSR, 1991), p. 62; Gregory Guroff, "The Legacy of Pre-Revolutionary Economic Education: St. Petersburg Polytechnic Institute," *Russian Review* 31, no. 3 (1972): 278.

6. See Ivanov, *Vysshaia shkola Rossii*, pp. 81, 93; V. V. Koropov, *Istoriia veterinarii v SSSR* (Moscow: Gosudarstvennoe izdatel'stvo sel'skohoziaistvennoi literatury, 1954), pp. 170–78.

7. In 1898–1916 only 3,136 railway engineers and 4,400 officers graduated, in contrast to 9,240 agricultural specialists. See Ivanov, *Vysshaia shkola Rossii*, pp. 318–19.

8. Based on Ivanov, *Vysshaia shkola Rossii*, Table 42, pp. 318–19.

9. Ibid., p. 198.

10. Ia. Nekludov, "Sel'skohoziaistvennye uchebnye zavedeniia v 1913 godu," *Sel'skohoziaistvennoe obrazovanie*, no. 10 (1913), p. 501.

11. A. E. Ivanov, *Vysshaia shkola Rossii*, pp. 81, 151.

12. Ibid., p. 320.

13. While a relatively small faction of students completed the full course of studies in private ("commercial") colleges, they managed to find employment having completed four and even three years of studies. Even more important, the overall enrollment (rather than the number of people who got diplomas) indicates the actual popularity of agriculture in the society, representing people who at some stage of their lives seriously considered joining the ranks of rural professionals. The total number of people who went through the private agricultural schools is hard to calculate. Between 1910 and 1916, the Evening Agricultural Courses run by the Society of People's Universities in St. Petersburg were attended by 575 students. See TsGIA, F. 451, Op. 1. The famous Stebut Women's Agricultural Courses registered up to 716 female students a year. RGIA, F. 450, Op. 3, D. 8, L. 16. Between 1906 and 1921, 3,753 students attended the St. Petersburg "Kamennoostrovskie" Agricultural Courses alone (females comprised 20 percent of the student body). TsGIA St. Petersburg, F. 449, Op. 1. Hence, the total number of students enrolled in the private agricultural schools should have been 5–10 times higher than the number of those who graduated with diplomas. This means that 10–20,000 Russians received full or partial training as agricultural specialists in the nongovernment institutions of higher learning.

14. According to the newspaper *Russkoe slovo*, in 1910 Moscow Agricultural Institute was 400 graduates short of meeting all the requests of local agronomy agencies (that year a total of 125 students got their diplomas, including Ekaterina Sakharova and Alexander Chaianov). Quoted in Iu. Larin, *Ekonomika dosovetskoi derevni* (Moscow and Leningrad: GIZ, 1926), p. 113. In 1910, the experimental stations of the Department of Agriculture were staffed with only 11 specialists, while at least 100 specialists were necessary. See Morachevskii, *Agronomicheskaia pomoshch'v Rossii*, pp. 451–52. The annual increase in agronomists' employment in 1909–12 fluctuated between 31.5 percent and 55.2 percent. See *Agronomicheskii zhurnal*, no. 8 (1913): 171. This trend persisted until the Revolution of 1917.

In 1916, 1,000 applications were submitted for 200 available seats at the Moscow Agricultural Institute, while the private Golitsyn Agricultural Courses could admit only 200 out of 1,600 applicants. In order to accommodate the high demand, the government planned to open nine new agricultural colleges and institutes. See D. Prianishnikov, "K osushchestvleniiu seti novykh agronomicheskikh institutov," *Zemledel'cheskaia gazeta*, no. 5 (February 4, 1917): 109.

15. For instance, the course of education in the Kazan mid-level agricultural college (*Kazanskoe srednee sel'skohoziaistvennoe uchilishe*) spanned six years, with only a short Christmas break during the academic year. Boys aged 14–16 years were admitted to the first year of the college with a two-year rural school education. The four-year course of municipal schools or *realschule* enabled graduates to be admitted directly to the third year of the agricultural college. A year of education cost only 40 rubles (five times less than at the agricultural institutes), and children of any social background and religion were admitted (although Jews were required to present resident permits from the local police). See *Svedeniia o Kazanskom srednem sel'skohoziaistvennom uchilishe i pravila dlia postuplenia v nego* (Kazan, 1913), pp. 3–7.

16. Thus in 1911 only 24 percent of all applicants were admitted, in 1912—25 percent, in 1913—24 percent, and in 1914—24 percent. See Ia. Nekludov, "Sel'skohoziaistvennye uchebnye zavedeniia v 1913 godu," *Sel'skohoziaistvennoe obrazovanie*, no. 10 (1913): 501, 502.

17. Ibid.

18. Based on Ivanov, *Vysshaia shkola Rossi*, Table 34, p. 274. These data reflect the situation at Moscow Agricultural Institute, Institute of Agriculture and Forestry in Novaia Alexandria, Forestry Institute, and Land Surveying Institute (*Mezhevoi institut*).

19. Thirty-six former students reported employment as agronomists in state service, and 31 had chosen a bohemian way of life. See A. Fortunatov, *K statistike rezultatov agronomicheskogo obrazovaniia* (St. Petersburg, 1894), pp. 4–5.

20. Based on data published in ibid.

21. Based on data published in *Zemskii agronom*, no. 10 (1915): 620.

22. This calculation is based on data derived from the following publications: "Diplomnaia rabota predstavlennaia okanchivaiushimi v techenii 1910 g.," in *Otchet o sostoianii Moskovskogo sel'skohoziaistvennogo instituta za 1910 god* (Moscow, 1911), pp. 43–50; *Spisok lits, sluzhashikh po vedomstvy Glavnogo upravlenia zemleustroiistva i zemledeliia na 1911* (St. Petersburg, 1911); *Spisok lits, sluzhashikh po vedomstvy Glavnogo upravlenia zemleustroiistva i zemledeliia na 1912. Ispravlen po 1 noiabria 1912* (St. Petersburg, 1912); N. A. Alexandrovskii, M. M. Glukhov, N. F. Shcherbakov, and V. N. Shtein, eds., *Mestnyi agronomicheskii personal, sostoiavshii na pravitel'stvennoi i obshchestvennoi sluzhbe 1 ianvaria 1914 g. Spravochnik* (Petrograd, 1914).

23. Based on data published in *Moskovskaia sel'skokhoziaistvennaia akademiia im. K. A. Timiriazeva: K 100-letiiu osnovaniia, 1865–1965* (Moscow, 1969), p. 107.

24. See Ivanov, *Vysshaia shkola Rossi*, p. 81.

25. Quoted from *Zemskii agronom*, no. 10 (1915): 620.

26. *Rech' i otchet chitannye v godichnom sobranii Moskovskogo sel'skohoziaistvennogo instituta 26 sentiabria 1900 goda* (Moscow, 1900), p. 58.

27. S. P. Fridolin, *Ispoved' agronoma* (Moscow: Novaia derevnia, 1925), pp. 18, 37–38.

28. Honorary citizens, merchants, and town commoners (*meshchane*) are united under the single denomination "middle-class representatives" because in real

social life these archaic estate categories were often applied to a rather distinctive conglomerate of protoprofessionals and mid-level entrepreneurs, most accurately defined as middle class.

29. *Rech' i otchet chitannye v godichnom sobranii Moskovskogo sel'skohoziaistvennogo instituta 26 sentiabria 1900 goda*, p. 59.

30. According to Seymour Becker, during the last half of the 1890s, when the would-be *petrovtsy* were at their previous stage of education, students from noble and bureaucratic families comprised 46–52 percent of the university contingent, 52–56 percent of that of the *gymnasia*, and about 36 percent of vocational schools. See Seymour Becker, *Nobility and Privilege in Late Imperial Russia* (Dekalb: Northern Illinois University Press, 1985), p. 125.

31. *Rech' i otchet chitannye v godichnom sobranii Moskovskogo sel'skohoziaistvennogo instituta 26 sentiabria 1906 goda* (Moscow, 1906), pp. 24–25.

32. *Otchet o sostoianii Moskovskogo sel'skohoziaistvennogo instituta za 1910 god* (Moscow, 1911), p. 35; *Otchet o sostoianii Moskovskogo sel'skohoziaistvennogo instituta za 1914 god* (Moscow, 1915), p. 96.

33. For an archetypal text, see Talcott Parsons, "The Professions and Social Structure," in Talcott Parsons, *Essays in Sociological Theory*, rev. edn. (New York: Free Press, 1954), pp. 34–49.

34. For a methodological outline of that approach, see Marvin Rintala, "Generations: Political Generations," in David L. Sills and Robert K. Merton, eds, *The International Encyclopedia of the Social Sciences*, vol. 6 (New York: Macmillan and Free Press, 1968), pp. 88–96. A characteristic example of practical application of this model can be found in Alan B. Spitzer, "The Historical Problem of Generations," *American Historical Review* 78 (1973): 1353–85.

35. Jane Pilcher, *Age and Generation in Modern Britain* (Oxford: Oxford University Press, 1995), pp. 29, et seq.

36. Karl Mannheim, "The Problem of Generations," in Karl Mannheim, *Essays on the Sociology of Knowledge* (London: Routledge and Kegan Paul, 1952), pp. 276–320.

37. Cf. Jane Pilcher, "Mannheim's Sociology of Generations: An Undervalued Legacy," *British Journal of Sociology* 45, no. 3 (September 1994): 481–95.

38. See Anthony Esler, " 'The Truest Community': Social Generations as Collective Mentalities," *Journal of Political and Military Sociology* 12 (Spring 1984): 99–112. We may recall that Pierre Nora has made this connection fundamental by introducing the concept of generation as a work of historical memory. Pierre Nora, "Generation," in Pierre Nora, ed., *The Realms of Memory*, vol. 1. *Conflicts and Divisions*, trans. A. Goldhammer (New York: Columbia University Press, 1996), pp. 499–613. Nora also notes the fundamental democratic implication of the concept of generation, stressing the horizontal ties and conceptualizing the difference in terms of time and experience.

39. For instance, many generational theorists tend to overplay the idea of generation as destiny. Even usually cautious Karl Mannheim did not escape this romantic position. Cf. Karl Mannheim, "The Problem of Generations," pp. 303, 306. Another popular cliché is present in the writings of sociologists who claimed to possess a universal key to the riddle of generation, which unequivocally translates the seemingly chaotic flow of time into a regular sequence of generations divided by constant intervals of, say, 15 years. Cf. Julián Marías, *Generations. A Historical Method* (Tuscaloosa: University of Alabama Press, 1970), pp. 164–65.

40. Annie Kriegel, "Generational Difference: The History of an Idea," in Stephen R. Graubard, ed., *Generations* (New York and London: Norton, 1978), p. 26.

41. As Annie Kriegel has put it, "Age is only a countable objective datum; seniority is the result of steps at least partly voluntary. It implies joining, personal affiliation. The date of birth is mere fate; but the date of entry into any sort of institution, particularly if the latter is meant to give a meaning to life, or even to transform it, represents the choice of one's destiny. No one can escape the reality of belonging to an age group. It is a fact established once and for all, but it yields little meaning, been absolutely general and passive. Seniority, on the other hand, forces us to look back to a significant selection between alternatives, choices the person had to make. It is all the more true that seniority binds generational communities more strongly than age because of the fact that, since the nineteenth century, both social and geographical mobility have vastly increased." Kriegel, "Generational Difference," p. 28.

42. "In social practice, it does not refer to an interval of time, but to an energy field that provides a framework for one or several experiences held to be crucial and worth remembering. A generation is only constituted when a system of references has retrospectively been set up and accepted as a system of collective identification." See Kriegel, "Generational Difference," p. 29.

43. Robert Wohl, *The Generation of 1914* (Cambridge, MA: Harvard University Press, 1979), p. 210.

44. Even populists who were traditionally viewed as champions of the peasant cause could not comprehend the nonideological concern for agriculture. One of the founding fathers of Russian populism, N. V. Chaikovskii, was imprisoned in the Petropavlovskaia Fortress in September 1908, when he learned about his son's plans to get married and engage in agriculture. Thanks to the commandant of the fortress, Colonel G. A. Ivanishin, who checked prisoners' letters, we know the advice that the old *narodnik* gave to his son: "It is necessary to be faithful to your calling by all means, and if you don't realize it clearly, to seek it…as a most precious shrine in your life…. Why not study agriculture? Only do not bury yourself in a voluntary prison, but find your true calling through it, to be useful not only for your family, but for the world." Colonel Ivanishin commented in his notebook on Chaikovskii's words with disapproval: "From this an old idealist is seen, a dreamer seeking something 'of world importance,' [thus] distracting [his] children from modest direct work, [and] forgetting that not all people were born talented, and that with such advice it is possible to multiply losers, dreamers, superfluous people unprepared for life." This episode demonstrates that the attitude toward the concept of professional service divided the representatives of "Undeground Russia" and "Legal Russia" no less than their political views. See G. A. Ivanishin, "Notebooks," in *Minuvshee: Istoricheskii al'manakh*, vol. 17 (Moscow and St. Petersburg: Atheneum; Phoenix, 1995), pp. 515–16.

45. Michael Confino, "On Intellectuals and Intellectual Tradition in Eighteenth- and Nineteenth-Century Russia," in S. N. Eisenstadt and S. R. Graubard, eds., *Intellectuals and Tradition* (New York: Humanities Press, 1973), p. 143. He wrote further: "This new intellectual stratum included a large and dynamic group of artists, poets, writers, and painters; a well-organized and effective group of urban professionals—lawyers, engineers, physicians, university professors, scientists; a large group of rural professionals—agronomists, statisticians, physicians, teachers." Ibid., pp. 140–41.

46. "Vnutrennee obozrenie," *Agronomicheskii zhurnal*, no. 8 (1913): 121.

47. "25 let obshchestvenno-agronomicheskoi raboty (k iubileiu Vladimira Nikolaevicha Vargina)," *Agronomicheskii zhurnal*, no. 5 (1913): 147–48.

48. S. F. Nikolaev, *Uchenyi-agronom V. N. Vargin* (Perm: Permskoe knizhnoe izdatel'stvo, 1966), pp. 7, 10, 20.

49. Ibid, p. 41.

50. Fridolin, *Ispoved' agronoma*, pp. 7–10, 13.

51. A classic example of this can be found in the story of A. J. Munby. See Leonore Davidoff, "Class and Gender in Victorian England: The Diaries of Arthur J. Munby and Hannah Cullwick," *Feminist Studies* 5, no. 1 (1979): 87–134.

52. Fridolin, *Ispoved' agronoma*, p. 12.

53. Alexandrovskii et al., *Mestnyi agronomicheskii personal, 1 ianvaria 1914 g.*, p. 197.

54. *Zhurnal zasedaniia gubernskogo agronomicheskogo soveshchaniia Kazanskogo gubernskogo zemstva 7 sentiabria 1911 goda* (Kazan, 1912), p. 171.

55. Fridolin, *Ispoved' agronoma*, p. 75.

56. V. N. Baliazin, *Professor Alexander Chaianov* (Moscow: Agropromizdat, 1990), pp. 10–12.

57. For instance, the Samara precinct agronomist Nikolai Dmitriev, who graduated from a mid-level agricultural college in 1908, found in the writings of Alexander Chaianov a blueprint for the project of precinct agronomy. N. L. Dmitriev, "K voprosu o rasprostranenii sel'skohoziaistvennykh znanii," in *Trudy 4-go Samarskogo gubernskogo agronomicheskogo soveshchania 16–20 dekabria 1913 goda* (Samara, 1914), p. 118.

58. See, for instance, S. P. Fridolin, "Nachalo kontsa," *Vestnik sel'skogo khoziaistva*, no. 3–4 (February 4, 1918): 3–4; A. V. Chaianov, "Letter to N. P. Makarov," published in V. A. Chaianov, *A. V. Chaianov—chelovek, uchenyi, grazhdanin* (Moscow: Izdatel'stvo MSKhA, 1998), pp. 122–23. The letter was written before December 4, 1917 (November 21, old style).

59. V. A. Chaianov, *A. V. Chaianov—chelovek, uchenyi, grazhdanin*, p. 16.

60. S. F. Nikolaev, *Uchenyi-agronom V. N. Vargin*, pp. 53, 56.

61. S. P. Fridolin, *Ispoved' agronoma*.

3 Transfer of the Italian Technology of Modernization and Birth of the Russian "Public Agronomy" Project

1. *Urozhaii 1905 goda*, vol. 1, *Ozimye khleba i seno* (St. Petersburg: Izdanie Tsentralnogo statisticheskogo komiteta MVD, 1906), p. v.

2. *Urozhaii 1906 goda*, vol. 1, *Ozimye khleba i seno* (St. Petersburg: Izdanie Tsentralnogo statisticheskogo komiteta MVD, 1907), p. xii.

3. *Urozhaii 1906 goda*, vol. 2, *IArovye khleba i kartofel'* (St. Petersburg: Izdanie Tsentralnogo statisticheskogo komiteta MVD, 1907), p. xxvii. A pood is a Russian unit of weight equal to about 36.11 pounds.

4. *Urozhaii 1907 goda*, vol. 2, *IArovye khleba i kartofel'* (St. Petersburg: Izdanie Tsentralnogo statisticheskogo komiteta MVD, 1908), p. xxi.

5. *Urozhaii 1908 goda*, vol. 2, *IArovye khleba i kartofel'* (St. Petersburg: Izdanie Tsentralnogo statisticheskogo komiteta MVD, 1909), pp. xviii, xxv.

6. *Urozhaii 1909 goda v Evropeiiskoi i Aziatskoi Rossii*, vol. 2, *Iarovye khleba i kartofel'* (St. Petersburg: Izdanie TSentralnogo statisticheskogo komiteta MVD, 1910), p. iv.

7. "Pechal'naia statistika," *Batumskii sel'skii khoziain*, no. 2 (February 1909): 84.

8. Cf. Alessandro Stanziani, "Russkie ekonomisty za granitsei v 1880–1914 gg.: Predstavleniia o rynke i tsirkuliatsii idei," in Iu. Sherrer and B. Anan'ich, eds., *Russkaia emigratsiia do 1917 goda—laboratoriia liberal'noi i revolutsionnoi mysli* (St. Petersburg: Evropeiskii Dom, 1997), pp. 164–66, et seq.; Kimitaka Matsuzato,

"Stolypinskaia reforma i rossiiskaia agrotekhnologicheskaia revolutsiia," *Otechestvennaia istoriia*, no. 6 (1992): 194–200.

9. V. I. Lenin, "Razvitie kapitalizma v Rossii," in V. I. Lenin, *Polnoe sobranie sochineniii*, 5th edn., vol. 3 (Moscow: Politizdat, 1963), pp. 39, 41, 55.

10. S. N. Bulgakov, *Kapitalizm i zemledelie*, vols. 1–2 (St. Petersburg: V. A. Tikhonov, 1900).

11. Ibid., vol. 1, p. 158.

12. As Bulgakov put it, "The major distinction of the peasant mode of production from the large-scale capitalist one consists in [the fact] that the peasant . . . strives for satisfaction of his needs and necessities, . . . while the capitalist strives for extraction of profit." See Bulgakov, *Kapitalizm i zemledelie*, vol. 1, p. 143.

13. A. I. Chuprov, *Melkoe zemledelie i ego osnovnye nuzhdy*, 4th edn. (Moscow: Kushnerev and Co., 1918; first published in 1907), p. 25.

14. On the problem of the conceptualization of peasant economy by zemstvo statisticians and the impact of their work on Russian economic theory, see David William Darrow, "The Politics of Numbers: Statistics and the Search for a Theory of Peasant Economy in Russia, 1861–1917" (Ph.D. diss., University of Iowa, 1996), esp. chapter VI; idem, "From Commune to Household: Statistics and and the Social Construction of Chaianov's Theory of Peasant Economy," *Comparative Studies in Society and History* 43, no. 4 (October 2001): 788–818. For a study of Russian zemstvo statisticians as professional corporation, see Martine Mespoulet, "Statisticiens des zemstva—Formation d'une nouvelle profession intellectuelle en Russie dans la période prérévolutionnaire (1880–1917). Le cas de Saratov," *Cahiers du Monde russe* 40, no. 4 (1999): 573–624; idem, *Statistique et révolution en Russie. Un compromis impossible (1880–1930)* (Rennes: Presses Universitaires de Rennes, 2001), chapters 1–3. On the interplay of scientific and ethical views of the statisticians, see Esther Kingston-Mann, "Statistics, Social Science, and Social Justice: The Zemstvo Statisticians of Pre-Revolutionary Russia," in Susan P. McCaffray and Michael Melancon, eds., *Russia in the European Context, 1789–1914* (Basingstoke: Palgrave, 2005), pp. 113–40.

15. In his dissertation, surprisingly little affected by the writings of Russian rural scholars of the early twentieth century, and dedicated to the agricultural, rather than agrarian history, David Kerans has arrived at a similar conclusion: "the peasant agrarian system—taken as a whole, with all its material, organizational and cultural components—was incapable of coping with the demographic and economic pressures of the era. Already by the turn of the century peasants were running out of ideas to improve grain yields." See Kerans, "Agricultural Evolution," p. 429. The already quoted statistics of labor/leisure balance in the Russian pre-1917 village collected by B. N. Mironov indicate that the average number of working days per year had decreased from 140 in the 1850s to 107 in 1902 in the Russian Orthodox communes. See B. N. Mironov, "Vsiakaia dusha prazdniku rada," pp. 200–10.

16. This figure, empirically established before the Revolution of 1917, was then theoretically validated by Alexander Chaianov, in A. V. Chaianov, "Optimal'nye razmery zemledel'cheskikh khoziaistv," *Trudy vysshego seminaria sel'skohoziaistvennoi ekonomii i politiki*, vol. 7 (Moscow, 1922), pp. 5–82.

17. Chuprov, *Melkoe zemledelie*, p. 112; A. V. Chaianov, *Organizatsiia krest'ianskogo khoziaistva* (Moscow: Kooperativnoe izdatel'stvo, 1925), pp. 53, 116.

18. F. Sev, "K voprosu o merakh uluchsheniia krest'ianskogo khoziaistva (Doklad 4-mu Samarskomu gubernskomu agronomicheskomu soveshchaniiu)," in *Otchet o deiatel'nosti i sostoianii sredstv Samarskogo obshchestva uluchsheniia krest'ianskogo khoziaistva za vtoroe trekhletie ego sushchestvovaniia (s 7/XI 1910 g. po XI 1913 g.)* (Samara, 1914), p. 183.

19. *Urozhaii 1906 goda*, vol. 1, p. xi; vol. 2, p. xiii.

20. *Urozhaii 1907 goda*, vol. 1, p. xii; vol. 2, p. xxv.

21. *Urozhaii 1908 goda*, vol. 2, p. xxv.

22. M. Frankfurt, "Zanimaites' skotovodstvom (Pis'mo k krest'ianam iuga-vostoka Rossii)," *Samarskii zemledelets*, no. 19 (October 1, 1916): 517.

23. Ferdinandov, "Neskol'ko slov o melkikh sel'skokhoziaistvennykh obshestvakh," 143, 144.

24. On Mikhailovsky (1842–1904), see the classic study James H. Billington, *Mikhailovsky and Russian Populism* (Oxford: Clarendon Press, 1958).

25. K. K. Sakovskii, "K podniatiiu i uluchsheniiu nashego zhivotnovodstva," *Vestnik obshchestvennoi veterinarii*, no. 7–8 (1907): 239.

26. P. B. Struve, "Ekonomicheskie programmy i 'neestestvennyi rezhim'," in P. B. Struve, *Patriotica: politika, kultura, religiia, sotsialism* (Moscow: Respublika, 1997), p. 96.

27. K. S. Ashin, "Obshchestvennaia agronomiia i zemleustroistvo," in K. S. Ashin, *Obshchestvenno-agronomicheskie etudy* (Kharkov: Izdatel'stvo Iuzhno-russkoi sel'skokhoziaistvennoi gazety, 1911), pp. 4–5. By "creative forces" Ashin meant the movement of public modernizers in countryside.

28. *Moskovskii oblastnoi s"ezd deiatelei agronomicheskoi pomoshch'i naseleniiu.*, pp. 50–51.

29. The Organization-Production School (OPS) in Russian economic thought attracted the attention of Western and Russian scholars at different points in time and for different reasons. The Western scholars who rediscovered the OPS in the 1960s were fascinated by the economic approaches of these scholars and their polemic with Marxist economists in the 1920s. These debates resembled disputes between the Marxists and nonclassical theorists of the 1970s. See Naum Jasny, *Soviet Economists of the Twenties: Names to Be Remembered* (Cambridge: Cambridge University Press, 1972), pp. 196–204; Teodor Shanin, *The Awkward Class: Political Sociology of Peasantry in a Developing SocietyRussia, 1910–1925* (Oxford: Oxford University Press, 1972); Susan Gross Solomon, *The Soviet Agrarian Debate: A Controversy in Social Science, 1923–1929* (Boulder, CO: Westview Press, 1977); idem, "Rural Scholars and the Cultural Revolution," in Sheila Fitzpatrick, ed., *Cultural Revolution in Russia, 1928–1931* (Bloomington: Indiana University Press, 1978), pp. 129–53; and so on.

    Soviet historians in 1987–1991 tried to interpret political declarations of the Chaianovists in support of NEP as grounds for perestroika's legitimization. (Vincent Barnett arrived at the same conclusion in "Recent Soviet Writings on Economic Theory and Policy from NEP," *Coexistence* 29, no. 3 (1992): 257–75.) As a rule, both historiographical currents ignored the prerevolutionary half of the OPS's activity and the social position and views of its members. This situation has been changing in historiography, as its main focus has shifted toward the genesis of the OSP. See, for instance, S. D. Domnikov, "Mirovozzrenie A. V. Chaianova" (Avtoreferat dissertatsii kandidata istoricheskikh nauk, Moscow: Institut istorii RAN, 1994); Stanziani, "Russkie ekonomisty za granitsei," pp. 157–75;

I. V. Gerasimov, "Tvorchestvo A. V. Chaianova v otechestvennoi i zarubezhnoi istoriographii" (Avtoreferat dissertatsii kandidata istoricheskikh nauk, Kazan: Kazan University, 1998); Alessandro Stanziani, *L'Économie en révolution: Le cas russe, 1870–1930* (Paris: Albin Michel S. A., 1998).

30. The first *Cattedra Ambulante di Agricoltura* was established in Rovigo in 1886. By the turn of the century, there were 30 mobile bureaus of agriculture in Italy, and by 1910, 112 bureaus with 79 additional branches. In 1910, mobile bureaus employed 309 specialists in agriculture, 95 percent of whom had received agricultural education in the institutions of higher learning. See V. Sazonov, "Populiarizatsiia sel'skokhoziaistvennykh znanii v Italii," *Sel'skokhoziaistvennoe obrazovanie*, no. 1 (1914): 10.

31. Bolton King and Thomas Okey, *Italy Today* (London: James Nisbet & Co., Limited, 1901), pp. 188–89.

32. Cf. Stanziani, "Russkie ekonomisty za granitsei," p. 165; idem, *L'Économie en révolution*, p. 136.

33. These articles were later included by Chuprov in a collection of his essays on the agrarian question. See A. I. Chuprov, "Reforma zemledeliia v Italii (1900)," in A. I Chuprov, *Krest'ianskii vopros: Stat'i 1900–1908 gg.* (Moscow: Izdatel'stvo br. Sabashnikovyh, 1909), pp. 1–43. See also A. I. Chuprov, "Agronomicheskaia pomoshch' naseleniiu v Italii (1901)," in A. I. Chuprov, *Krest'ianskii vopros*, pp. 44–79. In 1900, Alexander Chuprov wrote enthusiastically to the liberal juridical luminary A. F. Koni: "The most interesting part of my journeys in 1900 was a trip in May to Northern Italy, … by no means did I expect to encounter so much that was curious and to receive so many vivid impressions. In the person of professors of the so-called "mobile bureaus of agriculture" in Padova, Parma, and Verona I encountered true enthusiasts, saintly people who gave themselves entirely to serving the people. They are like our zemstvo agronomists, only with a much broader and more active role." Quoted in A. F. Koni, "Iz vospominanii ob A. I. Chuprove," in A. I. Chuprov, *Rechi i stat'i*, vol. 3 (Moscow: Izdatel'stvo br. Sabashnikovyh, 1909), pp. xxxii–xxxiii. Chuprov's publications on the Italian experience of public (i.e., nongovernment) agricultural assistance became truly archetypal for the new generation of Russian intelligentsia, who were literally brought up on these essays. As late as 1912, they were still recommended to a younger cohort of agricultural specialists as a sacred testament of the mass modernizing movement, as a "handbook that every agronomist must have." See *Zhurnal zasedaniia gubernskogo agronomicheskogo soveshchaniia Kazanskogo gubernskogo zemstva 7 sentiabria 1911 goda*, p. 242.

34. D. N. Prianishnikov, " 'Zemskaia agronomiia' v Italii," *Vestnik sel'skogo khoziaistva*, nos. 17–21 (1905).

35. A. V. Chaianov, "Stranstvuiushie kafedry v Italii," *Vestnik sel'skogo khoziaistva*, no. 33 (1908): 6.

36. V. Brunst, "Delo rasprostraneniia sel'skokhoziaistvennykh znanii v razlichnykh stranakh i vozmozhnye u nas vidy etogo roda agronomicheskoi raboty." I studied the typed manuscript of this article dated February 24, 1909, in TsDIAU, f. 2019, d. 352, ll. 1–10 ob.

37. Thus, I have to disagree with George Yaney, who believed that "the [agricultural] specialists set out to 'improve,' despite the fact that no one, least of all their supervisors, had any concrete, generally accepted idea of what the process of improvement entailed." George Yaney, *The Urge to Mobilize: Agrarian Reform in Russia, 1861–1930* (Urbana: University of Illinois Press, 1982), p. 347. His

otherwise fundamental study actually ignores the level of the public modernization movement, which, besides a tremendous potential of its own, was one of the major factors behind the government's "urge to mobilize" the countryside, as we shall see in the next chapters.

38. *Otchet o deiatel'nosti i sostoianii sredstv Samarskogo obshchestva*, pp. 1, 4.
39. Thus, on July 2, 1912, five peasants were brought to the Bezenchuk experimental agricultural station, and the next summer 11 more peasants visited the station under the auspices of the SOUKK. These excursions became popular, and the number of participating peasants continued growing. On June 26, 1916, 27 peasants spent a day at the station exploring new methods and techniques of agriculture. Society leaders concluded with satisfaction that peasants enjoyed the visit and were interested in what they had seen. The same impression is given by peasant letters published in the SOUKK's periodical the *Samara Agriculturist* but relating to the later period. See *Otchet o deiatel'nosti i sostoianii sredstv Samarskogo obshchestva*, p. 31; M. IA. Mordvinov, "Ekskursiia krest'ian na Bezenchukskuiu sel'skokhoziaistvennuiu opytnuiu stantsiu (pis'mo krest'ianina)," *Samarskii zemledelets*, no. 14 (July 15, 1916): 432–33; P. A. Nasonov, "Pis'mo krest'ianina s peredovykh pozitsii na Bezenchukskuiu opytnuiu stantsiu," *Samarskii zemledelets*, no. 15–16 (August 1 and 15, 1916): 467.
40. SOUKK kept records of all peasants who began cultivating corn, beets, or fodder grass as a part of an intensive crop rotation scheme. See *Otchet o deiatel'nosti i sostoianii sredstv Samarskogo obshchestva*, pp. 47–155.
41. Thus, in 1910–13 the society received a total of 16,340 rubles from the Department of Agriculture, a considerable sum at that time. A special grant of 500 rubles was given to subsidize the *Samara Agriculturist* in 1913 (which covered about 9 percent of the journal's expenses), 1,800 rubles were received to pay the Society's agronomist, separate sums were reserved for the purchase of sires and machines, and so on. See *Otchet o deiatel'nosti i sostoianii sredstv Samarskogo obshchestva*, p. 164; M. Vit-", "Sel'skohoziaistvennaia pechat' v Rossii," 80.
42. "Doklad Samarskoi gubernskoi zemskoi upravy 'K voprosu o sel'skokhoziaistvennom obrazovanii v sviazi s zaprosami obshchestvennoi agronomii'," in *Trudy Kazanskogo oblastnogo s"ezda predstavitelei gubernskikh zemstv (15–22 avgusta 1909 goda)* (Kazan, 1909), pp. 26, 27.
43. These calculations are based on data derived from E. Zaremba, "Uchastkovaia agronomiia v Rossii," *Agronomicheskii zhurnal*, no. 1 (1914): 143; Morachevskii, *Agronomicheskaia pomoshch'*, p. 168.
44. A. Iaroshevich, "Osnovnye printsipy uchastkovoi agronomicheskoi organizatsii," *Bessarabskoe sel'skoe khoziaistvo*, no. 8 (1912): 229; I. I. Shtutser, "Polozhenie uchastkovoi organizatsii i ee zadachi v Kazanskoi gubernii," in *Zhurnaly i postanovleniie predvaritel'nogo soveshchaniia agronomov pri Kazanskoi gubernskoi zemskoi uprave 26–27 iiulia 1911 g.* (Kazan, 1911), p. 17.
45. *Postanovleniia predvaritel'nogo soveshchaniia agronomov pri Kazanskoi gubernskoi zemskoi uprave 26 i 27 iiulia 1911 goda* (Kazan, 1911) pp. 3–5.
46. "Doklad agronoma D. S. Smirnova Alatyrskomu Uezdnomu agronomicheskomu soveshchaniiu 12-go sentiabria 1913 goda," in *Otchet ob agronomicheskoi deiatel'nosti Alatyrskoi uezdnoi zemskoi upravy za 1913 god* (Kazan, 1914), p. 309.
47. K. N. Mukhanov, "Samarskaia uezdnaia uchastkovaia agronomicheskaia organizatsiia za 1910 god," in *Uchastkovaia agronomicheskaia organizatsiia Samarskogo uezdnogo zemstva za 1910 god* (Samara, 1912), p. 3; *Obzor deiatel'nosti uchastkovoi*

agronomicheskoi organizatsii Samarskogo uezdnogo zemstva za 1913 god (Samara, 1914), pp. 3, 8–9.

48. "Otchet o deiatel'nosti Krasnoiarskogo uchastkovogo agronoma A. A. Kiseleva za 1909–1910 god," in *Uchastkovaia agronomicheskaia organizatsiia Samarskogo uezdnogo zemstva za 1910 god*, pp. 109–19, 156–57.

49. A. A. Ryshkin, "Otchet raionnogo agronoma po Zubovskomu uchastku," in *Uchastkovaia agronomicheskaia organizatsiia Samarskogo uezdnogo zemstva za 1910 god*, pp. 54, 55, 56.

50. Ibid., pp. 76–80, 89.

51. Mukhanov, "Samarskaia uezdnaia uchastkovaia agronomicheskaia organizatsiia za 1910 god," p. 10.

52. Alexandrovskii et al., *Mestnyi agronomicheskii personal, sostoiavshii na pravitel'stvennoi i obshchestvennoi sluzhbe 1 ianvaria 1914 g.*, pp. 86, 298.

53. In March 1914, an article published in the monthly *Agronomy Journal* claimed that Russian public agronomy was finally catching up with the Italian network of mobile bureaus. The personnel of the latter was better educated, while the staff of the former was better paid. See Sazonov, "Populiarizatsiia sel'skokhoziaistvennykh znanii v Italii," 53.

54. "Agronomicheskii personal zemskikh uchrezhdenii," in V. V. Morachevskii, ed., *Spravochnye svedenia o deiatel'nosti zemstv po sel'skomy khoziaistvu (po dannym na 1909 god)*, vol. 11 (St. Petersburg: GUZiZ, Department of Agriculture, 1911), p. 625; M. M. Glukhov, V. V. Zaretskii, and V. N. Shtein, eds., *Mestnyi agronomicheskii personal, sostoiavshii na pravitel'stvennoi i obshchestvennoi sluzhbe 1 ianvaria 1915 g. Spravochnik* (Petrograd, 1915) p. 386.

55. *Trudy III-go agronomicheskogo soveshchaniia pri Olonetskoi gubernskoi zemskoi uprave 1907 goda* (Petrozavodsk, 1907), pp. 1, 4.

56. Alexandrovskii et al., *Mestnyi agronomicheskii personal, 1 ianvaria 1914 g.*, p. 182.

57. *Trudy III-go agronomicheskogo soveshchaniia pri Olonetskoi gubernskoi zemskoi uprave 1907 goda*, pp. 1, 4.

58. Rat'kov was a physician born to a Vytegorsk district gentry family, he served as the board chairman from 1905 until the dissolution of zemstvos in 1918.

59. Rat'kov declared that this complex question was to be solved in the future, and that an agronomy conference could successfully cope only with specific questions. Provincial agronomist Veber added, that "the order of questions [on the agenda] is not accidental, there is a particular system [behind it]." *Trudy III-go agronomicheskogo soveshchaniia pri Olonetskoi gubernskoi zemskoi uprave 1907 goda*, p. 4.

60. Alexandrovskii et al., *Mestnyi agronomicheskii personal, 1 ianvaria 1914 g.*, p. 498.

61. See *Trudy III-go agronomicheskogo soveshchaniia pri Olonetskoi gubernskoi zemskoi uprave 1907 goda*, p. 71.

62. The deputy Vytegorsk district agronomist, G. P. Semenov, graduated from the Mariinsk mid-level agricultural college in 1905, simultaneously with I. D. Luke. See Alexandrovskii et al., *Mestnyi agronomicheskii personal, 1 ianvaria 1914 g.*, p. 222.

63. See *Trudy III-go agronomicheskogo soveshchaniia pri Olonetskoi gubernskoi zemskoi uprave 1907 goda*, p. 78.

64. As claimed by some of the participants in the conference. See *Trudy III-go agronomicheskogo soveshchaniia pri Olonetskoi gubernskoi zemskoi uprave 1907 goda*, p. 78.

65. *Zhurnal agronomicheskogo soveshchaniia pri Kazanskoi gubernskoi zemskoi uprave 17–24 avgusta 1907 goda* (Kazan, n.d.), p. 3.

66. Alexandrovskii et al., *Mestnyi agronomicheskii personal, 1 ianvaria 1914 g.*, p. 90.

67. *Zhurnal agronomicheskogo soveshchaniia pri Kazanskoi gubernskoi zemskoi uprave 17–24 avgusta 1907 goda*, p. 4.

68. See *Trudy tret'ego vserossiiskogo s"ezda veterinarnykh vrachei v Khar'kove*, vol. V, zhurnaly zasedanii (Kharkov, 1914), pp. vii, 1–3, 70, 75.

69. In 1913, about 81 percent of those who had graduated from veterinary institutes applied their knowledge professionally: as private practitioners (681), as zemstvo veterinarians (1,275), serving in the army (646) or in the agencies of the Ministry of Interior (1,408), occupying positions of city veterinarians (344), slaughterhouse inspectors (about 500), and even agronomists (12). The rest were employed elsewhere; there were a prison inspector and a priest with veterinary diplomas, 2 pharmacists, 26 land captains, and so on. See *Vestnik obshchestvennoi veterinarii*, no. 4 (1914): 225.

70. On September 28, 1908, the conference of veterinarians in Akmolinsk discussed the law requiring "mandatory slaughtering of those who are sick or suspected of being sick or infected." See "Khronika," *Veterinarnoe obozrenie*, no. 11–12 (1909): 368. Apparently, veterinarians were talking about animals, habitually forgetting to name their troubled "clients" directly. It is impossible to imagine physicians or agronomists debating such actions toward the objects of their professional expertise, or excersising such complete control over their clients. In 1910, veterinarian V. K. Rozovskii complained that in regard to veterinarians even zemstvos did not "separate a pure medical activity from police duties, which doubtless harms the good relations and trust of the population toward all veterinary personnel. It is time to release us from police duties." See *Zhurnaly i doklady Buinskoi uezdnoi zemskoi upravy ocherednoi i chrezvychainykh sessii 12 fevralia i 12 noiabria 1910 goda* (Simbirsk, 1911), p. 268.

71. Sakovskii, "K podniatiiu i uluchsheniiu nashego zhivotnovodstva," 239–41; I. M. Krasnopol'skii, "K voprosu o deiatel'nosti veterinarnykh vrachei dlia razvitiia produktivnogo skotovodstva," *Vestnik obshchestvennoi veterinarii*, no. 19 (1907): 726–28. This was his report delivered during the March 17, 1907, session of the Society of Tersk Region Veterinarians.

72. Quoted from S. F-v, "Po povodu stat'i 'Perezhivaemyi krizis'," *Vestnik obshchestvennoi veterinarii*, no. 3 (1912): 133.

73. Ibid., 133–34.

74. *Protokoly soveshchaniia zemskikh veterinarnykh vrachei Kazanskoi gubernii 24–26 aprelia 1912 goda* (Kazan, 1912), p. 3.

75. Ibid., 16–18, 20.

76. V. I. Simagin, "K voprosu o vzaimootnoshenii zemskikh agronomov i veterinarov v rabote po uluchsheniiu zhivotnovodstva," *Vestnik obshchestvennoi veterinarii*, no. 13–14 (1912): 612.

77. *Protokoly soveshchaniia zemskikh veterinarnykh vrachei Kazanskoi gubernii 24–26 aprelia 1912 goda*, pp. 23–24.

78. The very first suggestion during the agronomy conference of the Chernigov provincial zemstvo in December 1912 was one by precinct agronomist N. Balevich-Iavorskii, who argued for the necessity of veterinarians' participation in work for improving husbandry, and even in the agronomist conferences. See *Trudy gubernskogo agronomicheskogo soveshchaniia i ekonomicheskogo*

soveta Chernigovskogo gubernskogo zemstva 11–16 dekabria 1912 goda (Chernigov, 1914), p. 5.

79. See S. A. Gamaiunov, "Mysli i fakty," *Vestnik obshchestvennoi veterinarii*, no. 20 (1914): 965. "Agronomists say that it does not matter, let veterinarians work along with agronomists in the sphere of stock-breeding, it is their common cause, and at the same time a national cause. Yes, this is true,"—insisted another veterinarian—but veterinarians are "the junior partner and even the servant" in that alliance, and this cannot be tolerated. See Pavpertov, "Agronomiia i veterinariia," pp. 23, 25.

80. Pavpertov, "Agronomiia i veterinariia," p. 21.

81. Indeed, before 1905, veterinarians were not only more numerous but also better organized than agronomists. Between 1871 and 1905, 158 district and provincial conferences of veterinarians took place, but only 87 conferences of agronomists and/or statisticians. See Boris Veselovskii, *Istoriia zemstva*, vol. 2, p. 472.

82. Pavpertov, "Agronomiia i veterinariia," pp. 21, 26.

83. S., "Uezdnaia vystavka zhivotnovodstva v pos. Melekess, Samarskoi gub.," *Vestnik obshchestvennoi veterinarii*, no. 1 (1914): 63–65.

84. Ibid., p. 65.

85. Ibid., p. 64.

4   The Ambivalent Role of the State: A Conservative Patron and a "Progressive" Rival

1. Morachevskii, *Agronomicheskaia pomoshch'*, p. 136; "Agronomicheskii personal zemskikh uchrezhdenii," pp. 621–39; Koropov, *Istoriia veterinarii v SSSR*, p. 170.

2. "Veterinarian Blagodatnyi (in Stavropol) on December 5 [1906] was sentenced to 3 years in prison for participation in the pillaging of Count Orlov-Davydov's estate. The senior veterinarian Bratchikov (in Viatka) was fired by the governor on November 18 after 20 years of service…. The senior veterinarian of the Don stud farm, I. P. Kuchumov,…was fired on the grounds of suspicion that he organized a strike of stablemen…. The zemstvo veterinarian Spasskii (Melitopol district) was arrested after being denounced by a *rotmister* [captain] of the Krym regiment Stavraki, whose horse was declined treatment, and whom Spasskii told that the government itself provoked peasant unrest by keeping troops in the countryside and compelling peasants to service them." See "Svedeniia o dal'neishei sud'be poterpevshikh administrativnye presledovaniia," *Veterinarnoe obozrenie*, no. 1 (1907): 26–27.

3. "Veterinarnye knigi i revolutsiia," *Russkie vedomosti*, no. 6 (1912): 3.

4. On January 3, 1910, the Second Congress of Veterinarians began its sessions in the Moscow Polytechnic Museum. As usual, a police informer was present, taking notes during the work of the congress and writing regular reports to his superiors. On January 14, he reported the speech of the military veterinarian P. D. Osipchuk, whom he presented as a troublemaker because "his words are doubtless oppositional in character, if not their content, then for their intonations and gestures." See GARF, f. 102 Departament politsii; D-4, op. 117, d. 176, l. 9.

5. W. B. Bizzell, *The Green Rising* (New York: Macmillan, 1926), p. 12.

6. The role of the government in setting the stage for broad public initiative is discussed in David A. J. Macey, *Government and Peasant in Russia, 1861–1906: The Prehistory of the Stolypin Reforms* (DeKalb: Northern Illinois University Press,

1987). Some discussion of the evolution of the government's attitude toward rural professionals in connection with the state agrarian policy can be found in Yaney, *The Urge to Mobilize*, and Wcislo, *Reforming Rural Russia*. On the connection of the Stolypin reforms and modernization activity of agricultural specialists, see Matsuzato, "Stolypinskaia reforma i rossiiskaia agrotekhnologicheskaia revolutsiia," pp. 194–200.

7. Morachevskii, *Agronomicheskaia pomoshch'*, pp. i, ii.

8. *Agronomicheskii zhurnal*, no. 8 (1913), p. 171; Alexandrovskii et al., *Mestnyi agronomicheskii personal, 1 ianvaria 1914 g.*, p. i.

9. Iu. Larin, *Ekonomika dosovetskoi derevni* (Moscow and Leningrad: GIZ, 1926), p. 113.

10. In 1910, the Department of Agriculture allocated only 24,000 rubles to subsidize zemstvo agronomists, but by 1911 this sum had grown to 220,000 rubles, by 1912—to 500,000 rubles (of which 350,000 were allotted specifically to the precinct agronomists), by 1913—to 1,000,000 rubles (of which 631,000 were allotted to the precinct agronomists). See Morachevskii, *Agronomicheskaia pomoshch'*, p. 167. With an average annual agronomist's salary about 1,500 rubles, the government grant helped to sustain only 16 zemstvo agronomist jobs in 1910, but approximately 500 jobs in 1913 (the average salary of agronomists had increased to over 2,000 rubles by that time).

11. Glukhov et al., *Mestnyi agronomicheskii personal, 1 ianvaria 1915 g.*, pp. 556–59; Koropov, *Istoriia veterinarii v SSSR*, pp. 172, 184.

12. For a critique of this trend, see Kimitaka Matsuzato, "The Fate of Agronomists in Russia: Their Quantitative Dynamics from 1911 to 1916," *The Russian Review* 55 (April 1996): 172–200, esp. 172–73.

13. "Doklad s predstavleniem proekta instruktsii uezdnomu zemskomu agronomu," in *Zhurnaly i doklady Buinskoi uezdnoi zemskoi upravy ocherednoi sessii 1908 goda i chrezvychainykh sessii 1 marta i 28 noiabria 1908 g. s prilozheniiami* (Simbirsk, 1909), p. 335.

14. Bonch-Osmolovskii argued against the foundation of an experimental field to test the better technologies of land cultivation. When the chairman of the zemstvo board, V. L. Persiianinov, supported the district agronomist on this matter, Bonch-Osmolovskii argued against building normal housing for the agricultural workers who would till the experimental field. He also objected to the introduction of sample plots showing peasants the results of "scientific methods" of land cultivation; to the establishment of grain-purifying centers; to the development of apiculture; to the opening of a seed department at the zemstvo agricultural warehouse; to the hiring of a warehouse mechanic to assemble agricultural machines; to the decrease of fees for using the zemstvo mating centers; to a horse fair; to measures for improving peasant stock-breeding; to the opening of credit cooperatives; and to agricultural lectures to the peasants. See "Zasedaniia Buinskogo uezdnogo ekonomicheskogo soveta," in *Zhurnaly i doklady Buinskoi uezdnoi zemskoi upravy ocherednoi sessii 1908 goda i chrezvychainykh sessii 1 marta i 28 noiabria 1908 g.*, pp. 348, 349, 351, 354, 356, 360–365, 367.

15. He graduated after the Revolution of 1905 from the Mariinsk mid-level agricultural college, and for a while worked as manager of an agricultural warehouse in Samara province. See "Agronomicheskii personal zemskikh uchrezhdenii," p. 628. Apparently, a tight job market did not allow a professional without a higher education to compete with graduates from agricultural institutes for an agronomist position. The fact that he became the Buinsk district agronomist

indicated the appearance of a new trend in the job market in 1908. As discussed earlier, very soon the demand for specialists with mid-level education would catch up with the demand for better-educated professionals. However, for the Buinsk zemstvo board in 1908, the mid-level educational background of agronomist Ivanov only underlined his inferior status.

16. By unequivocally renouncing the zemstvo's efforts to modernize the peasantry after 1907 as reactionary, many influential Soviet and American scholars virtually blocked any possibility of studying public efforts to reform Russian agriculture within the traditional historiographical framework. See Leopold Haimson, ed., *The Politics of Rural Russia, 1905–1914* (Bloomington: Indiana University Press, 1979); Roberta Thompson Manning, *The Crisis of the Old Order in Russia. Gentry and Government* (Princeton, NJ: Princeton University Press, 1982); Terence Emmons and Wayne S. Vucinich, eds., *The Zemstvo in Russia: An Experiment in Local Self-Government* (Cambridge: Cambridge University Press, 1982); A. IA. Avrekh, *P.A. Stolypin i sud'by reform v Rossii* (Moscow: Izdatel'stvo politicheskoi literatury, 1991); P. N. Zyrianov, *Krest'ianskaia obshchina evropeiskoi Rossii 1907–1914 gg.* (Moscow: Nauka, 1992).

17. In 1911, M. F. Bonch-Osmolovskii was elected honorary justice of the peace. In 1913, he was unanimously elected a member of the zemstvo Medical Council, a member of the district School Council (20 votes for, 2 votes against), and one of four deputies to the Simbirsk province zemstvo board from Buinsk district (by 15 votes out of 22). See *Zhurnaly i doklady Buinskoi uezdnoi zemskoi upravy ocherednoi i chrezvychainykh sessii 12 fevralia i 12 noiabria 1910 goda*, p. 56; *Zhurnaly i doklady Buinskoi uezdnoi zemskoi upravy ocherednoi i chrezvychainykh sessii 16 fevralia, 15 iulia i 24 noiabria 1913 goda* (Simbirsk, 1914), pp. 155, 158.

18. A curious conversation preceded the approval of this decision by the board deputies. A board deputy, the wealthy peasant A. K. Susanin, declared that "Peasants look with hostility upon large expenses associated with the introduction of agronomy organization, because they cannot know what use it will bring." At this point, the zemstvo reaction was speaking not through a gentry conservative Bonch-Osmolovskii, but a peasant with the operatic name Susanin. The chairman of the board, V. L. Persianinov, responded with an argument no one on the board could challenge; he said that three-quarters of the expenses for agronomists would be paid by the government and the provincial zemstvo, "and if our district won't use the sums allotted for that purpose, they will go to other districts." See *Zhurnaly i doklady Buinskoi uezdnoi zemskoi upravy ocherednoi i chrezvychainykh sessii 12 fevralia i 12 noiabria 1910 goda*, p. 2.

19. Ibid., pp. 298–310.

20. *Zhurnaly i doklady Buinskoi uezdnoi zemskoi upravy ocherednoi i chrezvychainykh sessii 27 maia, 4, 22, 23 i 24 sentiabria 1914 goda* (Simbirsk, 1915), p. 303.

21. *Zhurnaly i doklady Buinskoi uezdnoi zemskoi upravy ocherednoi i chrezvychainykh sessii 16 fevralia, 15 iulia i 24 noiabria 1913 goda*, p. 355.

22. Ibid., p. 362.

23. A. N. Naumov, *Iz utselevshikh vospominanii*, vol. 6, pp. 1572–75, a manuscript in the Hoover Institution Archive, A. N. Naumov Collection, box no. 24. The published abridged version of Naumov's memoirs lacked an extensive characteristic of Slobodchikov, as well as some other important figures.

24. Characteristically, in 1916 a somewhat archaic "think tank"—the Academic Committee of the Ministry of Agriculture—was transformed into a modern

Institute of Experimental Agronomy. See T., "K ot"ezdu N. M. Tulaikova," *Samarskii zemledelets*, no. 15–16 (August 1 and 15, 1916): 457.

25. GARF, f. 102 Departament politsii, D-4, op. 117, d. 101 "O Moskovskom obshchestve sel'skogo khoziaistva"; GARF, f. 102 Departament politsii, D-4, op. 117, d. 114 "Kharkovskoe obshchestvo sel'skogo khoziaistva."

26. "Otnoshenie Kharkovskogo gubernatora v Departament Zemledeliia ot 19 dekabria 1911 goda," in GARF, f. 102 Departament politsii, D-4, op. 117, d. 114, l. 23.

27. "Raport Nachal'nika Kharkovskogo gubernskogo zhandarmskogo upravleniia g. Direktory Departamenta Politsii ot 4 dekabria 1911 goda," in GARF, f. 102 Departament politsii, D-4, op. 117, d. 114, ll. 14–20.

28. See GARF, f. 63 Moskovskoe okhrannoe otdelenie, op. 29, d. 1123 "Ob agronomicheskom s"ezde," l. 1.

29. Ibid. ll. 2–2 ob.

30. This was the case when the manager of the Kazan branch of the State Bank organized a conference on June 5, 1911, on the problem of building granaries in Kazan province. Among the 23 participants were inspectors of credit cooperatives in the State Bank's service and a few representatives of credit cooperatives. Despite the perfectly legal and even official nature of the conference, the Kazan governor sent a secret note to the head of the provincial Gendarme Administration requesting a check on the political profile of all the participants—including the provincial inspector of credit cooperatives. See "Otnoshenie Kazanskogo gubernatora k Nachal'niku Kazanskogo gubernskogo zhandarmskogo upravleniia," in NART, f. 199, Kazanskoe gubernskoe zhandarmskoe upravlenie, op. 2, d. 1240 "So spravkami o politicheskoi blagonadezhnosti raznykh lits," ll. 596–596 ob.

31. Z. I. Peregudova, *Politicheskii sysk Rossii. 1880–1917* (Moscow: ROSSPEN, 2000), pp. 48–49.

32. The State Examination Commission at Kazan University applied for such a reference concerning the student Vladimir Krotov before admitting him to the graduation exams; a peasant's son Sofronii Birelev was screened prior to being certified as a primary-school teacher; the merchant Pavel Kubasskii was first cleared of any possible suspicion by a reference of reliability, and only then started his mandatory military service. See NART, f. 199, op. 2, d. 1240, ll. 1, 2, 4. Kazan Provincial Gendarme Administration alone issued 2,700 references of reliability between April 5 and June 20, 1911, or an average of 35.5 references a day (counting business days as well as holidays and weekends). This suggests that political screening was an important factor in individual careers.

33. GARF, f. 102 Departament politsii, D-4, op. 119, d. 270 "Ob uchenom agronome, dvorianine Dmitrii Dmitrieve Korolevtsove."

34. Alexandrovskii et al., *Mestnyi agronomicheskii personal, 1 ianvaria 1914 g.*, p. 361.

35. *Agronomicheskii zhurnal*, no. 1 (1913): 135.

36. Alexandrovskii et al., *Mestnyi agronomicheskii personal, 1 ianvaria 1914 g.*, p. 124. One may speculate that networks of Ukrainian solidarity played a role in the promotion of Pokhodnia.

37. On July 29, 1914, the Samara precinct agronomist Georgii Meibom applied for a position as the Sviiazhsk district land settlement commission's agronomist (Kazan province). Four weeks later, on August 25, the Kazan governor ordered his hiring. Following some bizarre bureaucratic logic, on August 27, the immediate

supervisor of Meibom sent a written request to the governor's chancellery asking "whether there are any objections in terms of political reliability" to Meibom's appointment. The chancellery forwarded this request to the police. The reply came only on May 4, 1915, and it was favorable to Meibom. Still, even compromising information would hardly have hurt a specialist who had served for almost ten months by this time, if his supervisors were satisfied with his work. See NART, f. 580, Zaveduiushchii agronomicheskoi pomoshch'iu khoziaistvam edinolichnogo vladeniia Kazanskoi gubernii, op. 1, d. 1 "O sluzhbe agronoma Sviazhskoi zemleustroitel'noi komissii G. F. Meibom," ll. 1, 14, 26, 28.

38. NART, f. 199, Kazanskoe gubernskoe zhandarmskoe upravlenie, op. 2, d. 1028 "Po sobraniiu svedenii o politicheskoi blagonadezhnosti lits vkhodiashchikh v sostav kooperativnykh uchrezhdenii i upravlenii," ll. 2–5.

39. See GARF, f. 102 Departament politsii, D-4, op. 122, d. 128 "Ob ustroistve v poezde zheleznoi dorogi peredvizhnoi agronomicheskoi vystavki po glavnoi i vetviam Vladikavkazskoi zheleznoi dorogi," l. 1.

40. NART, f. 199, Kazanskoe gubernskoe zhandarmskoe upravlenie, Op. 2, d. 1384 "Po obvineniiu zaveduiushchego masterskimi Arskogo sel'skokhoziaistvennogo sklada Kazanskogo uezdnogo zemstva Aleksandra Egorova Amosova v prestuplenii predusmotrennom 2 p. 103 st. i 106 st. Ugolovnogo Ulozhenia," ll. 1–37; Alexandrovskii et al., *Mestnyi agronomicheskii personal, 1 ianvaria 1914 g.*, p. 110.

41. The leading role of the Kazan Society of Apiculture was reinforced in 1911, when it hosted the first All-Russian Congress of Beekeepers. See *Trudy Vserossiiskogo s"ezda pchelovodov v g. Kazani 17–23 avgusta 1911 g.* (Kazan, 1911).

42. S. Morozov, "Prizyv k pchelovodam Samarskoi gubernii," *Samarskii zemledelets*, no. 11–12 (June 1 and 15, 1916): 375–76.

43. "Donesenie Inspektora sel'skogo khoziaistva Kazanskoi gubernii A. P. Gortalova Nachal'niku Kazanskogo upravleniia zemleustroistva i gosudarstvennykh imushchestv ot 28 dekabria 1915 g.," in NART, f. 199, Kazanskoe gubernskoe zhandarmskoe upravlenie, Op. 2, d. 1526 "Po obvineniiu predsedatelia Kazanskogo obshchestva pchelovodstva V. I. Loginova po priznakam prestupleniia, predusmotrennogo 2 p. 103 st. Ugolovnogo Ulozheniia," ll. 22–27.

44. NART, f. 199, Kazanskoe gubernskoe zhandarmskoe upravlenie, Op. 2, d. 1526 "Po obvineniiu predsedatelia Kazanskogo obshchestva pchelovodstva V. I. Loginova po priznakam prestupleniia, predusmotrennogo 2 p. 103 st. Ugolovnogo Ulozheniia," ll. 3, 5, 13–14, 16, 28, 29, 34–36.

45. This insignia had three degrees and could be awarded upon the recommendation of a local or central office of the GUZiZ to anyone, regardless of social status or gender. See "Khronika," *Vestnik obshchestvennoi veterinarii*, no. 9 (1914), p. 499. The Romanov badge of honor was established soon after the celebration of the three hundredth anniversary of the Romanov dynasty in 1913. It was made in the shape of the cross of the Order of St. Vladimir, but was covered with green enamel. In the center was a garland around a silver griffin—the Romanov coat of arms. See I. V. Vsevolodov, *Besedy o faleristike* (Moscow: Nauka, 1990), p. 229.

46. See the editorial "Nashi zadachi," *Agronomicheskii zhurnal*, no. 1 (1913): 5.

47. On the emergence of culture and practices of social engineering in imperial Russia, see Ilya Gerasimov, "Redefining Empire: Social Engineering in Late Imperial Russia," in Ilya Gerasimov et al., eds., *Empire Speaks Out: Languages of Rationalization and Self-Description in the Russian Empire* (Leiden: Brill, 2009), forthcoming.

48. Ibid., pp. 169, 170–72.

49. David A. J. Macey, "Agricultural Reform and Political Change: The Case of Stolypin," in Taranovski, ed., *Reform in Modern Russian History: Progress or Cycle?*, p. 168.

50. The most perceptive studies of Stolypin reform have discussed the component of social engineering embedded in the seemingly administrative intervention; see, for example, Matsuzato, "Stolypinskaia reforma i rossiiskaia agrotekhnologicheskaia revolutsiia," pp. 194–200; David A. J. Macey, " 'A Wager on History': The Stolypin Agrarian Reforms as Process," in Judith Pallot, ed., *Transforming Peasants: Society, State and the Peasantry, 1861–1930. Selected Papers from the Fifth World Congress of Central and East European Studies* (New York: St. Martin's Press, 1998), pp. 149–73.

51. For an extensive treatment of the topic of the creation of a government "version of peasantry," see Macey, *Government and Peasant in Russia*, pp. 41–118, et seq.

52. C. A. Koefoed, *My Share in the Stolypin Agrarian Reforms*, ed., Bent Jensen, trans. Alison Borch-Johansen (Odense: Odense University Press, 1985), pp. 36–37.

53. The same can be said about the historiography of the Russian land commune, as was shown by P. N. Zyrianov in "Poltora veka sporov o russkoi sel'skoi obshchine," in Tsamutali, *Problemy sotsial'no-ekonomicheskoi i politicheskoi istorii Rossii XIX–XX veka*, pp. 87–100.

54. Many historians agree that the land commune did not hamper the spread of innovative technologies of land cultivation and machinery among the peasantry. See Esther Kingston-Mann, "Peasant Communes and Economic Innovation: A Preliminary Inquiry," in Esther Kingston-Mann and Timothy Mixter, eds., *Peasant Economy, Culture, and Politics of Imperial Russia, 1800–1921* (Princeton, NJ: Princeton University Press, 1991), pp. 23–51; Kimitaka Matsuzato, "Individualistskie kollektivisty ili kollektivistskie individualisty? Noveishaia istoriografiia po rossiiskim krest'ianskim obshchinam," in Gennadii Bordiugov, Norie Isia, and Takesi Tomita, eds., *Novyi mir istorii Rossii* (Moscow: AIRO-XX, 2001), pp. 189–201.

55. As we mentioned, the entire project of precinct agronomy in Kazan province was based on the principle of equal services provided to all zemstvo taxpayers. See *Postanovleniie predvaritel'nogo soveshchaniia agronomov pri Kazanskoi gubernskoi zemskoi uprave 26 i 27 iiulia 1911 goda*, p. 3. In its 1910 report, the Birsk district zemstvo board (Ufa province) specifically emphasized that it was against any discrimination in agronomy assistance on the grounds of form of land tenure. See "Doklad Upravy ob okazanii agronomicheskoi pomoshch'i khoziaistvam edinolichnogo vladeniia," in *Sbornik dokladov Birskoi uezdnoi zemskoi upravy i postanovlenii zemskogo sobraniia XXXVI ocherednoi sessii 1910 goda* (Ufa, 1911), p. 437.

56. The critique of government agronomy escalated to the point that a police officer present at the congress had to warn that he would shut down the whole congress if that critique continued. See GARF, f. 102 Departament politsii, D-4, op. 120, d. 181 "Ob oblastnom s"ezde v g. Moskve deiiatelei po okazaniiu agronomicheskoi pomoshch'i naseleniiu," l. 1.

57. *Zhurnaly osobogo agronomicheskogo soveshchaniia pri Kazanskoi gubernskoi zemleustroitel'noi komissii i Doklady inspektora sel'skogo khoziaistva za 1909 i 1910 gody*, (Kazan, 1911), pp. iv–vi.

58. Ibid., pp. xiii–xiv.

59. In December 1910, the annual zemstvo agronomist conference of Samara province specifically emphasized that agronomists could only assist a natural

evolution of the peasantry, hence any "templates, recipes, sample plans must be...condemned as contradicting the very nature of the economy. Sample farms...must be rejected for their inconsistency." See "Rezolutivnaia chast' trudov 2-go Samarskogo Gubernskogo agronomicheskogo soveshchaniia," *Samarskii zemledelets*, no. 8 (December 15, 1910): 244.

60. See "Prilozhenia k Ustavu o sluzhbe po opredeleniiu ot Pravitel'stva," in *Svod zakonov Rossiiskoi Imperii*, vol. 3, col. 150–65.

61. *Zhurnaly i protokoly osobogo agronomicheskogo soveshchaniia pri Kazanskoi gubernskoi zemleustroitel'noi komissii i Doklady inspektora sel'skogo khoziaistva za vtoruiu polovinu 1910 goda i za 1911 god* (Kazan, 1912), p. 138.

62. Ibid., p. 141.

63. "Rezultaty intsidenta s agronomom Freitagom," *Russkie vedomosti*, no. 114 (1912), p. 3. Agronomist Zalog managed to find a less important job in southwestern Podoliia province where, true to his public conscience, he became a precinct agronomist; agronomist Freitag kept his position until 1914, when he, also in line with his earlier preferences, was promoted to the position of acting government agronomist of the entire Erivan province in the Caucasus. See Alexandrovskii et al., *Mestnyi agronomicheskii personal, 1 ianvaria 1914 g.*, pp. 199, 262, 451.

64. P. Gr-v, "K voprosu o likvidatsii zemleustroitel'noi agronomii," *Agronomicheskii zhurnal*, no. 8 (1913): 133.

65. See A. Minin, "Agronomiia i zemleustroistvo v ikh otnoshenii k derevenskoi bednote," *Agronomicheskii zhurnal*, no. 1 (1914): 30–43.

66. Ibid., p. 41.

67. Cf. M. Aleksandrovich, "Voprosy organizatsionnogo stroitel'stva obshchestvennoi agronomii na gubernskikh agronomicheskikh soveshchaniiakh," *Agronomicheskii zhurnal*, no. 2 (1914): 71. In 1914, Aleksei Diomidov, the Iaroslavl district zemstvo agronomist, who graduated from the Moscow Agricultural Institute in the class of 1910, shared with his Kostroma colleagues the Iaroslavl third element's experience of taking over the government agronomy organization. He blamed government agronomy for all the problems of the third element and encouraged the Kostroma zemstvo to follow the Iaroslavl example: "it is hardly possible to reproach a zemstvo that relies on the opinion of the third element and trust its, generally speaking, competent knowledge of the situation." See A. Diomidov, "Iaroslavskoe i Kostromskoe zemstvo v ikh otnoshenii k agronomicheskoi organizatsii," *Kostromskoi kooperator*, no. 19–21 (December 19, 1914): 9.

68. See P. Gr-v, "K voprosu o likvidatsii zemleustroitel'noi agronomii," *Agronomicheskii zhurnal*, no. 8 (1913): 133–34. In 1914, out of 12 districts in Kazan province, only Sviiazhsk district still had two parallel agronomy structures representing the government and the zemstvo. See NART, f. 256, Kazanskoi gubernskoi zemleustroitel'noi komissii, op. 7, d. 43, l. 28 ob.

69. It took the GUZiZ two years of persuasion and pressure to compel the Birsk district zemstvo of Ufa province to accept the land settlement agronomy organization under its auspices in August 1914. See *Sbornik dokladov Birskoi uezdnoi zemskoi upravy i postanovlenii zemskikh sobranii ocherednoi sessii 20 sentiabria 1914 goda i chrezvychainyh zemskikh sobranii XXII sessii 19–20 fevralia i XXIII sessii 2 avgusta 1914 goda*, pp. 394–99. The government agronomy organization was completely turned over to the zemstvo on June 3, 1917. See "Khronika," *Zemledel'cheskaia gazeta*, no. 25–26 (Saturday, July 1, 1917): 529. See also

A. V. Efremenko, "Agronomicheskii aspekt stolypinskoi zemel'noi reformy," *Voprosy istorii*, no. 12 (1996): 3–15.

70. The story of trade unions created and supervised by the Russian secret service (*zubatovshchina*) is a good case in point. See Jonathan W. Daly, *Autocracy under Siege. Security Police and Opposition in Russia, 1866–1905* (DeKalb: Northern Illinois University Press, 1998), esp. chap. 5; Madhavan K. Palat, "Casting Workers as an Estate in Late Imperial Russia," *Kritika* 8, no. 2 (2007), esp. 337–38.

71. Cf. I. V. Lukoianov, *Rossiiskie konservatory (konets XVIII–nachalo XX vv.)* (St. Petersburg: Nestor, 2003); Richard Pipes, *Russian Conservatism and Its Critics: A Study in Political Culture* (New Haven, CT: Yale University Press, 2005).

72. RGIA, f. 403 "Russkoe zerno," op. 2, d. 1 "Obshchestvo sodeistviia pod"emy zemledeliia, sel'skogo khoziaistva i narodnoi trudosposobnosti 'Russkoe zerno.' "

73. E. P. Serapionova, "Kul'turno-ekonomicheskoe obshchestvo 'Russkoe zerno' v nachale XX veka," *Slavianskii al'manakh* (2000): 178.

74. RGIA, f. 403, op. 2, d. 2.

75. For a standard work on the topic, see Paul Vyšný, *Neo-Slavism and the Czechs, 1898–1914* (Cambridge: Cambridge University Press, 1977).

76. Ibid., pp. 72–73, et seq.

77. E. P. Kovalevsky (1865–1941) served as a high-ranking official in the Ministry of Education and was a member of multiple commissions on reforming primary and secondary schooling in Russia. In the Third and Fourth Dumas he was deputy head of the Commission for Public Education, then chairman of the permanent conference elaborating the legislation on universal education. In 1900, he organized educational excursions to the World Exhibition in Paris for Russian schoolteachers.

78. Cf. *"Russkoe zerno"* (Perm, 1910), p. 8; Serapionova, "Kul'turno-ekonomicheskoe obshchestvo," p. 169.

79. Vyšný, *Neo-Slavism and the Czechs*, pp. 107, 108.

80. Serapionova quotes excited remarks on this subject by two of the Russian Congress attendants, Count Bobrinsky and journalist G. Komarov (Serapionova, "Kul'turno-ekonomicheskoe obshchestvo," pp. 168–69).

81. RGIA, f. 403, op. 2, d. 2. Article 11 of the society's charter stated that one ruble was a standard annual membership fee; those who paid at least 25 rubles were exempt from any subsequent membership fees for life.

82. Serapionova, "Kul'turno-ekonomicheskoe obshchestvo," p. 177.

83. The government encouraged provincial governors to support the activities of Russian Grain. Local official periodicals published press releases on the activities of that formally nongovernmental association, sometimes revealing the paper's ignorance about the actual goals of the Russian Grain despite the utterly enthusiastic tone of a given article. That was the case with the official *Perm News*, which in its first article dedicated to the new society suggested that its goal was to teach Russian peasants how to produce complex agricultural machines themselves by studying abroad. "Russkoe zerno," *Permskie vedomosti*, no. 110 (1909), reprinted in *"Russkoe zerno"* (Perm, 1910), p. 5.

84. In 1909 alone, 13 extensive (up to 3,200-word) articles were published. The provincial administration's attempt to use the zemstvo-run *Zemskaia nedelia* for the same purposes was not supported, to the dismay of the administration. See A. V. "V 'Russkom zerne,' " in *"Russkoe zerno,"* p. 75.

85. "Otkrytie Permskogo gubernskogo otdela obshchestva 'Russkoe zerno,' " in *"Russkoe zerno,"* pp. 56–57, 59.

86. "Kak podniat' russkoe zemledelie," in *"Russkoe zerno,"* p. 20.
87. "Russkoe zerno," in *"Russkoe zerno,"* p. 7.
88. The Provincial Zemstvo Board and the annual provincial agronomist conference in July 1909 decided to abstain from supporting the Russian Grain's initiatives until the actual results and direction of its actions became clear. See "Kak podniat' russkoe zemledelie," in *"Russkoe zerno,"* p. 19.
89. A. V. Bolotov was only 39 when he became governor of Perm province in the stormy December of 1905. He curtailed political extremism in the province regardless of its ideological orientation, which led to a bitter conflict with the chief sponsor of the local Union of Russian People and charismatic preacher, high-ranked Orthodox monk Seraphim (Georgii Kuznetsov). Tatiana Bystrykh, "Gubernator ushel v monastyr'," *Permskie novosti*, May 10 (2001). Similarly, Bolotov did not subscribe to the Slavophile rhetoric of the Russian Grain. "I am not a supporter of politics' interference with purely economic matters, and therefore do not quite share the Society's view that preference should be given to studying agriculture in Slavic lands. I think that one should learn everywhere and from whomever." He was also against any discrimination in the treatment of peasants who had not left their communes in the activities of the Russian Grain. See "Otkrytie Permskogo gubernskogo otdela obshchestva 'Russkoe zerno,'" in *"Russkoe zerno,"* pp. 59, 66.
90. Ibid., pp. 62, 63. This was much less than the organizers of the Perm Russian Grain expected to receive from zemstvos: they estimated that each district zemstvo could contribute 400 rubles to its budget, and the provincial zemstvo would donate 1,000 rubles, thus accumulating about 6,000 rubles. "Kak podniat' russkoe zemledelie i 'Russkoe zerno.'" in *"Russkoe zerno,"* p. 27.
91. *Pis'ma krestian*, vol. 2, part 2 (Petrograd: Russkoe zerno, 1915), pp. 710–13.
92. Ibid., p. 748.
93. Ibid., pp. 674–77.
94. Ibid., pp. 568–92.
95. "Evrgaf Kovalevskii," in *Novyi entsiklopedicheskii slovar'*, vol. 22 (Petrograd: Izdatel'skoe delo byvshee Brokgauz-Efron, 1917), col. 35–36.
96. V. Iu. Rylov, "'My, voronezhskie natsionalisty'…Deiatel'nost' Vserossiiskogo natsional'nogo soiuza v Voronezhskoi gubernii (1908–1913)," unpublished paper, available at http://conservatism.narod.ru/juni/rylov.doc (accessed April 5, 2005).
97. Reportedly, he was more interested in the opening of a "national school" or "national credit union," and succeeded in founding the Union of Russian Women and a nationalist newspaper. See Rylov, "My, voronezhskie natsionalisty."
98. Serapionova, "Kul'turno-ekonomicheskoe obshchestvo," p. 170.
99. According to its chairman, A. Stolypin, the initial idea was to simply assist seasonal agricultural workers to go abroad without paying too much to recruiters and other intermediaries. See Alexander Stolypin, "Chto takoe 'Russkoe zerno'?" in *Pis'ma krestian* (St. Petersburg: Russkoe zerno, 1911), p. vi.
100. Cf. *Pis'ma krestian* (St. Petersburg: Russkoe zerno, 1911), p. iv.
101. A. V. "Velikoe nachinanie," in *"Russkoe zerno"* (Perm, 1910), pp. 49, 51, 52.
102. See RGIA, f. 403, op. 2, d. 70 "Perepiska s agentom GUZiZ v Amerike Kryshtofovichem," and d. 139 "Perepiska ob organizatsii i otpravke krestian v Argentinu na zarabotki."

5   The Economic Foundations of Social Mobilization

1. NART, f. 199, op. 1, d. 654, "So sluchainymi svedeniiami o prestupnoi deiiatel'nosti raznykh lits (anonimnye soobshcheniia)," ll. 24–24 ob.
2. For a general methodological interpretation of this process in respect to the urban intellectuals, revolutionaries-turned-workers–modernizers, see Leopold Haimson, "The Problem of Social Stability in Urban Russia, 1905–1917," *Slavic Review* 23, no. 4 (December 1964): 619–42; vol. 24, no. 1 (March 1965): 1–22.
3. Its local branches were no less active than the central organization; in 1910, the Kazan branch went so far as to discuss the possibility of its secession from the Moscow Council of the society and the establishment of an independent union. The professional self-awareness and solidarity of the members of the Kazan branch can be seen in the following episode. When, during a meeting of the Kazan land surveyors, a high-level union functionary proposed including land captains and members of land settlement commissions in the ranks of the society "for the success of the cause," the response was "a merry laugh," not even an angry protest. Although land surveyors worked for the land settlement commissions, unlike state officials, their primary loyalty was not on the side of the government but on that of their corporation and their comrades. See "Obshchestvo zemlemerov," *Samodeiatel'nost'*, no. 2–3 (February 1910): 36–37.
4. "Ustav o sluzhbe po opredeleniiu ot Pravitel'stva," in *Svod zakonov Rossiiskoi Imperii*, vol. 3, col. 1, article 4.
5. Ivanov, *Vysshaia shkola Rossii*, p. 347.
6. Ibid., p. 350.
7. Source: NART, f. 658, Chistopol'skaia uezdnaia zemskaia uprava, op. 21, d. 6 "Vedomost' na vydachu zhalovaniia sluzhashim v zemstve za ianvar'-mart mesiatsy 1910 g.," ll. 78–91.
8. Cf. Eklof, *Russian Peasant Schools*, pp. 218, 549 n. 12.
9. A. Diomidov, "K voprosu o material'nom polozhenii zemskogo agronomicheskogo personala," *Zemskii agronom*, no. 6–7 (1914): 2.
10. Thus, while in 1913, a precinct agronomist in the Alatyr district (Kazan province) earned 1,125 rubles a year, in Bessarabia province the norm was 2,400 rubles for a precinct agronomist. See *Zhurnaly Alatyrskogo uezdnogo zemskogo sobraniia sessii 1912 goda* (Alatyr, 1913), p. 66; *Doklady Bessarabskoi gubernskoi zemskoi upravy po agronomicheskomu otdelu 44-my ocherednomu gubernskomu Zemskomu sobraniiu* (Kishinev, 1913), p. 164.
11. Sazonov, "Populiarizatsiia sel'skokhoziaistvennykh znanii v Italii," 53.
12. The question of living standards of various social groups in Late Imperial Russia is one of the most understudied problems of both Russian and Western historiography. Ben Eklof is one of the few historians of the late imperial period who has actually put the abstract figures into historical perspective and tried to evaluate those figures in terms of the standard of living. See Eklof, *Russian Peasant School*, pp. 218–19, 550 n. 15. More research has been done concerning the post-1917 period, although the only special study on wages in Russia known to the author concerns the period of the 1930s–1950s. See J. Chapman, *Real Wages in Soviet Russia Since 1928* (Cambridge, MA: Harvard University Press, 1963). Figures of wages in industry since 1925 can be found in B. R. Mitchell, *European Historical Statistics* (New York: Columbia University Press, 1977). A small chapter "Consumption and Income Levels" describes in general terms the situation in all East

European countries since the 1950s, in Derek H. Aldcroft, *The European Economy 1914–1990*, 3rd edn. (London: Routledge, 1993), pp. 177–79. A brief but efficient analysis of fluctuations in industrial wages (their amount and composition) in the early 1920s is provided by Alec Nove in his *An Economic History of the USSR* (London: Allen Lane, 1969), p. 114.

13. Gladys Baker, *The County Agent* (Chicago: University of Chicago Press, 1939), pp. 41, 176.
14. P. P. Maslov, *Trudy Komissii po izucheniiu sovremennoi dorogovizny*, vol. 3 (Moscow: Obshchestvo im. A. I. Chuprova, 1915), pp. 20, 43.
15. For a discussion of the methodology for constructing historical price indices, see John J. McCusker, *How Much Is That in Real Money? A Historical Price Index for Use as a Deflator of Money Values in the Economy of the United States*, 2nd edn. (Worcester, MA: American Antiquarian Society, 2001). Up-to-date information is available at the online resource: *Six Ways to Compare the Relative Value of a U.S. Dollar Amount, 1774 to Present*; available at www.measuringworth.com/uscompare/ (accessed December 8, 2008).
16. For a classical outline of this method, see Sir Roy George Douglas Allen, *Index Numbers in Theory and Practice* (Chicago: Aldine Publishing, 1975).
17. F. A. Shcherbina, *Krest'ianskie biudzhety* (Voronezh: Izdanie Imperatorskogo vol'nogo ekonomicheskogo obshchestva, 1900); A. V. Chaianov, "Vliianie sostava i velichiny krest'ianskoi sem'i na eie khoziaistvennuiu deiatel'nost'," *Trudy Imperatorskogo Vol'nogo ekonomicheskogo obshchestva*, no. 1–2 (1912): 3–6.
18. Chaianov, *Organizatsiia krest'ianskogo khoziaistva*, p. 66.
19. In 1913, highly qualified master workmen, for instance, carpenters, earned up to 1.73 rubles a day. S. G. Strumilin, "Oplata truda v Rossii: istoricheskii ocherk," *Planovoe Khoziaistvo*, no. 4 (1930): 108. That could be as much as 450 rubles in annual income, while the wages of some joiners were even higher (Moscow city council paid joiners 1.88 rubles a day in 1908). See "Rabochii vopros," *Ekonomist Rossii*, no. 7–8 (1908): 22. The noted Soviet economist S. G. Strumilin based his calculations on the assumption that craftsmen worked 250–300 days a year. Hence, Moscow joiners earned 470 to 564 rubles annually.
20. P. P. Maslov, "Rabochii rynok i vzdorozhanie zhizni," in *Trudy Komissii po izucheniiu sovremennoi dorogovizny*, vol. 3, p. 161.
21. In 1913, an average worker's annual income in St. Petersburg province was 311 rubles, while in Kazan province it was only 201 rubles; and in Moscow province it was close to the empire's average—253 rubles. The lowest annual wage of 137 rubles was paid in the Moscow flax-spinning industry, and among the highest paid workers were the Petersburg power-plant employees, earning about 718 rubles a year. See *Statisticheskii sbornik za 1913–1917 gg.*, vol. 1 (Moscow: Trudy TsSU, T. Vii, vypusk 1, 1921), pp. 118–19.
22. Gymnasia teachers could receive a maximum salary of 1,260 rubles a year. See N. Orlov, *Sbornik rasporiazhenii i raz"iasnenii Ministerstva Narodnogo Prosveshcheniia* (Tomsk, 1910), p. 58. According to data published by Harley Balzer, in 1914 engineers received basic salaries usually in the range of 2,000–3,000 rubles. While some engineers earned even more, it was difficult to get a good job in their profession: unlike agronomists who enjoyed a booming job market, in 1913–14, nine engineers competed for one position. See Harley D. Balzer, "The Engineering Profession in Tsarist Russia," in Balzer, ed., *Russia's Missing Middle Class: The Professions in Russian History* (Armonk, NY: M. E. Sharpe, 1996), pp. 60, 74.

23. Diomidov, "K voprosu o material'nom polozhenii zemskogo agronomicheskogo personala," 2.
24. For instance, in Ufa province only 30 percent of the agronomists received a housing allowance, while the average rent paid by precinct agronomists was 128 rubles in 1912, and 140 rubles in 1913. One agronomist had to pay 240 rubles a year for his housing in 1912, and another agronomist paid even more, 300 rubles. At the same time, in the Birsk district of Ufa province a precinct agronomist earned only 1,440 rubles a year. See S. A. Gruzdev, "Polozhenie zemskoi uchastkovoi agronomii," *Zemskii agronom*, no. 10 (1914): 9, 10; *Sbornik dokladov Birskoi uezdnoi zemskoi upravy i postanovlenii zemskogo sobraniia XXXVI ocherednoi sessii 1910 goda* (Ufa, 1911), p. 601.
25. Gruzdev, "Polozhenie zemskoi uchastkovoi agronomii," pp. 10–11.
26. P. Rybalov, "Uchitel'skii domik," *Sel'skokhoziaistvennoe obrazovanie*, no. 6 (1916): 311–12. Rybalov criticized the zemstvos' practice of building apartment complexes instead of individual houses for teachers. While saving the zemstvos a few hundred rubles, this housing no longer satisfied the rural professionals.
27. Glukhov et al., *Mestnyi agronomicheskii personal, 1 ianvaria 1915 g.*, pp. 412–20.
28. *Trudy 11-go i 12-go gubernskikh agronomicheskikh soveshchanii pri Ufimskoi gubernskoi zemskoi uprave s 4 po 7 ianvaria i s 1 po 5 marta 1913 g.* (Ufa, 1913), p. 131.
29. Ibid.
30. Ibid., p. 132.
31. Ibid., p. 183.
32. Ibid., p. 75.
33. According to the data available for the 49 districts of the five central provinces (Moscow, Vladimir, Kostroma, Iaroslavl, and Tver), in 12 districts (one-quarter) there were no salary increases for agronomists at all. In 11 other districts (over one-fifth), an agronomist's salary could be increased no more than 20 percent in a number of small raises during the entire term of his or her service. Thirteen more districts (over one-quarter) limited the total of periodical increases to 30 percent of the initial salary. Eight districts (over one-sixth) provided raises for up to 50 percent of the initial salary, and the remaining five of the surveyed districts limited the scale of the periodical raises to 35–100 percent. See Diomidov, "K voprosu o material'nom polozhenii zemskogo agronomicheskogo personala," p. 4.
34. A telling incident occurred on the morning of October 3, 1912, during a session of the Alatyr district (Kazan province) zemstvo meeting. The district agronomist organization asked the zemstvo for 150 rubles for short-term academic training of agronomists in 1913, 100 rubles for an agricultural museum, and 50 rubles for pest control. The zemstvo meeting eagerly allocated the requested 300 rubles, and voted to give an additional 50 rubles to the museum of horticulture, in spite of the agronomist Krylov's declaration that this funding was excessive [sic!], for 100 rubles allocated to the agricultural museum would also support the museum of gardening. See *Zhurnaly Alatyrskogo uezdnogo zemskogo sobraniia sessii 1912 goda*, p. 129. Later on, a teacher of the Kuvakin school, Nikolai Vorotnikov, asked the zemstvo meeting for a small allowance. He had served for 19 years in the Alatyr zemstvo, but still earned 35 rubles a month, which after a pension fund deduction left him only 28.70 rubles a month (344.4 rubles a year). Meanwhile, his two children were studying in the district town of Alatyr, where the cheapest room and board cost 24 rubles a month. Teacher Vorotnikov asked the zemstvo

to cover at least some of his expenses for educating his children, but the zemstvo rejected his application on the grounds of scarcity of funds. Ibid., p. 155. To sum up: an agronomist who had held his position for two years and whose basic salary was 1,125 rubles a year (N. N. Krylov) did *not* ask the zemstvo to give extra money to some demonstrative facility, but the zemstvo allocated that extra sum anyway. A schoolteacher with 19 years of work experience and an annual salary of some 350 rubles (N. Vorotnikov) asked the zemstvo for support to help with the education of his children, and the zemstvo ignored his plea.

35. For instance, in February 1912, a lecturer at the fortnight courses in Elets (Orel province) was paid 5 rubles an hour. See *Nashe khoziaistvo*, no. 6–7 (1912): 22. Agronomist F. A. Kireevskii, who graduated from a middle-level agricultural college in 1908, had a teaching load of 14 hours at those courses. This means that in addition to his basic income he received at least 70 rubles, which could be over 50 percent of his monthly salary. See *Nashe khoziaistvo*, no. 8–9 (1912): 9.

36. Before the First Russian Revolution, the Porechensk district veterinarians (Smolensk province) received 700 rubles a year. Soon after the revolution, their salaries almost doubled reaching 1,200 rubles. See "Khronika," *Veterinarnoe obozrenie*, no. 15 (1909): 475.

37. In the Kazan district, veterinarians were still paid 1,200 rubles a year in 1914. See NART, f. 119, Kazanskaia uezdnaia zemskaia uprava, op. 2l, d. 47 "Vedomosti na vydachu zhalovaniia sluzhashchim v zemstve," l. 10.

38. The Stavropol veterinarian Sviatoslavskii was, apparently, an all-Russian champion of holding a number of positions simultaneously. He was paid: (1) a salary of 1,200 rubles as a precinct veterinarian; (2) a travel allowance of 300 rubles for the same job; (3) 420 rubles in housing allowance, by the railroad; (4) 400 rubles in housing and utilities, from the veterinary clinic; (5) a travel allowance of 200 rubles, from the city of Stavropol; (6) a half salary for working in another veterinary precinct, 500 rubles; (7) a travel allowance of 200 rubles for the same job; and (8) a salary for replacing the veterinary of the Osetin artillery division, 900 rubles. Instead of his basic salary of 1,200 rubles a year, this veterinarian actually received 4,120 rubles annually. See "Izvestiia i zametki," *Veterinarnoe obozrenie*, no. 3–4 (1907): 128.

39. In 1915, the Urzhum zemstvo (Viatka province) was willing to pay 420 rubles a year to veterinarian assistants with vocational-school training, and only 360 rubles to those with army-school training (with the housing and utilities provided by the zemstvo for free). While a private animal husbandry firm offered 552 rubles a year to an experienced veterinarian assistant (with free housing), the Simbirsk provincial zemstvo was willing to pay 600 rubles to an unspecified candidate (also provided with free housing). See *Veterinarnyi fel'dsher*, no. 4 (April 1915), inside cover. In the Orenburg district zemstvo, veterinarian assistants received a basic salary of 480 rubles a year, which could increase, after three periodic raises, to 600 rubles. See *Veterinarnyi fel'dsher*, no. 6 (June 1915): iii.

40. As late as January 1917, with an average inflation rate of 100 percent in comparison with the prewar level (cf.: *Statisticheskii sbornik za 1913–1917 gg.*, vol. 1, p. 230), the monthly salary of schoolteachers in the Kazan district was between 65 and 100 rubles, and the salary of veterinary assistants between 35 and 72 rubles, while agricultural elders were paid on average 60 rubles. See NART, f. 119, Kazanskaia uezdnaia zemskaia uprava, op. 2l, d. 81 "Spisok sluzhashchikh Kazanskogo uezdnogo zemstva," ll. 10, 10 ob., 18–41. In this respect, the case of the peasant son Stepan Burmistrov was truly symbolic. He was born in 1884, and

hence belonged to the same age group that shaped the New Generation of Russian intelligentsia. With a primary public-school education, in 1914, he took the job of secretary to the Kozmodem'iansk district land settlement commission (Kazan province), previously held by a certainly educated local nobleman, Boris Iumatov. This position provided a salary of 1,200 rubles a year. No disgrace to a gentleman, this salary made the young peasant Burmistrov almost three times better paid than the teachers who educated him at a zemstvo primary school. See NART, f. 256, Kazanskoi gubernskoi zemleustroitel'noi komissii, op. 9, d. 133 "O prokhozhdenii sluzhby Sekretaria Kozmodem'ianskoi uezdnoi zemleustroitel'noi komissii Burmistrova," ll. 5–14.

41. To quote a question addressed to the monthly *Agricultural Education* in 1914, "Please tell me, where can I take exams to receive the title of agronomist? I have a teacher's certificate allowing me to teach in two-grade schools." See "Voprosy i otvety," *Sel'skokhoziaistvennoe obrazovanie*, no. 3 (1914): 175. By becoming an agronomist, this teacher would have increased his annual income at least threefold. However, we do not know a single case of a teacher being converted into an agronomist, while hundreds of village teachers became cooperative instructors and managers. Holding the job of a teacher and working as an agricultural elder (for instance, at a zemstvo warehouse) was another, more realistic, possibility. Beginning in 1911, the Tver provincial zemstvo sponsored summer agricultural courses for the zemstvo school teachers at the Moscow Agricultural Institute (the zemstvo paid 2,000 rubles, and the Department of Agriculture paid 4,000 rubles for this program). Attending the courses were 100 teachers, half of whom were women. See *Russkie vedomosti*, no. 141 (1912): 3.

42. Source: Alexandrovskii et al., *Mestnyi agronomicheskii personal, 1 ianvaria 1914 g.*, pp. i–ii.

43. Morachevskii, *Agronomicheskaia pomoshch'*, p. 308.

44. Ia. I. Nekludov, "O voznagrazhdenii prepodavatel'skogo personala," *Sel'skokhoziaistvennoe obrazovanie*, no. 5 (1914): 251.

45. Ibid., pp. 252–54. There were seven types of agricultural vocational schools and colleges, each with a particular system of salaries. The quoted figures reflect the most typical salaries in the lowest and highest institutions in that hierarchy.

46. Cf. "Spros i predlozhenie truda," *Sel'skokhoziaistvennoe obrazovanie*, no. 1 (1914): 49.

47. The emperor signed the project on July 3, 1914, "So let it be," and the necessary funds were allocated. See "Odobrennyi Gosudarstvennym Sovetom i Gosudarstvennoi Dumoi zakon ob uchrezhdenii Uchitel'skogo sel'skokhoziaistvennogo instituta," *Sel'skokhoziaistvennoe obrazovanie*, no. 8 (1914): 401–8. The outbreak of the war put this plan on hold. Instead, on the Gora estate in Pskov province, purchased for the future institute, another primary agricultural school was open. See "Khronika," *Sel'skokhoziaistvennoe obrazovanie*, no. 9 (1914): 482.

48. See *Vestnik kooperativnykh soiuzov*, no. 12 (1915): 634, 642–43.

49. Morachevskii, *Agronomicheskaia pomoshch'*, p. 455.

50. By 1908, there were four women among 29 dairy masters in the service of the Department of Agriculture, earning more than many of their male colleagues. See "Spisok spetsialistov i instruktorov po zhivotnovodstvu i molochnomu khoziaistvu, masterov po molochnomu delu, zaveduiushchikh shkolami molochnogo khoziaistva i skotovodstva v Evropeiskoi i Aziatskoi Rossii (po svedeniiam na 1 iiulia 1908 g.)," in NART, f. 119, Kazanskaia uezdnaia zemskaia uprava, op. L,

d. 987 "Perepiska s gub. Upravoi o provedenii agronomicheskikh meropriiatii," ll. 8–12.

51. See "Otnoshenie Moskovskogo gradonachal'nika Tovarishu Ministra vnutrennikh del, P. G. Kurlovu," in GARF, f. 102 Departament politsii, D-4, op. 120 d. 181 "Ob oblastnom s"ezde v g. Moskve deiiatelei po okazaniiu agronomicheskoi pomoshch'i naseleniiu," l. 5.

52. Chaianov, *A. V. Chaianov—chelovek, uchenyi, grazhdanin*, p. 7.

53. "Diplomnaia rabota predstavlennaia okanchivaiushimi v techenii 1910 g.," pp. 43–50.

54. See McClelland, "Diversification in Russian-Soviet Education," p. 184.

55. "Khronika," Sel'skokhoziaistvennoe obrazovanie, no. 1 (1917): 39.

56. Ia. N., "Sel'skokhoziaistvennye uchebnye zavedeniia," *Sel'skokhoziaistvennoe obrazovanie*, no. 4 (1916): 203.

57. RGIA, f. 902, d. 86, l. 3 ob.

58. "Khronika," *Sel'skokhoziaistvennoe obrazovanie*, no. 3 (1916): 150.

59. Based on data published in "Khronika," *Sel'skokhoziaistvennoe obrazovanie*, no. 3 (1916): 150.

60. RGIA, f. 902, d. 86, l. 4.

61. RGIA, f. 450, op. 3, d. 8, l. 16 ob.

62. "Khronika," *Sel'skokhoziaistvennoe obrazovanie*, no. 1 (1917): 38–39.

63. A typical job offer read: "Seeking a candidate among graduates of the Stebut or Golitsin women's agricultural courses to take the position of manager [i.e., director] of the State Spiridonov primary women's agricultural and housekeeping school of the 1st degree…. This school is located six versts [four miles] from the Dno railway station in Pskov province. The manager of the school receives salary of 1,000 rubles a year, with free housing and utilities, in addition to payment for lessons and practical training." See "Spros i predlozhenie truda," *Sel'skokhoziaistvennoe obrazovanie*, no. 2 (1916): 112. The value of this package of benefits exceeded 1,500 rubles (with a minimum teaching load of 15 lessons a week), and apparently included periodical raises. This was a really good offer for a first job.

64. The standards of the contemporary job market were reflected in the following request published in the section "job wanted": "A graduate of the Stebut Women Agricultural Courses is seeking the position of director of a state, zemstvo, or public women's school of agriculture and housekeeping or dairy at a salary of no less than 1,200 rubles a year, with free housing and utilities." See "Spros i predlozhenie truda," *Sel'skokhoziaistvennoe obrazovanie*, no. 5 (1915): 288.

65. In the 1910s, the basic salary of a gymnasium teacher was 840 rubles a year (with a teaching load of 12 lessons a week; teachers of calligraphy earned several times less). After five years of service, a teacher received a first salary increase, and earned 900 rubles a year. After 15 years of teaching, she would receive the salary of 1,260 rubles. See Orlov, *Sbornik rasporiazhenii i raz"iasnenii*, pp. 49, 58.

66. In the early spring of 1915, two out of nine agronomists of the Okhan district zemstvo (Perm province) were women graduates of the Golitsin and St. Petersburg agricultural courses. In addition, three out of seven agricultural instructors were female (graduates of vocational agricultural schools and courses). See *Otchet o meropriiatiiakh po uluchsheniiu sel'skogo khoziaistva v Okhanskom uezde za vremia s 1-go aprelia 1914 g. po 1-e aprelia 1915 g.* (Okhansk, 1915), pp. 9–10. In general, different provinces revealed different patterns of women's employment. Only a few provinces (e.g., Astrakhan) had no female agricultural sppecialists employed

whatsoever. Other provinces could boast from 1–3 to 14–15 female instructors of agriculture and household, masters of diary, agronomists, and so on. See Glukhov et al., *Mestnyi agronomicheskii personal, 1 ianvaria 1915 g.* In addition, a growing number of women became cooperative instructors and inspectors.

67. One of the top officials in the Department of Agriculture, Nadezhda Dolgova (1853–1926), was married. She joined state service in 1907, and within a few years was a senior specialist in agricultural education. See *Spisok lits, sluzhashikh po vedomstvy Glavnogo upravlenia zemleustroiistva i zemledeliia na 1912*, col. 81. Dolgova was also the chair of the Society for Assistance to Women's Agricultural Education, the primary sponsor of the Stebut courses. The wife of the Povenets district agronomist A. K. Gagman (Olonets province), Mrs. Gagman, born Chaikina, worked as a creamery master in the Povenets zemstvo, after completing bacteriologist and creamery courses. See "Agronomicheskii personal zemskikh uchrezhdenii," pp. 626.

68. See *Sbornik dokladov Birskoi uezdnoi zemskoi upravy i postanovlenii uezdnogo zemskogo sobraniia XXXIX ocherednoi sessii 30 sentiabria 1913 goda* (Ufa, 1914), pp. 761–62.

69. During an annual agronomist conference, zemstvo agronomists of Samara province spent the entire day of December 17, 1913, discussing the problem of the frequent rotation of agronomists. Some tended to blame the conservative second element and the inert peasants, but others pointed to a new type of "agronomists-adventurers" constantly seeking a higher salary, apparently speaking about the careerists. See *Trudy 4-go Samarskogo gubernskogo agronomicheskogo soveshchania 16-20 dekabria 1913 goda* (Samara, 1914), pp. 30–34. In Ufa province, until 1914, the average tenure of a precinct agronomist had been 16.8 months. The majority of agronomists explained this high turnover of cadres by the inadequate material conditions. It is characteristic that with the improvement of the agronomists' situation in the early 1910s, their average tenure increased to 25 months. See Gruzdev, "Polozhenie zemskoi uchastkovoi agronomii," pp. 7–8.

70. Ivan Barkhatov, who graduated from the Moscow Agricultural Institute in 1910 together with Alexander Chaianov, became the chief agronomist of the Kazan Provincial Land Settlement Commission just 17 months after his graduation. See NART, f. 256, Kazanskoi gubernskoi zemleustroitel'noi komissii, op. 7, d. 43, l. 11. Even more impressive was the career of a close friend of Chaianov, A. N. Minin, who graduated from the Moscow Agricultural Institute in 1911, and by 1912 was the Chernigov provincial zemstvo agronomist. See Alexandrovskii et al., *Mestnyi agronomicheskii personal, 1 ianvaria 1914 g.*, p. 407.

71. Based on data derived from Alexandrovskii et al., *Mestnyi agronomicheskii personal, 1 ianvaria 1914 g.*

72. Cf. a quite explicit statement in the letter sent by the director of Kharkov Technological Institute to a professor of that institute, the renowned expert on agricultural cooperation, Aleksei Antsyferov, in 1908. The director asked Antsyferov to report whether he belonged, "openly or in secret, to any antigovernment organization, for instance, to the party of Constitutional Democrats, Social Democrats, Socialist Revolutionaries, Labor [*trudovikov*], Academic Union, and so on." TsDIAU, f. 2019, d. 9, l. 1. Similarly, in December, 1906, the former mayor of Orel, veterinarian Semen Zhivopistsev was prohibited by the governor from participating in the State Duma elections as a member of the Socialist Revolutionaries Party, while he was in fact a member of the Constitutional Democrats Party. "Khronika," *Veterinarnoe obozrenie*, no. 1 (1907): 23–24.

73. Katerina Clark, *The Soviet Novel. History as Ritual* (Chicago: University of Chicago Press, 1981); Irina Paperno, *Chernyshevskii and the Age of Realism. A Study in the Semiotics of Behavior* (Stanford, CA: Stanford University Press, 1988); Marina Mogilner, *Mifologiia podpol'nogo cheloveka* (Moscow: New Literary Review, 1999); idem, "The Russian Radical Mythology (1881–1914): From Myth to History" (Ph.D. diss., Rutgers University, 1999).

6   From Knowledge to Influence: Building a Bridge to the New Peasant

1. Ilya Gerasimov, " 'Vse vliianie znaiushchim liudiam:' Novaia generatsiia rossiiskoi intelligentsii kak modernizatory," in *Vlast' i nauka, uchenye i vlast': 1880-e–nachalo 1920-kh godov* (St. Petersburg: Dm. Bulanin, 2003), pp. 278–97.
2. A manifesto of this approach can be found in Kotsonis, *Making Peasants Backward.*
3. A. Minin, "Chto takoe zemskaia agronomiia?" *Krest'ianskoe delo*, no. 1 (November 15, 1909): 7.
4. The peasants of Buinsk district were not unique in their initial mistrust of agricultural specialists. A vivid illustration of this reaction can be found in the following account: "I remember one incident that happened in the depth of Izium district [Kharkov province]. We had scarcely interviewed the first peasants when the rumors started: 'The end of time has come, as was written in the Bible, they will take down our names at once, and then convert us to another religion.' " See N. Kovalevskii, "Na podvornoi perepisi," *Agronomicheskii zhurnal*, no. 9–10 (1913), p. 97.
5. I discussed this topic at length in Ilya Gerasimov, "On the Limitations of a Discursive Analysis of 'Experts and Peasants,' " *Jahrbucher fur Geschichte Osteuropas* 52, no. 1 (2004): 261–73.
6. Matsuzato, "The Fate of Agronomists in Russia: Their Quantitative Dynamics form 1911 to 1916," p. 173.
7. Eklof, *Russian Peasant Schools*, pp. 429–30.
8. In the words of Scott Seregny, "Socially and culturally, these 'conscious' peasants remained more closely tied to the village than many teachers of peasant origin and as such occupied a strategic position as intermediaries between the rural community and outsiders." See Scott J. Seregny, "Peasant Unions During 1905," in Esther Kingston-Mann and Timothy Mixter, eds., *Peasant Economy, Culture, and Politics of Imperial Russia, 1800–1921* (Princeton, NJ: Princeton University Press, 1991), p. 353.
9. Morachevskii, *Spravochnye svedenia o deiatel'nosti zemstv po sel'skomy khoziaistvu (po dannym na 1909 god)*, p. xxx. This work provided information about the statistical organizations of 22 out of 34 zemstvos, which altogether had over 19,000 voluntary correspondents. The total number of correspondents must be somewhere beyond 30,000, for such "peasant-dominated" provinces as Viatka or Olonets must have had many hundreds of correspondents. There is indirect evidence that Pskov province alone had over a thousand correspondents, although the exact figure was not reported (ibid., p. 424).
10. In Moscow province, there were 374 voluntary correspondents including 240 peasants (64 percent), 63 clergymen (17 percent), and 39 teachers (10.5 percent). In Iaroslavl province, the figures were almost identical: 64 percent of 800 correspondents were peasants, 19 percent were members of the clergy. See

Morachevskii, *Spravochnye svedenia o deiatel'nosti zemstv po sel'skomy khoziaistvu (po dannym na 1909 god)*, pp. 252, 568. In 1913, in Samara province 66 percent of the 1,813 correspondents were peasants—members of the commune, and 8.4 percent were individual farmers. Altogether, peasants accounted for 74 percent of the Samara statistical bureau's voluntary correspondents. See A. V. Teitel, "Chto govorit naselenie Samarskoi gubernii o zemskoi agronomii," in *Trudy 4-go Samarskogo gubernskogo agronomicheskogo soveshchania 16-20 dekabria 1913 goda*, p. 172.

11. Morachevskii, *Spravochnye svedenia o deiatel'nosti zemstv po sel'skomy khoziaistvu (po dannym na 1909 god)*, pp. 35, 127, 175, 568. Some zemstvos provided correspondents with special notebooks containing questions to be answered during a year and a schedule for them. The Kazan zemstvo notebook for 1915, for instance, included 64 questions, which were more or less evenly distributed from January to December. See *Zapisnaia knizhka korrespondenta statisticheskogo otdeleniia na 1915 god* (Kazan: Statisticheskoe otdelenie Kazanskoi gubernskoi zemskoi upravy, 1914), pp. 6–8.

12. Each correspondent annually received agricultural periodicals and literature worth two rubles on average. Very often, zemstvos gave correspondents free subscriptions to their own publications (thus, 1,000 Kazan correspondents received free subscriptions to the *Kazan Gazette*), or to a major provincial agricultural periodical (in 1909, 1,300 Poltava correspondents were made subscribers to the weekly *Farmer* of the Poltava Agricultural Society). The peasants thus received free access to quality and relevant regional information, while the publishers increased the number of their subscribers.

13. *Otchet o deiatel'nosti i sostoianii sredstv Samarskogo obshchestva*, p. 25.

14. Ibid., p. 24.

15. Ibid., pp. 26, 28.

16. GARF, f. 102 Departament politsii, D-4, Op. 119, d. 237 "O proizvedeniiah povremennoi pechati, izdavaiemyh professional'nymi organizatsiiami", l. 16.

17. *Krest'ianskoe delo*, no. 14 (August 1, 1911), cover page.

18. "'Krest'ianskoe delo' i ego chitateli," *Krest'ianskoe delo*, no. 14 (August 1, 1911): 289.

19. In 1909, at the beginning of the public modernization campaign, a revolutionary-turned-cooperative ideologist, V. V. Khizhniakov, explained to the educators: "Writing popularly is not as easy as many authors of 'books for the people' think. Besides a complete knowledge of the question and general ability to express one's thoughts, it is also necessary to know the psychology of the audience; it is necessary to have much intuition to speak with this audience not only in a comprehensible language, but also using understandable images and examples, and an adequate system of thinking." See V. V. Khizhniakov, "Bibliografiia i literaturnoe obozrenie," *Vestnik kooperatsii*, no. 2 (1909): 167. Eight years later, on the eve of the February Revolution, the editorial board of the magazine the *New Spike* urged its contributors: "Our first plea—write simply, using foreign words and complex phrases only if absolutely necessary, and always with explanations. Remember that almost all our readers are genuine peasants, who will not understand a foreign word…. It is not so hard to write simply: there is no necessity to emulate folk speech…. The second plea—do not write at great length…. Write no more than four pages of a usual sheet [format] (always on one side of it). Thus, do not write in a partisan and lengthy manner, but *simply* and *briefly*—in this way you will help the *New Spike* to become a truly *popular*

magazine sooner." See Redaktsiia, "K sotrudnikam 'Novogo kolosa,'" *Novyi kolos*, no. 5–6 (February 15, 1917): 1.

20. For instance, in 1911–12, a special commission of Kazan agricultural specialists chaired by the provincial agronomist I. I. Shtutser reviewed 600 brochures on various aspects of agriculture, and found only 228 of them appropriate for distribution among the peasants of the province. Just a few examples of reviews will give an insight into the criteria of the Kazan commission. About the book *The Land, and How to Better Utilize It*: "An agronomist is not a land settlement officer, and therefore should not distribute this book." About *Preparing Pork*: "The brochure is written, apparently, by a veterinarian from a veterinarian's point of view, and not for the peasants. First, one must have good pigs, and know how to fatten them up." The founders of the SOUKK, the provincial agronomist Teitel and the agronomist Sev, earned extremely negative remarks from their Kazan colleagues: "The author [Teitel] always forgets that he is writing a popular interpretation for the peasants, and not a report for a provincial agronomist conference.... It is obvious that [the brochure] is written by an armchair specialist, alienated from real work." About Sev, they wrote, "There are 68 pages of small print in the book, and exclusively about the seeds and crops. One has to hit upon such a popularization!" In the commission's opinion, one should avoid such foreign words as "process," "period," and "factor." To sum up, Kazan agricultural specialists (predominantly, agronomists) wanted a popular book to be written in lively, simple language, from the point of view of zemstvo or public agronomists, not their "narrow-minded" rivals, the government agronomists and veterinarians. See *Zhurnal zasedania gubernskogo agronoicheskogo soveshchania Kazanskogo gubernskogo zemstva 7 sentiabria 1911 goda*, pp. 149–341.

21. Some of those lists were designed to keep local rural specialists posted on the latest publications available for distribution among peasants (such as the list edited by the provincial agronomist Shtutser), others targeted popular village libraries organized and patronized by the local "village intelligentsia." In the latter case, lists of suggested literature indiscriminately covered a broad spectrum of topics but differentiated by cost among the suggested sets of books. Thus, there were lists costing 10, 25, and even 100 rubles. See *Spisok knig dlia narodnykh sel'skokhoziaistvennykh bibliotek. Sostavlen Komissiei o merakh sodeistviia ustroistvu narodnykh chtenii po sel'skomu khoziaistvu* (Petrograd: GUZiZ, Departament Zemledeliia, 1914).

22. "'Krest'ianskoe delo' i ego chitateli," p. 288.

23. Ibid., p. 290.

24. *Otchet o deiatel'nosti i sostoianii sredstv Samarskogo obshchestva*, p. 29.

25. In 1913, a peasant of the village of Semenovka, Novouzensk district, commented on the publications in the *Samara Agriculturist*: "in general, we understand articles in *S.A.*, particularly by the agronomist Sev, but the thing is, nothing in those articles is about our district, and other localities are alien to us, the people of Semenovka." See *Otchet o deiatel'nosti i sostoianii sredstv Samarskogo obshchestva*, p. 29.

26. Ibid.

27. In 1911 and 1913, a few dozen respondents to the *Samara Agriculturist*'s survey speculated that while they perfectly understood the magazine's publications, their "content is unclear for a peasant, who graduated only from a primary school, because of the many incomprehensible foreign words and intellectual expressions"; "I understand, but for the peasant it should be written in a more

popular style." See *Otchet o deiatel'nosti i sostoianii sredstv Samarskogo obshchestva*, pp. 26, 28. It is worth noting that among those who were concerned about the common peasant's ability to grasp the sense of "overly intellectual" publications, there were at least four better-educated peasants in 1911. This makes the essence and configuration of the boundary between the modernizers and the peasants as their "objectified class par excellence" more complicated than the historians of Imperial Russia used to think. Apparently, discursive projections on and for traditional peasantry were not an exclusive privilege of the educated city elite; rather, every participant in a cultural dialogue sooner or later discovered that in their isolation from the outer world, traditional peasants were "semiotically invisible," and thus semantically marked by the others—until the moment when they would speak out for themselves.

28. According to the 1911 survey of the magazine the *Peasant Cause*, the majority of respondents shared every issue of the magazine with 5–50 people. Even more people read every single issue of the magazine when it was subscribed to by a teahouse, agricultural cooperative, or a civic-minded deacon. See " 'Krest'ianskoe delo' i ego chitateli," 289–90.

29. The emergence of reading circles among the peasants testified to the beginning of this process: 26 farmers of Mogilev province together subscribed to a number of magazines, including the *Peasant Cause*; in Tetushi district (Kazan province), the growing circle of 20 peasants subscribed to a few periodicals together. See " 'Krest'ianskoe delo' i ego chitateli," 289; N. Iakushkin, "Iz krest'ianskikh pisem," *Krest'ianskoe delo*, no. 20 (November 1, 1911), p. 451. According to the questionnaire of the Moscow Literacy Society, in 1915 under 6 percent of respondents (18) read newspapers alone, while 11 percent (35) read "with the entire village," 5 percent (15) shared a newspaper "with several villages." Sixty-four percent (205) answered that the same newspaper was read "many" people. Murinov, "Gazeta v derevne," p. 183. On the patterns of village reading practices, see also Reitblat, *Ot Bovy k Bal'montu*.

30. A. Lazarenko, "Rasprostranenie sel'skokhoziaistvennykh znanii vneshkol'nym putem," *Sel'skokhoziaistvennoe obrazovanie*, no. 10 (1915): 485.

31. Ibid., p. 487. If every peasant represented one household, then agricultural specialists directly contacted 6.5 percent of all farms.

32. Ibid., pp. 490, 493.

33. The most frequent attendees were two peasants under arrest, who were held in the same building of the county court. See K. G[udevich], "Korrespondentsia," *Bessarabskoe sel'skoe khoziaistvo*, no. 24 (1912): 735–36.

34. Jane Burbank, *Russian Peasants Go to Court. Legal Culture in the Countryside, 1905–1917* (Bloomington: Indiana University Press, 2004).

35. For instance, only 269 out of 360 applicants were admitted to the winter courses in Okhansk district (Perm province) in 1913–14, which meant a competition of 1.34 applicants for each available place. See *Otchet o kursakh, byvshikh v Okhanskom uezde s 1-go oktiabria 1913 goda po 1 oktiabria 1914 goda i o postoiannykh sel'skokhoziaistvennykh kursakh v Shalashakh*, n.d., p. 17.

36. The ten-month long agricultural courses organized by the Uman mid-level college of gardening and agriculture in 1912 dedicated 64 hours to agricultural economics and 14 hours to cooperation. These lessons happened to be the most popular among the 62 peasant students, of whom 90–95 percent attended every class on economics, demonstrating particular interest in the question of farm organization. See *Otchet o desiatimesiachnukh kursakh dlia*

vzroslykh krest'ian pri Umanskom srednem uchilishche sadovodstva i zemledeliia v 1912 godu (St. Petersburg: GUZiZ, Departament Zemledeliia, 1914), p. 78. Almost 81 percent of the students were actually engaged in agriculture before entering the courses, and half of them were members of land communes (ibid., pp. 9–10).

37. V. Nikolaev, "O predstoiiashchikh kursakh v g. El'tse po sel'skomu khoziaistvu," *Nashe khoziaistvo*, no. 1 (January 17, 1912): 21.

38. V. Nikolaev, "Eletskie kursy po sel'skomu khoziaistvu," *Nashe khoziaistvo*, no. 6–7 (April, 1912): 22.

39. Ibid., pp. 21–22.

40. There were, of course, transitional forms of peasant discourse, when peasants attempted to describe their current experience in a traditional epic form, or in the genre of folk song. See, for instance, a poem about economic and political issues of the day written by a peasant from Olonets province, quoted in Iakushkin, "Iz krest'ianskikh pisem," p. 451. However, for the majority of educated peasants (and workers), borrowing from the language of the modernized elite seemed to be more appropriate and, in a way more organic, than using traditional folklore forms to express realities they were never meant to express.

41. "Vpechatleniia kursistov o Eletskikh kursakh po sel'skomu khoziaistvu," *Nashe khoziaistvo*, no. 6–7 (April, 1912): 22–23.

42. Ibid., p. 23.

43. Ivan Demin, "Vpechatleniia o kursakh," *Nashe khoziaistvo*, no. 8–9 (1912): 9.

44. Peasants' comments after the Okhansk courses in the winter of 1913–14 repeated the response to the Elets courses, which were located a thousand miles away: "these courses are enlightenment. If not for the courses, we would know nothing, for now we've learned something and can think about changing our life finally"; "No doubt, courses are useful, exactly because I decided to change much on my farm. I will change [it] to extract more profits. But first, I would like to verify by my own experiments what I have heard during the lectures." See *Otchet o kursakh, byvshikh v Okhanskom uezde s 1-go oktiabria 1913 goda po 1 oktiabria 1914 goda*, p. 37.

45. Krest'ianin Vlasov, "Staroe i novoe," *Nashe khoziaistvo*, no. 36–37 (1912): 12, 13.

46. At the beginning of the twentieth century, it was still a common practice in many land communes to control the observance of the holidays by the commune members. Violators were subject to serious fines (ranging from 50 kopecks to 4 rubles) and even imprisonment. See Mironov, " 'Vsiakaia dusha prazdniku rada': trud i otdykh v russkoi derevne vtoroi poloviny XIX–nachala XX v.," p. 204. One village gathering in Riazan province in 1897 cynically ruled that any work on any holiday should be penalized by a bucket of vodka, provided by the violator to the whole commune. Tul'tseva, "Obshchina i agrarnaia obriadnost' riazanskikh krest'ian na rubezhe XIX–XX vv.," p. 53.

47. Krest'ianin I. Demin, "Delo za vami, gg. intelligenty," *Nashe khoziaistvo*, no. 34–35 (1912): 18.

48. After the peasant Aleksei Petrovich Aref'ev had consolidated his communal plots in 1910, he realized that the three-field system of crop rotation would not benefit him in his small individual field. He approached his district agronomist Ekimov, consulted with the director of an experimental field, Bogomolov (who graduated in 1905 from a mid-level agricultural college), and implemented a four-field system with tilled crops. In 1910, Aref'ev enjoyed a rich harvest, but what is more important, he conducted himself as a professional farmer rather than a traditional peasant. He made professionals actually work for him,

instead of explaining endlessly their potential usefulness. See A. P. Aref'ev, "Pis'mo v redaktsiiu," *Samarskii zemledelets*, no. 1 (January 15, 1911): 19. It took six years for the peasant Shmakov to realize that "what was applicable in a commune with its enormous pasture lands, cannot be squeezed onto an individual farm." Finally, he consulted with a local agronomist and customized the system of land cultivation on his farm. Once he had abandoned traditional routine, Shmakov revealed the ingenuity of a qualified farmer. He even constructed a simple weeding mechanism, 10 times cheaper than the factory-built device (something he had never attempted to do before). See M. Shmakov, "V bor'be s sorom," *Samarskii zemledelets*, no. 11–12 (June 1 and 15, 1916): 361–67.

49. To quote Robert Bideleux, by 1913 "[t]here occurred a steady diffusion of inexpensive, unexciting, ... mainly small-scale advances which were accommodated without intolerable risk by prudently cautious and pragmatic smallholders: new seed strains; new or newish crops (especially potatoes, maize, new wheats, sunflowers and tobacco); improved livestock breeds; poultry and pig rearing; iron parts of wooden ploughs, carts, wheelbarrows." See Robert Bideleux, "Agricultural Advance Under the Russian Village Commune System," in Roger Bartlett, ed., *Land Commune and Peasant Community in Russia: Communal Forms in Imperial and Early Soviet Society* (New York: St. Martin's Press, 1990), p. 202. Cf. also Matsuzato, "Stolypinskaia reforma i rossiiskaia agrotekhnologicheskaia revolutsiia," pp. 194–200.

50. Thus, while in 1912, the Poltava experimental agricultural station proudly reported that the crop yield of the neighboring peasant farms was 10–15 percent higher than the average, in 1916, the staff of the Bezenchuk experimental station (Samara province) was touched by a letter received from the Field Forces. Peasant Nasonov, who visited the station before the war, asked for recent station reports to be sent to the trenches, "for entertainment ... and persuasion of comrades from other provinces." See N., "O deiatel'nosti Poltavskogo Obshchestva sel'skogo khoziaistva v sviazi s voprosami o zadachakh deiatel'nosti sel'skokhoziaistvennykh obshchestv bolshogo i malogo raiona Orlovskoi gubernii," *Nashe khoziaistvo*, no. 32–33 (1912): 15; P. A. Nasonov, "Pis'mo krest'ianina s peredovykh pozitsii na Bezenchukskuiu opytnuiu stantsiiu," *Samarskii zemledelets*, no. 15–16 (August 1 and 15, 1916): 467.

51. Cf. V. A. Kosinskii, "Neskol'ko slov ob osnovnykh nuzhdakh derevni," *Agronomicheskii zhurnal*, no. 3 (1914): 5.

52. All records were probably beaten by the Samara Society for Improving the Peasant Economy, which in its 1913 account reserved over a hundred pages for detailed lists of peasants who had begun planting corn, beets, or fodder grass as part of an intensive crop rotation scheme. It is worth noting that one can find the same people on different lists, which means that those peasants were radically changing the pattern of farming. See *Otchet o deiatel'nosti i sostoianii sredstv Samarskogo obshchestva*, pp. 47–155.

53. See *Zadachi po matematike (arifmetika, algebra, geometriia, trigonometriia) s podrobnymi resheniiami* (N.p.: Izdanie zhurnala Samoobrazovanie, 1901), col. 8, 17, 27, 29, 33.

54. See S. Glazenap, *Zadachnik po sel'skomu khoziaistvu. Otdel VII i VIII. Sel'skokhoziaistvennye raschety i geometricheskie i fizicheskie zadachi po sel'skomu khoziaistvu* (Petrograd, 1915), pp. 41, 96, 138, 144.

55. In the obscure Birsk district of Ufa province, peasants of all nationalities and confessions submitted to the zemstvo board 21 appeals for new schools in 1910, while the zemstvo was able to open only 12 schools in 1911. See "Spisok khodataistv ob otkrytii nachal'nykh shkol, predstavliaemykh na rassmotrenie Birskogo uezdnogo zemskogo sobraniia, sessii 1910 goda," in *Sbornik dokladov Birskoi uezdnoi zemskoi upravy i postanovlenii zemskogo sobraniia XXXVI ocherednoi sessii*, pp. 199–206. On the popular demand for schooling in the countryside, also see Eklof, *Russian Peasant Schools*, pp. 300–303.
56. Nekludov, "Sel'skohoziaistvennye uchebnye zavedeniia v 1913 godu," p. 497.
57. "Ustav narodnykh sel'skokhoziaistvennykh shkol," in N. G. Kovalenko, *Ob organizatsii narodnykh sel'skokhoziaistvennykh shkol* (Petrograd, 1914), pp. 1–6.
58. Kovalenko, *Ob organizatsii narodnykh sel'skokhoziaistvennykh shkol*, p. 22.
59. Professional periodicals of the mid-1910s recorded many instances of peasant agitation in favor of women's education. According to one account, peasants of the Sudogod district (Vladimir province) attending a zemstvo session in 1914 argued that while "a new life is beginning to be established in the village, our women and girls are absolutely unprepared for it." See A. Novikov, "O shkolakh dlia krest'ian," *Sel'skokhoziaistvennoe obrazovanie*, no. 10 (1914): 512.
60. Eklof, *Russian Peasant Schools*, p. 313.
61. In 1914 a top GUZiZ education official, Nadezhda Dolgova, came out with a sample plan of lecturing and short-term courses on housekeeping for adult women, which would focus on such topics as the qualities and importance of a housewife; the organization of the dwelling; childbearing, and so on. See N. Dolgova, "Kursy i chteniia po domovodstvu," *Sel'skokhoziaistvennoe obrazovanie*, no. 10 (1914): 523–26. The subsequent courses on housekeeping throughout the country were a tremendous success. The first women's courses in Osinsk district (Perm province) on November 21–December 20, 1915, could admit only 80 students out of 153 applicants. Out of 59 women who graduated from the courses, only 39 percent had studied in rural schools before, while 22 percent were completely illiterate. These figures indicate that the peasants' demand for agriculture-related schooling for women had dynamics of its own, not necessarily tied to the demand for a general education. See V. G., "Kursy domovodstva," *Sel'skokhoziaistvennoe obrazovanie*, no. 4 (1916): 182–87. Zemstvos welcomed the emerging trend toward women's agricultural education, and established special stipends for girls in agricultural schools. See "Khronika," *Sel'skokhoziaistvennoe obrazovanie*, no. 2 (1915): 100.
62. Cf. Eklof, *Russian Peasant Schools*, pp. 464–65. Eklof used incomplete statistics, and limited the scope of his analysis to 1909.
63. "Vypusk uchenikov," *Sel'skokhoziaistvennoe obrazovanie*, no. 5 (1914): 286; *Sel'skokhoziaistvennoe obrazovanie*, no. 10 (1914): 502.
64. In 1906, 19 coevals of Alexander Chaianov and his fellow members of the new generation of Russian intelligentsia graduated from the Belebei three-grade agricultural school. Only half of them were of the peasant estate, while two of the students were nobles (apparently, quite dumb; one of them spent six years in a three-year school prior to graduation). Those who reported their subsequent occupation earned between 300 and 600 rubles a year. Some Belebei graduates filled the ranks of zemstvo clerks, thus changing the social balance in the zemstvo. However, more of them became managers of zemstvo warehouses and experimental agricultural farms, instructors of creameries,

and so on. See *Otchet po Belebeevskoi, Ufimskogo Gubernskogo zemstva, nizshei sel'skokhoziaistvennoi shkole 1-go razriada za 1907 god* (Ufa, 1908), pp. 186–89.

65. See S. V. Matveev, "V volostnykh starshinakh," *Russkoe bogatstvo*, no. 2 (1912): 76; idem, "Iz zhizni sovremennogo krest'ianskogo 'mira,'" *Russkoe bogatstvo*, no. 9 (1913): 117.

66. See "Otchet agronoma N. A. Pospelova po 2-mu agronomicheskomu uchastku, za iiul'-sentiabr' 1912 goda," in *Zhurnaly Ardatovskogo ocherednogo uezdnogo zemskogo sobraniia 1912 goda* (Ardatov, 1913), pp. 662–63.

67. S. L. Teitel, "Po voprosu o sviazi deiatel'nosti uchastkovykh agronomov s rabotoiu opytnykh uchrezhdenii," in Alexandrovskii et al., *Mestnyi agronomicheskii personal, 1 ianvaria 1914 g.*, pp. 1–2.

68. See M. Krasil'nikov, "Chto znaet Ufimskaia derevnia ob agronomakh," *Zemskii agronom*, no. 3 (1913): 14–16.

69. See Teitel, "Chto govorit naselenie Samarskoi gubernii o zemskoi agronomii," pp. 172–79.

70. See S. M. Mikhailov, "O kastratsii v zemskoi praktike," *Vestnik obshchestvennoi veterinarii*, no. 1 (January 1, 1914): 15.

71. For instance, we know that some 7,000 visitors paid for admission to the Alatyr exhibition of agriculture and crafts between September 1 and 4, 1913. Even though many peasants took notes, writing down useful information, it is still unclear how many people saw the exhibition as sheer entertainment in their dull provincial life. See *Otchet ob Alatyrskoi raionnoi sel'skokhoziaistvennoi vystavke s 1-go sentiabria po 5-e sentiabria 1913 g.* (Alatyr: Izdanie Alatyrskogo uezdnogo zemstva, 1915), p. 198.

72. On this issue, see M. L. Kheisin, *Istoriia kooperatsii v Rossii: vse vidy kooperatsii s nachala ee sushchestvovaniia do nastoiashchego vremeni* (Leningrad: Izdatel'stvo Vremia, 1926); L. E. Fain, *Otechestvennaia kooperatsiia: istoricheskii opyt* (Ivanovo: Ivanovskii gosudarstvennyi universitet, 1994).

73. Cf. Richard Wortman, *The Crisis of Russian Populism* (Cambridge: Cambridge University Press, 1967), pp. 137–56.

74. A general survey of the intelligentsia's experiments with agricultural cooperatives in the nineteenth century can be found in Kotsonis, *Agricultural Cooperatives*, pp. 21–63; Yoshio Imai, "The Artel' and the Beginnings of the Consumer Cooperative Movement in Russia," in Bartlett, ed., *Land Commune and Peasant Community in Russia*, pp. 363–75. See also Alexander Dillon, "The Rural Cooperative Movement and Problems of Modernizing in Tsarist and Post-Tsarist Southern Ukraine (New Russia), 1871–1920" (Ph.D. diss., Harvard University, 2003), esp. chap. 3, on Levitskii artels.

75. V. V. Khizhniakov, "K sovremennomy polozheniiu kooperativnogo dela v Rossii," *Vestnik kooperatsii*, no. 2 (1910): 20–21.

76. Cf. N. Gibner, "Kak ukrepit' nashi potrebitel'skie obshchestva," *Vestnik kooperatsii*, no. 2 (1909): 27. A retired colonel, Nikolai Gibner was one of the founding fathers of Russian consumer cooperatives and the founder of the Russian cooperative periodical press. For a biographical sketch of N. P. Gibner, see V. V. Kabanov, "Kooperatory Rossii: shtrikhi k portretam," in *Kooperatsiia: stranitsy istorii*, Vypusk 4 (Moscow: Institut ekonomiki RAN, 1994), p. 122.

77. Kotsonis, *Agricultural Cooperatives*, pp. 121, 159–66. See also A. P. Korelin, *Sel'skokhoziaistvennyi kredit v Rossii v kontse XIX-nachale XX v.* (Moscow: Nauka, 1988).

78. Based on data published in A. Merkulov, "Kooperativnoe dvizhenie v Rossii," *Vestnik kooperatsii*, no. 4 (1912): 130.
79. Ibid.
80. *Sel'skokhoziaistvennyi promysel v Rossii* (Petrograd: Izdanie Departamenta zemledeliia, 1914), p. 3.
81. See V. V. Kabanov, *Oktiabr'skaia revolutsiia i kooperatsiia: 1917 g.–mart 1919 g.* (Moscow: Nauka, 1973).
82. See *Vestnik kooperativnykh zoiuzov*, no. 12 (1915): 649.
83. Their opposite approaches toward industrial workers' political and socioeconomic initiative became a focal point of an influential historiographical tradition. See Leopold Haimson, "The Problem of Social Stability in Urban Russia, 1905–1917," *Slavic Review* 23, no. 4 (December 1964): 625, et seq.; R. Edwood, *Russian Social Democracy in the Underground: A Study of the RSDRP in the Ukraine* (Assen: Van Goorcum, 1974); Robert Service, *The Bolshevik Party in Revolution* (London: Macmillan, 1979), Part 1; Victoria E. Bonnel, *Roots of Rebellion, Workers' Politics and Organizations in St. Petersburg and Moscow, 1900–1914* (Berkeley: University of California Press, 1983).
84. "Ot redaktsii," *Vestnik kooperatsii*, no. 1 (1909): 4–5.
85. Torgashev, "Kooperativnoe dvizhenie v Rossii," *Vestnik kooperatsii*, no. 1 (1909), otdel II, 1.
86. See A. Evdokimov, "Kooperatsiia i obshchestvennaia agronomiia v Poltavskoi gubernii," *Vestnik kooperatsii*, no. 5 (1912): 129–30.
87. M. Kheisin, "Potrebitel'nye obshchestva v Rossii (ocherk ikh razvitiia i sovremennogo polozheniia)," *Vestnik kooperatsii*, no. 1 (1909), p. 41. At most, the masses would have been blamed for their ignorance and inertia—something that sounded like an excuse for them, and an additional reproach to their educators. Cf. A. Merkulov, "Ocherk razvitiia i sovremennogo polozheniia maslodel'nykh artelei," *Vestnik kooperatsii*, no. 1 (1909): 13.
88. For instance, in the winter of 1909–10, the Kazan magazine *Initiative* [Samodeiatel'nost'] published a very peculiar story about two young civil servants who back in 1882, allegedly, organized a small cooperative bank. This story was written in a documentary style but attributed very modern ideas and attitudes to the educated people living in the 1880s. The idea of establishing a credit association was born in the following dialogue:

> — You should buy watermelons and have your wife sell them out there, in the market. That would be great, a real, productive credit, you know what I mean? ... Well, take another business. Trade second-hand clothes.
> — What's with you, really? Am I a rag-and-bone man?
> — Every occupation, my dear fellow, is noble.

The story turned into a satire on the traditional mores of the intelligentsia, describing the moment when the initiative of two proto-cooperators became a success: "Everything was going somewhat prosaicly without splendid phrases, without particular élan. Ordinary! Nikolai Ivanovich and Vasilii Artem'evich had imagined that everything would have been different: high-minded people delivering good speeches and warm words; the crowd listening with reverence and becoming excited. Then, ideologists commanding in a deafening voice: comrades! Move it, act! Comrades rushing into ebullient work and acting. Meanwhile, hero-ideologists assembling a light stage, and benevolently accepting signs of gratitude. From those, who wish.... That is how, in Philistine [sic!] logic, everything should have happened. In fact, everything was different." See S. A. Kaimakov, "Kak oni ustroili bank," *Samodeiatel'nost'* (December 1909): 12–13; (January 1910): 13. This piece explicitly parodied Mikhailovskii's sociological model of the "heroes" and the "masses."

89. A tragedy happened when the chairman of the Konstantinograd agricultural society (Poltava province), Averkii Babich, engaged too deeply in a complicated grain-pawning business without possessing the necessary skills and resources. His agricultural society lost thousands of rubles, and Babich committed suicide. For contemporary observers, this was a result of the fatal mistrust of the peasant economic initiative by the intelligentsia modernizers, who were eager to work for the people, but did not know how to work with the people: "He wanted to enrich the Society against its own will, for all risky operations have been rejected by a resolution of the Society." See A. Evdokimov, "Pis'mo s Iuga," *Vestnik kooperatsii*, no. 1 (1912): 101–6. By itself, the grain-pawning cooperative business was an attempt to improve the position of the peasantry on the grain market, relieving them of the necessity to sell grain quickly in the autumn very cheaply to intermediate traders because of the peasants' desperate need for cash. While agitating for this type of commercial operation, the cooperative press warned about its complicated and risky character. Cf. F. I., "Khlebozalogovaia operatsiia," *Trudovoi soiuz* (September, 1912): 4–7.

90. "Vnutrennee obozrenie," *Agronomicheskii zhurnal*, no. 7 (1913): 127.

91. In an interview with the journal *Messenger of Cooperation*, Pavel Sadyrin argued that a cooperative must be, first of all, a vital economic organization; its educational and philanthropic agenda was a matter of secondary importance. Sadyrin thought that all human motivations, including a strive for profit, could be accommodated by the cooperative movement. See "Anketa. Opros mnenii o kooperatsii, proizvedennyi V. F. Totomiantsem," *Vestnik kooperatsii*, no. 2 (1912): 69–71.

92. A brief episode in a theoretical polemic between M. Tugan-Baranovskii and M. Kheisin gives a good insight into the type of discourse represented by Tugan-Baranovskii. In 1911–12, Tugan and Kheisin engaged in a discussion on the economic nature of cooperatives. Kheisin argued that cooperatives were a type of capitalist enterprise, with all its major attributes, including the category of capitalist profit. This approach contradicted Tugan's theoretical views, yet his most powerful argument did not come from the arsenal of political economy. He suggested that as a cooperative activist, Kheisin must have been against the taxation of cooperatives. "And still, M. L. Kheisin tries to prove to me that the income of cooperatives is their profit, without fearing that his considerations will be used to harm cooperatives," that is, to tax them as capitalist firms. See M. Tugan-Baranovskii, "Ekonomicheskaia priroda kooperativov," *Vestnik kooperatsii*, no. 1 (1912): 6. This argument betrays the rather unscrupulous thinking of a politician rather than the logical analysis of a social scientist. For more information about the views of this patriarch of the Russian cooperative utopia, see M. I. Tugan-Baranovskii, *Obshchestvenno-ekonomicheskie idealy nashego vremeni* (St. Petersburg: Izdatel'stvo Viestnika Znaniia, 1913); idem., *Kursy po kooperatsii: ekonomicheskaia priroda kooperativov i ikh klassifikatsiia*, 2nd edn. (Moscow: Moskovskii gorodskoi nar. universitet im. A. L. Shaniavskago, 1914); *Sotsial'nye osnovy kooperatsii* (Moscow: Ekonomika, 1989).

93. The problem of the cooperativist project is too complex to be examined here in any detail. For a discussion of Russian cooperativism as a protofascist ideology of

corporate society, as presented in the writings of one of its most prominent ideologists, Alexander Chaianov, see Ilya Gerasimov, *Dusha cheloveka perekhodnogo vremeni: sluchai Aleksandra Chaianova* (Kazan: ANNA, 1997), pp. 135–62.

94. During the first two decades of the twentieth century, almost 20 titles of works by Charles Gide were published in Russia, some of them in more than a dozen editions. See Iurii V. Latov, "Knigi zapadnykh ekonomistov XVIII–nachala XX veka, izdannye v Rossii," *THESIS* 1 (Winter 1993): 242.

95. Quoted from E. Kachner, "Obzor periodicheskoi kooperativnoi pechati," *Vestnik kooperatsii*, no. 5 (1912): 120.

96. Ibid., p. 121.

97. Quoted from Kachner, "Obzor periodicheskoi kooperativnoi pechati," pp. 122–23.

98. The decisive takeoff of the cooperative periodical press took place in the middle of 1909. See "Kooperativnoe dvizhenie i literatura," *Vestnik kooperatsii*, no. 4 (1909): 120. In May 1915, various types of cooperative associations published 41.3 percent of all agricultural periodicals in Russia, while during only one year, between 1911 and 1912, the number of special periodicals dedicated to the problems of cooperative movement had increased 1.6 times. See Vit-", "Sel'skohoziaistvennaia pechat' v Rossii," p. 82; Vol'fson, *Gazetnyi mir na 1911 god*, col. 329–30; Vol'fson, *Gazetnyi mir: Adresnaia i spravochnaia kniga*, col. 522–23.

99. See "Ot redaktsii," *Trudovoi soiuz*, no. 1 (August 1912): 3.

100. See Z. S. Bondurovskii, "Pis'mo v redaktsiiu," *Iuzhnyi kooperator*, no. 4 (February 28, 1913): 78–79.

101. See V. F. Shvets, "K tovarishcham," *Iuzhnyi kooperator*, no. 1 (January 13, 1913): 3–4.

102. *Kooperativnyi prazdnik. 26-oe noiabria 1912 g.* (Ekaterinburg: Ekaterinburgskoe obshchestvo potrebitelei, 1913), p. 2.

103. See "Kratkii ocherk deiatel'nosti Ekaterinburgskogo O-va potrebitelei za 5 let," in *Kooperativnyi prazdnik*, pp. 4–5.

104. See M. D. Popov, "Rol' zhenshchiny v kooperatsii," in *Kooperativnyi prazdnik*, pp. 17–20.

105. N. Ch., "Prazdnik kostromskoi kooperatsii," *Kostromskoi kooperator*, no. 4 (February 23, 1914): 3.

106. Ibid., pp. 3, 4.

107. Quoted from D. Koleno, "Trudovye idealy kooperatsii," in *Kooperativnyi prazdnik*, p. 12.

108. See *Trudy vtorogo Ekaterinburgskogo uezdnogo obshchekooooperativnogo s"ezda v g. Ekaterinburge 18–20 iiunia 1914 goda* (Ekaterinburg: Izdanie Ekaterinburgskogo uezdnogo zemstva, 1914), p. 9.

109. Ordinary members of cooperatives did not read the theoretical treatises of Tugan-Baranovskii or Charles Gide, but tens of thousand of them read the *Cooperative Desk Calendar* published by the influential Moscow Union of Consumer Societies. The 1914 Calendar edition for 1915 featured articles denouncing commerce as a source of unfair profits, blaming the right of private propriety for all economic problems, and proclaiming the coming of a just regime based on cooperatives. See *Kooperativnyi nastol'nyi kalendar' na 1915 god* (Moscow: Izdanie Moskovskogo soiiuza potrebitel'nykh obshchestv, 1914), pp. 4, 6, 73.

110. The peasant Tupitsin greeted the opening of a cooperative teahouse as an event resulting in no less than "regeneration" of the local people and revealing in

them "a healthy unity of interests." See I. Tupitsin, "Otkrytie kooperativnoi chainoi," *Kostromskoi kooperator*, no. 19–21 (December 19, 1914): 13. Another peasant, Evrgafov, had a vision of the cooperative movement, which "creates more new threads that connect and unite together large regions. But those threads should be developed further and further, so that finally, all nations will be united in one harmonious and peaceful family." See M. Evgrafov, "Mechta kooperatora," *Kostromskoi kooperator*, no. 19–21 (December 19, 1914): 13.

111. The Kazan consumer society Labor Union was founded in 1908 on the initiative of a group of Kazan intelligentsia, and the majority of its 1,300 members were also representatives of educated society (professors, physicians, lawyers, journalists, etc.) See V. N[ikol'skii], "Obshchestvo potrebitelei 'Trudovoi Soiiuz' v g. Kazani," *Samodeiatel'nost'*, no. 2–3 (February 1910): 18; Ar-skii, "Chleny i obshchestvo," *Trudovoi soiuz*, no. 7 (May 1913): 6. Such a social composition seemed to promise a particular responsiveness of the cooperative's members to the ideological side of the cooperative movement, but that proved not to be the case. While leaders of the Labor Union did their best to ignite enthusiasm for cooperativism and hostility toward the "class of producers and middlemen," ordinary members of the cooperative ignored annual meetings, campaign rallies, and even ordinary surveys. At first, only some 10 percent of all members displayed interest in cooperative public activity, and a few years later this figure fluctuated between 3 and 6 percent. See P. Z[lokazov], "Kooperatsiia i politika," *Trudovoi soiuz*, no. 5 (March 1913): 6. The oldest cooperative society in Kazan (founded in the early 1870s) was a special mutual fund established by the Society of Sales Assistants with the sole purpose of purchasing lottery tickets. Despite annual losses of about 400 rubles, this "lottery cooperative" had a steady membership, as the dream of hitting a jackpot kept attracting new "cooperators." N., "Kazanskoe obshchestvo prikazchikov," *Samodeiatel'nost'* (January, 1910): 30.

112. For instance, the Tarlashin consumer society (Kazan district) appealed to high cooperative principles explaining its unusual and complicated practice of servicing its members. The effect of that practice was quite unambiguous and pragmatic: "this way the society completely smokes out village traders, attaching all residents to the cooperative." See "Doklad Pravleniia Tarlashinskogo O-va potrebitelei Kazanskogo uezda sobraniiu predstavitelei uchrezhdenii melkogo kredita, imeiushchego byt' v Kazani 19–20 maia tekushchego 1914 goda," in *Trudy raionnykh kooperativnykh soveshchanii organizovannykh v 1913–1914 gg. kassoi melkogo kredita Kazanskogo gubernskogo zemstva* (Kazan: Izdanie Kazanskoi gubernskoi kassy melkogo kredita, 1915), p. 364.

113. Neither education nor social background predetermined one's views on this question (ibid., pp. 48–49); *Trudy vtorogo koooperativnogo s"ezda Shadrinskogo uezda v iiule 1912 goda* (Shadrinsk: Izdanie Shadrinskogo uezdnogo zemstva, 1912), p. 16.

114. "Pochtovyi iashchik," *Nashe khoziaistvo*, no. 1 (January 17, 1912): 31.

115. Ibid., no. 2–3 (January 31, 1912): 31.

116. See L. V-on, *Vserossiiskii kooperativnyi s"ezd v S.Petersburge* (Kiev: Izdanie zhurnala Nashe delo, 1912), pp. 50, 51.

117. Cf. "Russian cooperatives were, are, and, hopefully, will be politically neutral. They unite people of very different views." See Merkulov, "Kooperativnoe dvizhenie v Rossii," 118. See also A. A. Isaev, "Arteli i obshchestvennaia bor'ba," *Vestnik kooperatsii*, no. 2 (1912): 3–16.

118. See NART, f. 199, op. 2, d. 1028, ll. 90 ob., et seq.

119. See Zviagintsev, "Obshchie godovye sobraniia kreditnykh tovarishchestv Lebedianskogo uezda," *Nashe khoziaistvo*, no. 4–5 (February 18, 1912): 15–16.
120. Cf. Zyrianov, *Krest'ianskaia obshchina evropeiskoi Rossii 1907–1914 gg.*, pp. 234–35.
121. See Matveev, "V volostnykh starshinakh," p. 77.
122. See Mikhail Butov, "Otkrytie potrebitel'nogo obshchestva v s. Cherkasakh," *Nashe khoziaistvo*, no. 16–17 (1912): 21–22.
123. This attitude was revealed by the Kazan provincial zemstvo board during the Kazan regional congress of representatives of 12 provincial zemstvos in 1909. Kazan zemstvo bosses declared that the Kazan zemstvo was against "compulsory introduction of cooperatives," advocating their volunteer development and the population's freedom of choice between the zemstvo small credit funds and cooperative credit associations. They went as far as to proclaim the potential dominance of cooperatives as "the second serfdom," advocating instead the zemstvo as a "compulsory cooperative." Such a mix of liberal ideology and protectionist policy provoked a heated discussion, ending in the passing of a resolution sponsored by the Kazan second element (20 votes for, 16 against). See *Trudy Kazanskogo oblastnogo s"ezda predstavitelei gubernskikh zemstv*, pp. 204, 214, 222–31, 242.
124. Cf. "As a matter of fact, we have here the two major and most important branches of the same socioeconomic activity, directed toward the development of production forces, only divided between two organizations." See K. A. Matseevich, "Iz istorii otnoshenii obshchestvennoi agronomii i kreditnoi kooperatsii," *Vestnik kooperatsii*, no. 2 (1912): 69.
125. See P. B. Shimanovskii, "Zemstvo, agronomiia i kooperatsiia," *Zemskii agronom*, no. 6 (1913): 9.
126. V. Brunst, "O neobkhodimosti peresmotra zemskoi agronomicheskoi deiatel'nosti s vydeleniem iz nee voprosov ekonomicheskikh," *Agronomicheskii zhurnal*, no. 9–10 (1913): 15.
127. V-on, *Vserossiiskii kooperativnyi s"ezd v S. Petersburge*, pp. 7, 37; E. K., "Sel'skokhoziaistvennaia kooperatsiia na Vserossiiskom s"ezde deiatelei po melkomu kreditu i sel'skokhoziaistvennoi kooperatsii," *Vestnik kooperatsii*, no. 2 (1912): 123.
128. The cooperative ideologists instructed their followers to "treat the zemstvo with significant attention, to be interested in the zemstvo's activities," for the zemstvo assisted many cooperative initiatives and employed "many conscious people, eager to help peasants in all their needs." See "Kooperatory i zemstvo," in *Kooperativnyi nastol'nyi kalendar' na 1915 god*, pp. 101, 102. In Riazan province, cooperatives and the zemstvo found a way to settle their mutual pretensions (cooperatives demanded the democratization of zemstvo activities, the zemstvo wanted greater control over the cooperatives). For instance, in 1916, in Mikhailov district, a local credit association allocated 606 rubles to build a zemstvo medical assistant's precinct, something which only a few years before was impossible. Previously, it was the zemstvos who subsidized the cooperatives, and not vice-versa. See *Pervoe gubernskoe kooperativnoe soveshchanie pri Riazanskoi gubernskoi zemskoi uprave. Trudy soveshchaniia* (Riazan: Ekonomicheskoe otdelenie Riazanskoi gubernskoi zemskoi upravy, 1916), p. 21.
129. As the available sources suggest, it was often the wartime crisis and conflicts over procurement orders that reinforced the existing old animosities between

cooperatives and zemstvos. Thus, the conflict between cooperatives and zemstvos transcended ideological boundaries and took the form of institutional rivalry over economic and political influence in the countryside. See I. Kudriashev, "Otnoshenie kooperativov k zemstvu," *Kostromskoi kooperator*, no. 20 (November, 1915): 8–9; Ivanov, "Zemstvo, prodovolstvennyi vopros i samodeiatel'nost' naseleniia," *Kostromskoi kooperator*, no. 1 (January 1917): 2–6; Nika, "Melochi dnia," *Kostromskoi kooperator*, no. 4 (January 1917): 10; S. Pichkurov, "Kooperativnyi s"ezd v Odesse 21–25 oktiabria (vpechatleniia uchastnika)," *Iuzhnyi kooperator*, no. 2 (January 31, 1916): 43–46; M. Khristodor, "K kooperativnomu sbytu khleba pri posredstve Odesskogo soiuza kred. i ss. sb. t-v," *Iuzhnyi kooperator*, no. 2 (January 31, 1916): 510–12; Pravlenie Odesskogo Soiuza, "Po povodu stat'i agronoma M. Khristodora," *Iuzhnyi kooperator*, no. 2 (January 31, 1916): 513–14.

130. *Rezolutsii Pervogo Vserossiiskogo s"ezda deiatelei po melkomu kreditu i sel'skokhoziaistvennoi kooperatsii v S. Petersburge 11–16 marta 1912 g.* (St. Petersburg, 1912), p. 26.

131. See A. Golubev, "Dukhovenstvo v kooperativnom kredite," *Vestnik kooperatsii*, no. 2 (1909): 175–76.

132. See "Episkop i kooperativy," *Samarskii zemledelets*, no. 21 (November 1, 1916): 597–98.

133. A point extensively discussed by Yanni Kotsonis, for example, in *Making Peasants Backward*.

134. In 1910, at the second congress of credit cooperatives of Belebei district (Ufa province), a representative of the Troitsk credit associations asked the district instructor of credit cooperatives whether he would oppose the participation of local agronomists in cooperative activities. Inspector P. V. Kamkin replied that he would welcome "that promising influx of intellectual forces." Characteristically, none of the agronomists present reacted to the initiative of the Troitsk cooperator or to the response of inspector Kamkin. See *Trudy 2go s"ezda predstavitelei kreditnykh tovarishchestv Belebeevskogo uezda, 15–17 iiunia 1910 goda v g. Belebee* (Belebei, 1910), p. 19. In 1914, working now in Kazan province, inspector Kamkin demanded that all agronomists should attend all cooperative meetings in their precincts. This time, agronomists explicitly refused to sacrifice their professional duties to the cooperative cause. "Soveshchanie predstavitelei uchrezhdenii melkogo kredita, proiskhodivshee v g. Cheboksarakh, Kazanskoi gub., 9 i 10 iiunia 1914 goda," in *Trudy raionnykh kooperativnykh soveshchanii organizovannykh v 1913–1914 gg. kassoi melkogo kredita Kazanskogo gubernskogo zemstva*, pp. 184, 185.

135. This precise collision took place during the agronomist conference held by the Viatka provincial zemstvo board (February 27 to March 6, 1914). On two days, March 4 and 6, agronomists and other agricultural specialists debated the problem of concentration on cooperative activity, raised by the director of the zemstvo small credit fund, A. A. Valaev. See *Trudy agronomicheskogo soveshchaniia pri Viatskoi gubernskoi zemskoi uprave 27 fevralia–6 marta 1914 goda* (Viatka, 1914), pp. 73–89. Other conferences did not witness such passionate discussions. The Chernigov provincial agronomist conference even showed a skeptical attitude toward agronomists' involvement in the tutoring of cooperatives. See *Trudy gubernskogo agronomicheskogo soveshchaniia i ekonomicheskogo soveta Chernigovskogo gubernskogo zemstva 11–16 dekabria 1912 goda* (Chernigov, 1914), pp. 10–11. In July 1913, the Kazan provincial agronomist conference

displayed a most practical approach toward the question of cooperative activity. The precinct agronomist, Ivan Kopysov, who graduated from a mid-level college in 1909, delivered a report on the topic. Agronomists discussed some exremely problematic points of that report and decided to start with the organization of cooperative courses for peasants: "These courses will allow the establishment of personal relationships between the agronomist and future cooperative activists from the peasant milieu." See *Zhurnaly zasedanii gubernskogo agronomicheskogo soveshchaniia 12–14 iulia 1913 goda* (Kazan: Kazanskoe gubernskoe zemstvo, 1914), pp. 20–23, 98–101.

136. See K. V. Pyzhenkov, "O vzaimootnosheniiakh kreditnoi i sel'skokhoziaistvennoi kooperatsii," in *Trudy raionnykh kooperativnykh soveshchanii organizovannykh v 1913–1914 gg. kassoi melkogo kredita Kazanskogo gubernskogo zemstva,* pp. 357–60.

137. According to a 1912 survey in Moscow province, 57.5 percent of the existing cooperatives were organized with the assistance of local agronomists, and 54.3 percent enjoyed regular consultations with agronomists. The survey indicated that agronomists mostly assisted dairy associations (100 percent of these were controlled by agronomists) and agricultural societies and associations (85 percent were assisted on a regular basis, 15 percent occasionally). At the same time, only 27.7 percent of the consumer societies were regularly inspected by agronomists. These figures indicate that by assisting cooperatives, agronomists fulfilled their program of peasant modernization, while not being absorbed by the cooperative movement with its particular goals. See N-skii, "Kooperativnaia deiatel'nost' agronomov," *Zemskii agronom,* no. 5 (1914): 48–49.

138. See Sevkovskii, "Novye formy sel'skokhoziaistvennoi kooperatsii," *Vestnik kooperatsii,* no. 6 (1912): 34.

139. A. B-ov, "Fakty i itogi agronomicheskoi praktiki," *Zemskii agronom,* no. 3 (1916): 216.

140. See A. Malyshev, "Iz deiatel'nosti Terskogo soiuza," *Vestnik kooperativnykh soiuzov,* no. 6 (June 1916): 346.

141. See A. Gusakov, "Zapiski inspektora melkogo kredita," *Viatskii kooperator,* no. 3 (March 15, 1918): 11–17.

142. See B-ov, "Fakty i itogi agronomicheskoi praktiki," 210.

143. Cf. Dm. Uspenskii, "Agronomiia i kooperatsiia," *Zemskii agronom,* no. 7–8 (1916): 409–17.

144. A concept used by James Scott in his influential critique of modern social engineering projects, in *Seeing Like a State: How Certain Schemes to Improve the Human Condition Have Failed* (New Haven, CT: Yale University Press, 1998), 89–90, and passim.

7 At the Crossroads: Coping with Modernization as Routine

1. See V. Onufriev, "Usloviia, tormoziashchie normal'noe razvitie agronomicheskoi raboty," *Zemskii agronom,* no. 6–7 (1914): 6–15.

2. See "Agronomicheskii personal zemskikh uchrezhdenii," pp. 630; Alexandrovskii et al., *Mestnyi agronomicheskii personal, 1 ianvaria 1914 g.,* p. 100; "Diplomnaia rabota predstavlennaia okanchivaiushimi v techenii 1910 g.," pp. 43–50.

3. See A. I. D'iakov, "Pessimisty i optimisty," *Zemskii agronom,* no. 2 (1915): 129–33.

4. K. S. Ashin, "Ob agronomicheskom krizise," *Agronomicheskii zhurnal,* no. 2 (1915): 4–5.

5. "Since the measures adopted hitherto proved to be insufficient, it is urgently necessary to undertake some other, new, measures." See Agronom E., "Obshchestvennaia agronomiia i poslednie techeniia politicheskoi ekonomii," *Zemskii agronom*, no. 10 (1916): 583. For similar assesments of the situation, see Iu. Eremeeva, "Chem zatrudniaetsia rabota zemskoi agronomii?" *Agronomicheskii zhurnal*, no. 2 (1915): 67–69; I. Soloviev, "Krizis agronomii," *Zemskii agronom*, no. 6 (1915): 370.

6. Soloviev, "Krizis agronomii," p. 371.

7. See examples of the discussion of a new balance between public activism and professional expertise in the *Agronomy Journal*: I. Iakushkin, "Prepodavatel' sel'skogo khoziaistva ili sel'skii khoziain?" *Agronomicheskii zhurnal*, no. 1 (1913): 19; N. Kostrov, "Opyt issledovaniia krest'ianskogo khoziaistva v agronomicheskom otnoshenii," *Agronomicheskii zhurnal*, no. 5 (1913): 51; N. M. Kataev, "Bol'noe mesto v nashei sisteme sel'skokhoziaistvennogo obrazovaniia," *Agronomicheskii zhurnal*, no. 6 (1913): 22; V. Benzin, "Zhenskoe sel'skokhoziaistvennoe obrazovanie v Amerike," *Sel'skokhoziaistvennoe obrazovanie*, no. 2 (1915): 73; P. Shimanovskii, "Krizis zemskoi agronomii," *Agronomicheskii zhurnal*, no. 3–4 (1915): 111–114.

8. Cf. the diary entry by Sakharova quoted in the Introduction. For the fundamental dualism of Chaianov's worldview, see Gerasimov, *Dusha cheloveka perekhodnogo vremeni*.

9. P. Lunegov, "Agronomicheskaia deistvitel'nost'," *Zemskii agronom*, no. 10 (1915): 589. Pavel Lunegov was not a veteran of the zemstvo agronomist movemenet but a young man, who had just graduated from a mid-level agricultural college in 1914 and at the time of publication worked as the Okhan district (Perm province) agronomist of the government land settlement commission (Glukhov et al., *Mestnyi agronomicheskii personal, 1 ianvaria 1915 g.*, p. 279). His young age and government service did not spare Lunegov from inheriting the missionary ethos of the early activists of the public modernization movement.

10. "Aleksei Fedorovich Fortunatov. K tridtsatiletiiu ego professorskoi deiatel'nosti," *Zemskii agronom*, no. 9 (1916): 502.

11. Lecturing in the circle of public agronomy in the Moscow Agricultural Institute on November 21, 1912, Professor Fortunatov expressed the same ideas in prose: "We think, that … [the agronomist] must be, first of all, a scientifically educated preacher." See A. Fortunatov, "Kto on?" *Zemskii agronom*, no. 1 (1913): 18. In July 1913, Fortunatov addressed the entire community of Russian agricultural specialists with a most explicit sermon on the necessity of antiprofessional education of rural professionals: "For us, the agronomist is a synonym for the agricultural *intelligent* …. What [type of] school is preferable for the education of a future local agronomist? We will answer: a … 'special one,' kindly asking to that special be distinguished from professional." To Fortunatov, an ideal student possessed the special knowledge of theoretical agronomy and the skills of an independent researcher, while also being a conscious citizen. Somehow, professional qualities were reduced in the mind of Fortunatov to bazaar experience: "We think, that the bazaar will teach commercial skills better than the laboratory." See A. Fortunatov, "O podgotovke mestnogo agronoma," *Agronomicheskii zhurnal*, no. 7 (1913): 5, 6, 13. Fortunatov, a founding father of the public agronomy project, was the adamant keeper of the original testament. In 1916 he was literally repeating words he had said a decade before: "in our opinion, the *scientific* school in general is hostile to any professional orientation. See A. F[ortuna]-tov, "Review of L. Sokal'skii,

Obshchestvennaia agronomiia i nashi agronomicheskie uchebnye zavedeniia (Odessa, 1905)," *Russkie vedomosti* XVIII, no. 255 (September 19, 1905).

12. See N. G. Kovalenko, "S"ezd deiatelei po sel'skokhoziaistvennomy obrazovaniiu v Moskve," *Sel'skokhoziaistvennoe obrazovanie*, no. 2 (1914): 52.

13. For instance, the teacher at the Viatka technical secondary school, I. P. Lashkevich, argued in favor of syncretism of the agronomists' training: "The agronomist is a symbiosis of a public man, plants, and animals .... Agronomy cannot be split into parts. There can be some specialists in agronomy, but the leading role should belong to agronomists." See *Trudy agronomicheskogo soveshchaniia pri Viatskoi gubernskoi zemskoi uprave 27 fevralia–6 marta 1914 goda*, p. 5.

14. See Fridolin, *Ispoved' agronoma*, pp. 46–53, 70, 75; *Trudy Viatskogo raionnogo agronomicheskogo soveshchaniia pri Viatskoi gubernskoi zemskoi uprave 25–28 avgusta 1913 goda* (Viatka, 1914), pp. 4, 26, 18.

15. K. F. Golovin, "Budto ne za chto priniatsia," *Ekonomist Rossii*, no. 10 (April 18, 1909): 8.

16. See N. P. Il'inskii, "Neskol'ko myslei o zhelatel'noi uchebnoi obstanovke Kazanskogo zemledel'cheskogo uchilishcha," in *Trudy Kazanskogo oblastnogo s"ezda predstavitelei gubernskikh zemstv*, pp. 30, 35–37.

17. See N. V. Utekhin, "Po voprosu o shkol'nom sel'skokhoziaistvennom obrazovanii," in *Trudy Kazanskogo oblastnogo s"ezda predstavitelei gubernskikh zemstv*, pp. 74–76, 82.

18. See I. Dolgikh, "Iz praktiki prepodavaniia obshchestvennoi agronomii," *Sel'skokhoziaistvennoe obrazovanie*, no. 3 (1915): 192–201; no. 4 (1915): 196–97.

19. See Iaroshevich, "Osnovnye printsipy uchastkovoi agronomicheskoi organizatsii," 134, 135. The precinct agronomist Iaroshevich had just begun his professional career (in 1911), and apparently referred to his own experience claiming that agricultural colleges produced specialists "with negligible practical training."

20. On the traditional studenthood ethos and radical subculture, see Susan K. Morrissey, *Heralds of Revolution: Russian Students and the Mythologies of Radicalism* (Oxford: Oxford University Press, 1998).

21. See GARF, f. 102, D-4, op. 121, d. 122 "O S.-Peterburgskikh sel'skokhoziaistvennykh kursakh," ll. 2–3 ob.

22. "Iz deiatel'nosti kruzhkov," *Vestnik zhizni slushatelei i slushatel'nits SPb. Sel'skokhoziaistvennykh kursov*, no. 3 (November 1913): 22.

23. Less unanimously, members of the circle criticized the idea of priority treatment of small-size farms by agronomists, and only one of them advocated the necessity for agronomists to concentrate entirely on the cooperative activity. See "Iz deiatel'nosti kruzhkov," 23–24.

24. After 1912, the circle of public agronomy in the institute, which had 120 members in 1913, was actually run by Alexander Chaianov. He was, apparently, the author of the circle's report for the 1913–14 academic year, which admitted the inadequacy of practical training at the institute. See "Otchet pravleniia," in *Otchet o sostoianii Moskovskogo sel'skohoziaistvennogo instituta za 1914 god. Kruzhok obshchestvennoi agronomii pri Moskovskom sel'skokhoziaistvennom institute: Otchet o deiatel'nosti za 1913–1914 uch. g.* (Moscow, 1915), pp. 6, 8. Presentations made by members of the circle emphasized the necessity of acquiring as many practical skills as possible while at the institute. See S. Elenevskii, "K voprosu o tekhnike i tekhnicheskoi podgotovke agronomov," in *Otchet o sostoianii Moskovskogo sel'skohoziaistvennogo instituta za 1914 god. Kruzhok obshchestvennoi agronomii pri Moskovskom sel'skokhoziaistvennom institute*, pp. 48–53.

25. On March 5, 1901, during a search in the home of the Kazan Veterinarian Institute student Boris Riurikov, the gendarme lieutenant-colonel Burago confiscated the statutes of a veterinarian circle. According to the statutes, the goal of this veterinarian circle consisted in no less than "developing a conscious, intellectually and morally cultivated individual personality—a public activist-veterinarian." Among the means accepted by the circle members, section 2D of the statutes mentioned "a possibly extensive acquaintance with social literature, political, economic, historical, geographical, and ethnographic conditions of the Russian people." However noble the intentions of the members of that veterinarian circle, professional self-improvement was the least of their concerns. See NART, f. 199, op. 1, d. 125 "Protokoly ob obvinenii lits i postanovleniia ob osvobozhdenii iz pod strazhi i zakliucheniia pod osobyi nadzor politsii," ll. 33, 37.

26. For instance, the circle of public agronomy at the Moscow Agricultural Institute was established on November 6, 1907. It had 65 members in the 1907–08 academic year, 83 members in 1908–09, 120 in 1909–10, and about 200 members annually as of 1911, which accounted for about 17 percent of all students. See Gerken, "Kruzhok obshchestvennoi agronomii pri Moskovskom sel'skokhoziaistvennom institute," *Studencheskoe delo*, no. 2 (February 1912): 54. Between 1907 and 1910, 28 student circles and associations emerged in the Moscow Agricultural Institute, and in 1914, there were altogether 30 student circles. See *Otchet o sostoianii Moskovskogo sel'skohoziaistvennogo instituta za 1910 god*, pp. 37–38; *Otchet o sostoianii Moskovskogo sel'skohoziaistvennogo instituta za 1914 god*, p. 125. In April 1909, a student circle of agronomy was organized even at St. Petersburg University, where, from the very beginning, it attracted two-thirds of all agricultural students. See "Studencheskii agronomicheskii kruzhok pri Sanktpetersburgskom universitete," *Sel'skokhoziaistvennoe obrazovanie*, no. 5 (1914): 287.

27. Cf. "Circles…finish and deepen the kind of training, which is required outside the school walls." See "Moskovskii sel'skokhoziaistvennyi institut. Iz zhizni kruzhkov," *Vestnik zhizni slushatelei i slushatel'nits SPb. Sel'skokhoziaistvennykh kursov*, no. 3 (November 1913): 17. Student circles also played an important role as career counseling agencies. See Gerken, "Kruzhok obshchestvennoi agronomii pri Moskovskom sel'skokhoziaistvennom institute," pp. 54–55.

28. See "Vnutrennee obozrenie," *Agronomicheskii zhurnal*, no. 7 (1913): 112, 127.

29. "Vnutrennee obozrenie," *Agronomicheskii zhurnal*, no. 2 (1913): 84–85.

30. For instance, see Minin, "Agronomiia i zemleustroistvo v ikh otnoshenii k derevenskoi bednote"; V. Brunst, "O liniiakh naimen'shego soprotivlenia v agronomii," *Zemskii agronom*, no. 7 (1913): 3–5.

31. By 1914, only one Samara precinct agronomist worked exclusively with peasants living in communes, 44 percent of agronomists assisted predominantly individual farmers, while 51 percent equally consulted individual farmers and commune members. At the same time, 89 percent of Samara agronomists considered individual farmers to be the most suitable clients of the agronomists, and only 5 percent of agronomists believed that commune members were a better milieu for agronomist improvement measures. See A. V. Teitel, "Zemskaia agronomicheskaia organizatsiia Samarskoi gubernii, sovremennoe sostoianie i zaprosy," in *Trudy 4-go Samarskogo gubernskogo agronomicheskogo soveshchania 16-20 dekabria 1913 goda*, p. 55.

32. Cf. "Vnutrennee obozrenie," *Agronomicheskii zhurnal*, no. 7 (1913): 117; Iu. S. Eremeeva, "O nedelimosti melkogo zemlevladeniia," *Agronomicheskii zhurnal*,

no. 2 (1914): 18–19; A. Kaufman, "Zemskaia sel'skokhoziaistvennaia statistika i agronomicheskii personal," *Zemskii agronom*, no. 1 (1916): 7.

33. See E. N. Sakharova-Vavilova, "Dnevnikovye zapisi," in RGAE, f. 328, E. N. Sakharova-Vavilova, op. 1, ed. khr. 8, l. 35.
34. Louis Guy Michael, "Russian Experience, 1910–1917," vol. 1, pp. 1–2, a manuscript in the Hoover Institution Archive, Louis Guy Michael Collection.
35. See True, *A History of Agricultural Extension Work in the United States*, pp. 28–30.
36. See "Khronika," *Agronomicheskii zhurnal*, no. 8 (1913): 173.
37. During the first half of 1914 alone, the Siberian train attracted 14,484 people, and the Vladikavlaz train about 20,000 people, while the Moscow agronomist train ran through 7 provinces during that year, and was attended by 45,000 peasants. See *Sel'skokhoziaistvennoe obrazovanie*, no. 10 (1915): 490.
38. See "Khronika," *Vestnik obshchestvennoi veterinarii*, no. 8 (1914): 440.
39. See "Chteniia v poezdakh," *Sel'skokhoziaistvennoe obrazovanie*, no. 4 (1914): 239–40.
40. See GARF, f. 102, D-4, op. 122, d. 128 "Ob ustroistve v poezde zheleznoi dorogi peredvizhnoi agronomicheskoi vystavki po glavnoi i vetviam Vladikavkazskoi zheleznoi dorogi," l. 1.
41. See *Veterinarnyi fel'dsher*, no. 8 (August 1915): 142.
42. See "Khronika," *Sel'skokhoziaistvennoe obrazovanie*, no. 5–6 (1917): 220.
43. V. Benzin, "Osennii reis 1916 goda agronomicheskogo poezda Vladikavkazskoi zh. d.," *Zemledel'cheskaia gazeta*, no. 5 (February 4, 1917): 124–25.
44. GARF, f. 102, D-4, op. 124, d. 133 "O deiatel'nosti agronomicheskogo poezda Obshchestva Moskovsko-Kazanskoi zheleznoi dorogi," ll. 4 ob., 6–6 ob.
45. For example, in 1914, a special train car was employed by the Educational Department of the Ministry of Communications to host free mobile courses in apiculture, gardening, and horticulture. The Pavlovgrad district zemstvo (Ekaterinoslav province) organized two mobile agronomist precincts in train cars. See *Veterinarnyi fel'dsher*, no. 4 (April 1915): 63, 64. In the summer of 1914, the Don Agricultural Society launched a floating agricultural exhibition on a boat without an engine, provided by the private company E. T. Paramonov's Sons. The exhibition covered a broad range of topics, from veterinary to cooperation, and in the 35 days ending August 11, it was attended by 12,000 people. See "Khronika," *Sel'skokhoziaistvennoe obrazovanie*, no. 4 (1915): 222–23.
46. In the early summer of 1917, the board of the Moscow–Kazan Railway Company informed the Department of Agriculture that it had to discontinue the agronomist train under the current circumstances on the road. See *Sel'skokhoziaistvennoe obrazovanie*, no. 7–8 (1917): 289. At the same time, the Vladikavkaz train continued its activity. Its last trip was scheduled to start on January 15, 1918. It was planned that the train would run through 36 stations, and the main emphasis of the crew's activity would be on selling seeds, agricultural tools, and spare parts. See "Khronika," *Iugo-vostochnyi khoziain*, no. 1 (1918): 14.
47. The infamous "Trotsky train" was perhaps a replica of the train of Prime Minister Petr Stolypin during his trip to Siberia, and even more directly of the Grand Duke Nikolai Nikolaevich's train of the war period. The agricultural trains became a model for a different type of mobile expertise. For example, the Lenin literature-instructor train was used during forceful procurement campaigns. During its trip in the autumn of 1919 from Simbirsk to Cheliabinsk, it was even equipped with documentary films on agricultural topics, as all prerevolutionary

agronomist trains had been. See "Poezd im. Lenina," in *Vestnik VTsIK*, no. 190 (August 28, 1919): 3. At the same time, the special train of the VChK (the first name of the infamous Soviet secret service) ran in the front area, punishing deserters. See "Deiiatel'nost' poezda VChK," *Vestnik VTsIK*, no. 184 (August 21, 1919): 3. The train of the VChK was followed by the train of the October Revolution with Soviet dignitary Mikhail Kalinin on board, representing the central government and releasing from jails those who had not been shot by the "specialists" of the VChK train. See "Poezd Oktiabr'skoi revolutsii VTsIK (beseda s tov. Kalininym)," *Vestnik VTsIK*, no. 194 (September 3, 1919): 1. Still, the final profanation of the idea of agricultural trains occurred a few years later, when at the peak of the 1921 famine, an "agronomist train" was running through the countryside of Kazan province. Its few cars did not contain any food for the starving peasants, only agricultural expositions and museums. See *Kazanskii zemledelets*, no. 5–8 (May–August, 1921): 94.

48. *Otchet o deiatel'nosti i sostoianii sredstv Samarskogo obshchestva*, p. 30.
49. See *Kalendar'-spravochnik zemskogo deiatelia na 1915 god* (Petrograd, 1914), p. 234.
50. See *Sel'skokhoziaistvennoe obrazovanie*, no. 4 (1914): 241–42.
51. See GARF, f. 102, D-4, op. 124, d. 133, l. 4 ob.; "Khronika," *Agronomicheskii zhurnal*, no. 8 (1913): 173.
52. The compact and safe products by Cock were particularly popular. A 1.5 sq. m. screen cost 27 rubles, a film projector cost 175 rubles, an amateur camera was 260 rubles, while a professional camera cost 900 rubles—half of a precinct agronomist's annual salary. Pathé company sold new films at a price of 22 kopecks per meter, which made an average 100-meter film cost some 22 rubles. There was also the option of renting a film for a few rubles, as we saw in the case of the Romen district zemstvo. See P. I. Surov, "Kinematograf," *Sel'skokhoziaistvennoe obrazovanie*, no. 5 (1914): 267, 269, 271, 272.
53. See Surov, "Kinematograf," 266–67.
54. *Katalog nesgoraemykh lent Kinematografa dlia vsekh. General'naia kompaniia Br. Pate. Otdel Kok* (Moscow, n.d.), pp. 117–24. This catalogue was published in the second half of 1916, definitely after July 14.
55. See *Sel'skokhoziaistvennoe obrazovanie*, no. 4 (1915): 223; "Zemskie kinematografy," in *Kalendar'-spravochnik zemskogo deiatelia na 1915 god*, p. 234.
56. The provincial zemstvo allocated a huge sum of 16,000 rubles for interest-free credits to the villages wishing to purchase movie projectors and films. The zemstvo reserved the right to control the repertoire of cinemas established with its financial aid. See "V zemstvakh," *Volostnoe zemstvo*, no. 5 (1917): 167.
57. The Kineshma zemstvo (Kostroma province), the Cherdynsk zemstvo (Perm province), and the Voronezh zemstvo (Voronezh province). See *Sel'skokhoziaistvennoe obrazovanie*, no. 12 (1915): 651; "Zemskie kinematografy," p. 234.
58. For example, in 1916, the Department of Agriculture granted 700 rubles to the Kazan Society of Apiculture to purchase a Cock film projector and six films about apiculture. The society's film showings in hospitals were a success. See V. I. Loginov, "O chteniiakh v lazaretakh," *Zhurnal Kazanskogo obshchestva pchelovodstva*, no. 1 (1917): 25.
59. A consumer cooperative in the small township of Kellerovo, Vladimir province (1,800 inhabitants in 1897) opened a cinema of its own in 1915, and although the tickets were sold very cheaply the annual profit totaled 984 rubles. See A. M. Popov, "O kul'turno-proizvodstvennoi rabote kooperativov," in *IV Gubernskoe soveshchanie po voprosam kooperatsii. 4–5 sentiabria 1916 g.*

Protokoly, doklady, postanovleniia (Ufa: Ufa provincial zemstvo small credit bank, 1916), p. 52.

60. See *Sel'skokhoziaistvennoe obrazovanie*, no. 2 (1917): 95; B. A., "Zametka o demonstratsii traktorov na Bezenchukskoi sel'skokhoziaistvennoi stantsii," *Samarskii zemledelets*, no. 22 (November 15, 1916): 613–15.

61. When in the winter of 1908–09, the Ekaterinoslav Zemstvo Agency in the United States received a request from Russia to provide details about the self-propelling plowing machines and tractors, the agency could find information only about the first experimental models. Apparently, the expectations were far ahead of reality. See *Otchet starshego agenta Ekaterinoslavskogo Zemstva v Soedinennykh Shtatakh I. B. Rozena za vremia s 1-go oktiabria 1908 g. po 1-e marta 1909 g* (Ekaterinoslav: Tipografiia Gubernskogo Zemstva, 1909), pp. 54–55. In the early 1910s, professional periodicals sympathetically described the newest agricultural machines, such as tractors or grain combines, without much hope of their introduction in Russia in the near future. For example, see "Khronika," *Samarskii zemledelets*, no. 12 (July 1, 1911): 339.

62. The government agricultural agency in the United States purchased and sent to Russia two tractors in 1911, accompanied by a mechanic. *Obzor deiiatel'nosti Sel'skokhoziaistvennogo agentstva v Severo-Amerikanskikh Soedinennykh Shtatakh s 1 iiulia 1909 g. Po 1 ianvaria 1912 g.* (St. Petersburg: GUZiZ, Department of Agriculture, 1913), p. 206.

63. In 1915, in Kursk province alone, 53 zemstvo renting stations (a few per agronomist precinct) could not meet the peasants' demand for tools and machines. See K. U., "Prokatnye stantsii v Kurskoi gubernii," *Mashina v sel'skom khoziaistve*, no. 3 (February 15, 1917): 115–17. Unlike the southern provinces of Samara and Poltava, the extensive system of land cultivation, which presupposed vast landholdings even by peasants, was not characteristic of Kursk province. Still, peasants experienced an urgent need to intensify their agricultural production by means of special tools and machines. Between 1906 and 1912, the import of agricultural machines had grown fourfold, while domestic production was also on the rise. See V. Kostrovskii, "Kursy remonta i postroiki sel'skokhoziaistvennykh mashin pri Saratovskom Aleksandrovskom remeslennom uchilishche," *Sel'skokhoziaostvennoe obrazovanie*, no. 1 (1917): 24. See also George S. Queen, "The McCormick Harvesting Machine Company in Russia," *Russian Review* 23, no. 2 (1964): 164–81; S. V. Kalmykov, "Amerikanskoe predprinimatel'stvo v Rossii," in V. I. Bovykin, ed., *Inostrannoe predprinimatel'stvo i zagranichnye investitsii v Rossii: Ocherki* (Moscow: ROSSPEN, 1997), pp. 257–65.

64. Cf. N. Blagoveshchenskii, "Tri derevenskie zadachi (Iz Shchigrovskogo uezda)," *Zemledel'cheskaia gazeta*, no. 6 (February 11, 1917): 139–40. According to the detailed calculations of one landowner, the cost of one *desiatina* plowed by a tractor was 16 rubles (including all service costs plus housing, utilities, and salaries of two mechanics), while the same land plowed by oxen was 19 rubles. See T. Ruzskii, "Eshe o traktorakh," *Mashina v sel'skom khoziaistve*, no. 6–7 (April 15, 1917): 253–56. According to another landowner, the cost of one *desiatina* plowed by a light tractor was 3.3 rubles, or 5–6 rubles a *desiatina* counting the workers' wages. See *Zemledel'cheskaia gazeta*, no. 8 (February 25, 1917): 180.

65. See *Zemledel'cheskaia gazeta*, no. 4 (January 28, 1917): 103.

66. Ibid., 104; N., "K voprosu o vypiske Depatamentom Zemledeliia iz Ameriki traktorov," *Samarskii zemledelets*, no. 24 (December 15, 1916): 673.

67. See *Zemledel'cheskaia gazeta*, no. 2 (January 14, 1917): 17; N., "K voprosu o vypiske Depatamentom Zemledeliia iz Ameriki traktorov," *Samarskii zemledelets*, no. 24 (December 15, 1916): 673–74.

68. Besides 378 tractors ordered by the Department of Agriculture and the Western Zemstvos, the War Ministry transferred 82 tractors for agricultural purposes: 27 tractors to the Kuban procurement committee, 17 to the Ekaterinoslav committee, and so on. The tractors were organized into units that were employed in accordance with a committee plan. See "Snabzhenie sel'skokhoziaistvennymi traktorami," *Zemledel'cheskaia gazeta*, no. 32–34 (September 9, 1917): 612.

69. See G. A. Shcherbinin, "Neotlozhnye nuzhdy sel'skogo khoziaistva v sviazi s oboronoi strany," *Novyi Ekonomist*, no. 49 (December 3, 1916): 8–11. Grigorii Scherbinin graduated from the Higher School of Agriculture in Vienna around 1910, and took a job within the Department of Agriculture in 1911. See *Spisok lits, sluzhashikh po vedomstvy Glavnogo upravlenia zemleustroiistva i zemledeliia na 1912*, col. 76.

70. The General Meeting of the All-Russian Chamber of Agriculture (November 28–December 1, 1916) accepted a special plan for the "immediate development" of countryside tractorization. The plan included an ambitious import program, education of mechanics, building of tractor plants in Russia and, contrary to the Stolypinist legacy, it also envisioned "studying the question of the artel [communal] land cultivating by tractors." See "Postanovlenie Chrezvychainogo obshchego sobraniia Vserossiiskoi sel'skokhoziaistvennoi palaty 28 noiabria–1 dekabria 1916 g.," *Vestnik sel'skogo khoziaistva*, no. 3 (January 22, 1917): 13.

71. A typical course for army privates in early 1917 trained 57 men in the basics of arithmetic, geometry, and physics, and gave them a good practical knowledge of three or four types of tractors. Each of the students received 25 hours of actual work experience as a tractor operator. See Belianchikov, "Neskolko slov o podgotovke traktornykh motoristov," *Mashina v sel'skom khoziaistve*, no. 6–7 (April 15, 1917): 226–33.

72. See *Zemledel'cheskaia gazeta*, no. 8 (February 25, 1917): 180; Belianchikov, "Neskolko slov o podgotovke traktornykh motoristov," p. 230.

73. See Fedor Kamenskii, "Pis'mo v redaktsiiu," *Samarskii zemledelets*, no. 24 (December 15, 1916): 693–94.

74. See Kostrovskii, "Kursy remonta i postroiki sel'skokhoziaistvennykh mashin pri Saratovskom Aleksandrovskom remeslennom uchilishche," pp. 24–30.

75. See "Plany ustroistva trudovykh druzhin uchashchikhsia," in *Ttrudovye druzhiny uchashchikhsia i ikh trudoustroistvo. Sbornik spravochnykh svedenii, sostavlennykh Komissiei po vneshnemu sel'skokhoziaistvennomu obrazovaniiu, sostoiashchei pri Departamente Zemledeliia* (Petrograd: Departament Zemledeliia, 1916), pp. 8–10.

76. I. Meshcherskii, "Trudovye druzhiny uchashchikhsia," in *Ttrudovye druzhiny uchashchikhsia i ikh trudoustroistvo*, p. 2.

77. "Pravila o trudovykh druzhinakh uchashchikhsia," in *Ttrudovye druzhiny uchashchikhsia i ikh trudoustroistvo*, p. 22.

78. See M., "Sel'skokhoziaistvennye trudovye druzhiny uchashchikhsia," *Sel'skokhoziaistvennoe obrazovanie*, no. 2 (1917): 58–59.

79. See K., "Viazovskaia uchenicheskaia trudovaia druzhina," *Sel'skokhoziaistvennoe obrazovanie*, no. 3–4 (1917): 140–42.

80. See *Sel'skokhoziaistvennoe obrazovanie*, no. 2 (1917): 93–94.

81. See "Rabota shkol'nykh druzhin," *Iugo-vostochnyi khoziain*, no. 4 (1918): 11.

82. See *Sel'skokhoziaistvennoe obrazovanie*, no. 2 (1917): 93–94.

83. There are only indirect indicators of the actual scale of this movement. For example, in 1916, the Department of Agriculture spent 43,650 rubles in subsidies to the agricultural school brigades. The detailed budget of the Don region school brigades mentioned 8,000 rubles of subsidies from the Department of Agriculture to support 229 school students, or about 35 rubles per student. If this rate was typical of other regions, the total sum spent by the Department of Agriculture could sustain 1,250 students, organized in many dozens of agricultural brigades. See "Khronika," *Sel'skokhoziaistvennoe obrazovanie*, no. 11 (1916): 569; "Rabota shkol'nykh druzhin," *Iugo-vostochnyi khoziain*, no. 4 (1918): 10.
84. See *Sel'skokhoziaistvennoe obrazovanie*, no. 7–8 (1917): 290.

8 Nation as Motherland

1. "Vnutrennee obozrenie," *Agronomicheskii zhurnal*, no. 9–10 (1913): 137.
2. On this current in Russian studies, see "In Search of a New Imperial History," *Ab Imperio*, no. 1 (2005): 33–56.
3. See Alexander Semyonov, "Empire as a Context Setting Category," *Ab Imperio*, no. 1 (2008): 193–204.
4. See Ernest Gellner, *Language and Solitude: Wittgenstein, Malinowski, and the Habsburg Dilemma* (Cambridge: Cambridge University Press, 1998).
5. See A. V. Chaianov, "Letter to A. A. Rybnikov, A. N. Chelintsev, and N. P. Makarov," in Chaianov, *A. V. Chaianov—chelovek, uchenyi, grazhdanin*, p. 120.
6. See Dillon, "The Rural Cooperative Movement," p. 466.
7. See "Pis'mo I. V. Emel'ianova I. K. Okulichu v Boston ot 6 sentiabria 1920 g.," in Museum of Russian Culture Archival Collection, I. V. Emel'ianov Collection, box no. 2.
8. See Oleg Budnitskii, ed., *Evrei i russkaia revolutsiia* (Moscow: Gesharim, 1999), p. 293.
9. For instance, Alexander Chaianov was a famous collector of Russian antique art; his outstanding expertise is reflected in the survey "Sobiratel'stvo v staroi Moskve," *Sredi kollektsionerov*, no. 1 (1922): 26–29. Chaianov was among the founders of that journal [Among the Collectors].
10. See *Otchet o deiatel'nosti i sostoianii sredstv Samarskogo obshchestva . . .* p. 29.
11. Cf. Marina Loskutova, "A Motherland with a Radius of 300 Miles: Regional Identity in Russian Secondary and Post-Elementary Education from the Early Nineteenth Century to the War and Revolution," *European Review of History* 9, no. 1 (January 2002): 7–22.
12. Studies of attempts to forge a pan-imperial political community by the institutes of officialdom are not numerous. See Heinz-Dietrich Löwe, "Russian Nationalism and Tsarist Nationalities Policies in Semi-Constitutional Russia, 1905–1914," in Robert B. McKean, ed., *New Perspectives in Modern Russian History* (Basingstoke, UK: Macmillan, 1992), pp. 250–77; Geoffrey Hosking, *Russia: People and Empire, 1552–1917*, enlarged ed. (Cambridge, MA: Harvard University Press, 1998).
13. For instance, the government modernizers decided to "Russify" the rising cooperative movement in an attempt to secure control over it and limit its democratic political potential. Instead of the Western-sounding term "rich" in subversive connotations, a pseudo-Russian word "splotchina" was coined, allegedly "more comprehensible" to the common folk and "genuinely Russian." Publishers of the new government periodical *Splotchina* suggested calling members of

cooperatives "splotniks," and cooperative managers "splotchiks." See "Ot redaktsii," *Splotchina*, no. 1 (1909): 1. Needless to say, not a single cooperative activist actually used this "new Russian" cooperative language.

14. See *Doklad Komissii po peresmotru torgovogo dogovora s Germaniei 49-my ocherednomy Kazanskomu gubernskomu zemskomu sobraniiu* (Kazan, 1913[?]), p. 1.

15. Ibid., p. 6.

16. On the Russo-German trade treaty and the public discussion of its conditions in 1911–14, see Merle Wesley Shoemaker, "Russo-German Economic Relations, 1850–1914" (Ph.D. diss., Syracuse University, 1979), pp. 274, 276–78; Iu. F. Subbotin, *Rossiia i Germaniia: partnery i protivniki: torgovye otnosheniia v kontse XIX v.–1914 g.* (Moscow: Institut rossiiskoi istorii RAN, 1996), 196–208, et seq.

17. "Ob assignovanii Komissii po peresmotru torgovogo dogovora s Germaniei i Avstro-Vengriei," in *Doklady Bessarabskoi gubernskoi zemskoi upravy po agronomicheskomu otdelu 44-my ocherednomu gubernskomu zemskomu sobraniiu*, p. 196.

18. For instance, see M. N. Sobolev, "Ekonomicheskoe znachenie russko-germanskogo dogovora," *Agronomicheskii zhurnal*, no. 2 (1913): 24–30; D. Piskunov, "Russkaia pechat' o novom torgovom dogovore s Germaniei," *Agronomicheskii zhurnal*, no. 2 (1913): 69–82.

19. Cf. N. D. Beliaev, "Rasshirenie tekhniki v sel'skom khoziaistve," *Ekonomist Rossii*, no. 7 (February 20/March 5, 1910): 7; Sobolev, "Ekonomicheskoe znachenie russko-germanskogo dogovora," pp. 28–29.

20. See Subbotin, *Rossiia i Germaniia*, p. 208.

21. On the impact of Western, and first of all German, economic theory on Russian economic thought, see Stanziani, *L'Économie en révolution*, pp. 45–57, esp. 45, 47.

22. See "Ot redaktora," in N. K. Dokuchaev, ed., *Materialy po izucheniiu uslovii khraneniia i razlozheniia navoza*, vol. 1 (Petrograd: Kabinet obshchego i chastnogo rastenievodstva Stebutovskikh vysshikh zhenskikh sel'skokhoziaistvennykh kursov, 1915), p. 3.

23. For example, see K. M-ch [Konstantin Matseevich], "Metody izucheniia ekonomicheskikh problem v sel'skom khoziaistve," *Agronomicheskii zhurnal*, no. 4 (1913): 106.

24. "Raznye soobshcheniia i izvestiia," *Veterinarnyi fel'dsher*, no. 5 (May 25, 1909): 143. The example of the United States, apparently, was highly appealing to rural "technical specialists." In 1911, the magazine *Veterinarian Assistant* even introduced a regular section "Letters from America," meant to satisfy the interest of veterinarian assistants in the United States where "technicians" played a far more prominent role than in Russia.

25. Viktor Viktorovich Talanov (1871–1936) graduated from the St. Petersburg Forestry Institute and later from the New Alexandria Institute of Agriculture and Forestry, the first choice of agronomists interested in the scientific aspect of the profession. Indeed, Talanov became prominent as an organizer of experimental agricultural stations and research institutes, and he selected new sorts of grain and corn. The zemstvo agricultural agency in the United States was one of his favorite creations; Talanov coordinated the Agency's research activity and edited dozens of voluminous issues of the *Izvestiia zemskoi sel'skokhoziaistvennoi agentury v SASh*. See N. P. Goncharov, "Organizator sistemy gosudarstvennogo sortoispytaniia i vydaiiushchiisia selektsioner (130 let so dnia rozhdeniia V. V. Talanova)," *Informatsionnyi vestnik VOGiS*, no. 20 (2002); available at www.bionet.nsc.ru/vogis/vestnik.php?f=2002&p=20_5 (accessed February 7, 2007).

26. Joseph (Iosif Borisovich) Rosen (1877–1949) was born in Moscow, studied agronomy in Russia and Germany, and joined the Russian Social Democratic Party. As a Menshevik, he was arrested and exiled to Siberia. In 1903 he emigrated to the United States, earned a degree in Michigan, and in 1908 accepted an invitation from Talanov to become the zemstvo agricultural agent in United States. He spent two years in this capacity, fulfilling zemstvo orders and cooperating with Talanov in botanical experiments. They would meet again 15 years later, when Rosen returned to Russia as director of the Agro-Joint (American Jewish Agricultural Corporation) and offered Talanov a position as head of the Bureau for Introduction and Dissemination of New Sorts of Field Plants. See Joseph A. Rosen papers in the archive of the YIVO Institute for Jewish Research in New York. After Rosen, from 1910 to May 1912, Ivan Emel'ianov, whom we already mentioned in this book, served as senior specialist in the Ekaterinoslav Zemstvo Agency in the United States. Before accepting this position, Emel'ianov held the office of Kharkov provincial zemstvo agronomist. See "Curriculum Vitae uchenogo agronoma i razriada Ivana Vasil'evicha Emel'ianova, 1921," in the Museum of Russian Culture Archival Collection, I. V. Emel'ianov Collection, box no. 2.

27. Morachevskii, *Spravochnye svedenia o deiatel'nosti zemstv po sel'skomy khoziaistvu (po dannym na 1909 god)*, p. 98. Vengerovskii died four months after his arrival in the United States in October 1908. See *Otchet starshego agenta Ekaterinoslavskogo Zemstva v Soedinennykh Shtatakh*, p. 2.

28. Morachevskii, *Spravochnye svedenia o deiatel'nosti zemstv po sel'skomy khoziaistvu*, vol. 11, p. 99.

29. *Otchet starshego agenta Ekaterinoslavskogo Zemstva v Soedinennykh Shtatakh*, p. 38.

30. "Khronika," *Bessarabskoe sel'skoe khoziaistvo*, no. 5 (1912): 150.

31. Morachevskii, *Agronomicheskaia pomoshch' v Rossii*, p. 141. Government plans to establish an agricultural agency in the United States went back to 1901–02, but it was not until the beginning of the Stolypin reforms, and, most important, the emergence of the zemstvo agency, that the government bureaucracy hurried to implement these plans. See Raymond A. Heider, "Russian-American Agricultural Agency," in Joseph L. Wieczynski, ed., *The Modern Encyclopedia of Russian and Soviet History*, vol. 32 (Gulf Breeze, FL: Academic International Press, 1983), p. 33.

32. See *Obzor deiatel'nosti sel'skokhoziaistvennogo agentstva v Severo-Amerikanskikh Soedinennykh Shtatakh s 1 iiunia 1909 g. po 1 ianvaria 1912 g.* (St. Petersburg: GUZiZ, Departament Zemledeliia, 1913), p. 8.

33. See Hoover Institution Archive, Russia. Posol'stvo (US), box no. 328.

34. *Obzor deiatel'nosti sel'skokhoziaistvennogo agentstva*, pp. 3–4.

35. Iosif Rosen of the zemstvo-founded Agency studied agronomy in three countries, and already in the United States developed a new variety of winter rye that was named after him. Officially registered in 1912, the "Rosen rye" was regarded as "the most productive rye for Michigan" even a decade later. Cf. H. C. Rather, "Setting a Standard for Seed," *Extension Bulletin*, no. 34 (March 1924): 7. When Ivan Emel'ianov became senior specialist at the zemstvo agency, he had already received the diploma of agronomist of the first degree from the Kiev Polytechnic in 1907, and five years of prior work experience. Fedor Kryshtofovich, senior specialist at the government agency, had graduated from Kiev University around 1907 with the diploma of physician, and before taking the job in the agency had only two years' experience as an agronomist in the steppe region. (Raymond A. Heider mistakenly identified Kryshtofovich as a former "governor general," in "Russian-American Agricultural Agency," p. 34.) At the same

time, Emel'ianov's salary was 3,500 rubles or $1,750 per annum, while Kryshto-fovich received 6,000 rubles or $3,000 (an American agricultural specialist of their level, a county agent, earned some $2,000 annually). Kryshtofovich was probably chosen for the job because of his prior experience in America: he came to the United States in 1894 with his wife and six children, spent over ten years on a farm near Pomona, California, and returned to Russia sometime after 1905. His first encounter with the United States was described in F. F. Kryshtofovich, *Amerikanskie pis'ma* (St. Petersburg: Izdanie zhurnala *Nauka i zhizn'*, 1905). Thus, the generational parallelism between Emel'ianov and Kryshtofovich only underlined the difference between their employer agencies. See "Curriculum Vitae uchenogo agronoma i razriada Ivana Vasil'evicha Emel'ianova, 1921;" Morachevskii, *Spravochnye svedenia o deiatel'nosti zemstv po sel'skomy khozi-aistvu*, vol. 11, p. 99; *Obzor deiatel'nosti sel'skokhoziaistvennogo agentstva*, p. 4. In 1913–14, new and much more experienced specialists were appointed to the Government Agricultural Agency, while their salaries were tremendously increased: in 1916, the senior agent, agronomist William Anderson, received 7,000 rubles per year. His assistant, agronomist Viacheslav Kochetkov, was paid 5,000 rubles. "Vedomost' lichnogo sostava Otdela Ministerstva zemledeliia pri Russkom zagotovitel'nom komitete v Amerike," Hoover Institution Archive; Russia. Posol'stvo (US), box no. 328, folder 1.

36. Cf. *Obzor deiatel'nosti sel'skokhoziaistvennogo agentstva*, p. 204.
37. Louis Guy Michael, "Russian Experience, 1910–1917," vol. 1, 2, manuscript in the Hoover Institution Archive, Louis Guy Michael Collection.
38. Ibid. The memoirs of L. G. Michael are written in the style of Mark Twain's *A Connecticut Yankee in King Arthur's Court*, depicting with sympathy and irony the adventures of an American agricultural specialist in the strange world of prewar rural Southern Russia. The memoirs provide a rare opportunity to look at the machinery of a local zemstvo and the work of rural professionals through the eyes of a foreign, yet very well-informed, observer.
39. The interest of the Russian agricultural public toward America in the 1910s was part of the broader phenomenon of the "Americanization" of Russian society, which was reflected, for instance, in popular fiction. See Jeffrey Brooks, *When Russia Learned to Read: Literacy and Popular Literature, 1861–1917* (Princeton, NJ: Princeton University Press, 1985), pp. 144–46. For a general assessment of the impact of the American model on Russian economic moderniza-tion, see H. Rogger, "*Amerikanizm* and the Economic Development of Russia," *Comparative Studies in Society and History*, no. 23 (1981): 382–420. Characteristi-cally, the rise of public interest in America coincided with a period of severing the official relations between the two countries, which once again underlines the degree of independence achieved by Russian civic society after 1905–07. See Jeanette E. Tuve, "Changing Directions in Russian-American Economic Relations, 1912–1917," *Slavic Review* 31, no. 1 (1972): 52–70.
40. Vadim Kukushkin, *From Peasants to Labourers: Ukrainian and Belarusan Immi-gration from the Russian Empire to Canada* (Montreal: McGill-Queen's University Press, 2007), pp. 33, 40–41.
41. Chernigov province became the "largest and earlier emigration cluster" in Ukrainian lands due to its poor quality of soil and long tradition of migrant labor (ibid., p. 39).
42. See A. Mukhin, "Otkhod krest'ian na zarabotki v Ameriku," *Agronomicheskii zhurnal*, no. 9–10 (1913): 148–54.

43. *Pis'ma krestian* (St. Petersburg: Russkoe zerno, 1911); *Pis'ma krestian*, vol. 2, parts 1–2 (Petrograd: Russkoe zerno, 1914, 1915).

44. For instance, in the 1913 manuscript selected for publication, the editor criss-crossed the formal rhetorical preamble (commenting in the margins: "no need") and also, along the text, some excessive or overly personal details. Nothing was added to the original text. See the notebook (of A. Tsyganov) in RGIA, f. 403, op. 2, d. 252, ll. 4–43; "Letter no. 36, Tsyganov, A. M.," in *Pis'ma krestian*, vol. 2, part 1 (Petrograd: Russkoe zerno, 1914), pp. 231–46.

45. In Denmark, Russian peasant-"interns" were generally treated as cheap laborers, and the language barrier hampered their active contact with the local population, while in Southern Europe (in Bulgaria, Serbia, and Croatia) Russian visitors did not see a dramatic difference from Russian realities.

46. Peasant Mikhail Alekseev (27 years old, from Viatka Province), in *Pis'ma krestian*, vol. 2, part 1, p. 41.

47. Peasant Alexander Ivlev (26 years old, from Novgorod province), ibid., p. 95.

48. Peasant Vasilii Stepanov (21 years old, Pskov province), in *Pis'ma krestian* (St. Petersburg: Russkoe zerno, 1911), p. 81.

49. Peasant Mikhail Alekseev (27 years old, from Viatka Province), in *Pis'ma krestian*, vol. 2, part 1, pp. 42–43.

50. Peasant Andrei Kriuchkov (19 years old, Smolensk province), ibid., p. 297.

51. Peasant Aleksei Reva (no data), in RGIA, f. 403, op. 2, d. 252, l. 44.

52. Peasant Vasilii Remezov (27 years old, Saratov province) felt himself among the Latvians "as in the home country" speaking in his "native tongue," as "all the Latvian intelligentsia and youth speak Russian." In *Pis'ma krestian*, vol. 2, part 2, p. 438. Peasant Vasilii Shmutin (22 years old, Tver province) found another formula: Baltic peoples were "stepsons" of "Mother-Rus'," ibid., p. 483. Another Saratov native, the village teacher and farmer, Ivan Filin, was not sure about the "Russianness" of Latvia but his experience in the Baltics made him aware of the complexity of the Russian Empire: "Far away from us to the west, a small people of Latvians live. The region populated by Latvians is usually called the 'Baltic region,' but Latvians call it 'Latvia.' For a long time nothing was known about Latvians, only it is being written and said that Germans lived there. But in fact there is not a single farm of a German peasant" (see *Pis'ma krestian*, vol. 2, part 2, p. 458).

53. Peasant Sergei Khazreev (26 years old, Novgorod province), *Pis'ma krestian* (St. Petersburg: Russkoe zerno, 1911), p. 48.

54. Peasant Stefan Zenchenko (28 years old, Vitebsk province, probably a Belorussian), ibid., p. 107.

55. Peasant Nikita Marusich (32 years old, from Kherson province), *Pis'ma krestian*, vol. 2, part 1, p. 144.

56. Peasant Ivan Staritsky (18 years old, Novgorod province), *Pis'ma krestian* (St. Petersburg: Russkoe zerno, 1911), p. 114.

57. Peasant Mikhail Prokofiev (30 years old, Tver province), *Pis'ma krestian*, vol. 2, part 1 (Petrograd: Russkoe zerno, 1914), p. 387.

58. Peasant Stephan Zenchenko, *Pis'ma krestian*, vol. 2, part 2, p. 505.

59. As was the case with Deacon Alexander Maslov (20 years old, from Tver province), who spent a year in Moravia and considered marrying the daughter of a local village elder. See RGIA, f. 403, op. 2, d. 252, l. 2 ob.

60. After a year or even two spent abroad, those young, usually single peasants spoke Czech well and felt themselves quite at home in the new world that fascinated

them so much. A few peasants mentioned being asked by their Czech hosts to stay for good.

61. As can be seen in the case of Stefan Zenchenko, who combined a local Byelorussian identification with a sense of belonging to "Russia." Cf. *Pis'ma krestian* (St. Petersburg: Russkoe zerno, 1911), pp. 91, 92, 103. While Alexander Karkuzaka (19 years old, native of Poltava province) could be regarded as a Ukrainian, he identified himself with "Russia" and "Russian peasants." See *Pis'ma krestian*, vol. 2, part 1, p. 110.

62. As peasant Andrei Kriuchkov (19 years old , Smolensk province) put it in a letter to his parents, not meant for publication. See *Pis'ma krestian*, vol. 2, part 1, p. 299.

63. Between 1911 and 1914, 15,000 copies of the first volume of *Pis'ma krestian* were sold (*Pis'ma krestian*, vol. 2, part 1, p. xx). Individual letters were published in thin brochures, in print runs of up to 4,000 copies each (cf. RGIA, f. 403, op. 2, d. 253, l. 1). Many peasants admitted that they had read various editions of *Pis'ma krestian* and were impressed by what they read.

64. See Holquist, *Making War, Forging Revolution*, and especially Sanborn, *Drafting the Russian Nation*.

65. "Voina ili mir?" *Novyi ekonomist*, no. 29–30 (July 27, 1913): 1.

66. Cf. "The question of international politics never enjoyed the particular attention of our [intelligentsia] society. Interests of this category did not exist at all in very broad circles.... The shift of concepts and ideas that took place in the consciousness of our society...can be compared to what had happened during the epoch of 1905–6 in the sphere of internal politics." See K. M-kii, "Voina i derevnia," *Agronomicheskii zhurnal*, no. 7–8 (1914): 4, 5. This issue was published sometime after June 1915.

67. This perspective gradually penetrated all aspects of professional discourse. In 1915, the Samara-based *Zemstvo Agronomist* juxtaposed the spontaneous drift of agriculturists toward "intensive West European farming" to the national "Russian agronomy," which was supposedly oriented toward extensive agriculture. "Vnutrennee obozrenie," *Zemskii agronom*, no. 2 (1915): 69.

68. See, for example, "Vnutrennee obozrenie," *Agronomicheskii zhurnal*, no. 4–5 (1914): 133, 144; K. S. Ashin, "Voina i sel'skoe khoziaistvo," *Agronomicheskii zhurnal*, no. 4–5 (1914): 148. This first wartime issue of *Agronomicheskii zhurnal* was published after October 6, 1914.

69. For instance, see N. Chernyshev, "Voina, trezvost' i kooperatsiia," *Kostromskoi kooperator*, no. 13–15 (August 29, 1914): 2–3. The Bugulma district congress of credit cooperatives (Samara Province) was opened on August 6, 1914, in an atmosphere of patriotic élan. After a flamboyant speech by an inspector of credit cooperatives A. Il'in, the audience shouted "hurrah," then the local battalion orchestra played the national anthem, and the cooperative activists decided to send a telegram of greeting to the emperor and General Brusilov—something very untypical of a cooperative forum hitherto. See N. B., "Bugul'minskii s"ezd predstavitelei kreditnykh kooperativov uezda," *Samarskii zemledelets*, no. 15–16 (August 15, 1914): 461.

70. As a matter of fact, the anti-alcohol campaign took over the cooperative movement before the war. See, for instance, W. Arthur McKee, "Sobering Up the Soul of the People: The Politics of Popular Temperance in Late Imperial Russia," in *Russian Review* 58 (April 1999): 212–33. "New peasants" and many cooperative managers supported the prohibition of alcohol, but they could only censure

the cooperatives that were engaged in sales of alcohol. See N. Litakov, "Vino i kooperatsiia (pis'mo krest'ianina)," *Kostromskoi kooperator*, no. 3 (February 1, 1914): 4–5; *Trudy tret'ego kooperativnogo s"ezda Okhanskogo uezda, byvshego v g. Okhanske 28-29 maia 1914 g.* (Okhansk, 1914), pp. 37–38. The war made urgent the economic aspect of the prohibition: how to compensate for approximately one-third of state budget revenues previously received from the sale of alcohol. It was comprehensible, when a cooperative activist enthusiastically claimed that "the sober nation" would find a way to cope with the loss of the "drunk money" (Chernyshev, "Voina, trezvost' i kooperatsiia," p. 3). But when a leading rural economist, A. N. Chelintsev, declared that "from the point of view of agricultural production, we should unequivocally say that the great cause of sobering up the Russian people coincides with the road to general growth of agricultural productivity," he was at least not completely honest with his readers about the complexity of the problem. See A. Chelintsev, "Sel'skoe khoziaistvo i veterinariia," *Agronomicheskii zhurnal*, no. 1 (1915): 60. By mid-1915, the vine-growing industry of Bessarabia experienced a severe crisis because of the prohibition of alcohol sales. See "Khronika," *Bessarabskoe sel'skoe khoziaistvo*, no. 15 (August 1, 1915): 406–7. By 1916, it became apparent that the prohibition of alcohol resulted in severe economic problems. See Z. Ioralova, "Alkogolizm vo Frantsii i sredstva bor'by s nim," *Agronomicheskii zhurnal*, no. 1 (1916): 95.

71. The magazine's section "Chronicle" was especially biased, featuring numerous anecdotes about the humiliation of German and Austrian troops by farm and draught animals. For example, see "Khronika," *Vestnik obshchestvennoi veterenarii*, no. 18 (September 15, 1914): 900–10. While publishing a purely analytical article "How the Separator Affected German Agriculture" in August 1914, the editorial board of the cooperative magazine *Kostroma Cooperative Activist* was so confused that they added a special footnote: "This article was sent to us before the war." See *Kostromskoi kooperator*, no. 13–15 (August 29, 1914): 16. For the cooperative press in general, professional expertise without a certain political or ideological message was a strange animal.

72. See "Vnutrennee obozrenie," *Zemskii agronom*, no. 2 (1915), 109; "Vnutrennee obozrenie," *Agronomicheskii zhurnal*, no. 3–4 (1915): 102–7. Without any excuses, the *Agronomist Journal* published a very analytical and correct article "The War and German Agriculture," in which the author went as far as an appeal to learn the methods of wartime management of agriculture from the Germans, something almost unthinkable at the beginning of the war. See Boris Frommet, "Voina i sel'skoe khoziaistvo Germanii," *Agronomicheskii zhurnal*, no. 7–8 (1914): 110–34.

73. M. Zubrilov, "Geografiia v srednikh sel'skokhoziaistvennykh uchilishchakh," *Sel'skokhoziaistvennoe obrazovanie*, no. 6 (1915): 308.

74. For example, in 1915, the agronomist of the Sviiazhsk land settlement commission (Kazan province), G. F. Meibom, changed his middle name, very important in Russia, from Ferdinand to Fedor. See NART, f. 580, Zaveduiushchii agronomicheskoi pomoshch'iu khoziaistvam edinolichnogo vladeniia Kazanskoi gubernii, op. 1, d. 1, l. 22.

75. This was the case when Alexander Chaianov explained to his friend N. P. Makarov his attitude to the prospect of conscription in 1915: "[I]n August, I will have to exchange the rostrum for the rifle. I do not think this will be socially efficient." See Chaianov, *A. V. Chaianov—chelovek, uchenyi, grazhdanin*, p. 118. The letter was written between February and May 1915.

76. For instance, 15 out of 35 precinct veterinarians of Ufa province had been called up only two months after the beginning of the war. See "Khronika," *Vestnik obshchestvennoi veterinarii*, no. 19 (October 1, 1914): 951. Alexander Chaianov's fear of being drafted was quite realistic: in 1914, seven faculty members and graduate students of the Moscow Agricultural Institute were called up, one of whom was soon killed. In addition, 150 students were drafted (about 11 percent of all), and by 1915, 16 of them had been wounded and four were killed. See *Otchet o sostoianii Moskovskogo sel'skohoziaistvennogo instituta za 1914 god*, pp. 14–15. During the first year of the war, altogether 80 teachers in agriculture schools were conscripted, while almost a quarter of their male students were 19 and older and hence expected imminent conscription. See Ia. Nekludov, "Voina i sel'skokhoziaistvennye shkoly," *Sel'skokhoziaistvennoe obrazovanie*, no. 11 (1915): 541, 542.

77. Already on September 30, 1914, veterinarian students were granted the right of conscription deferment until the end of their studies (like medical students). See "Khronika," *Vestnik obshchestvennoi veterinarii*, no. 19 (October 1, 1914): 993–94. Circular no. 6 of the Main Committee for Granting Conscription Deferments listed the personnel of warehouses, machinery rental stations, and workshops of agricultural machines as enterprises crucial to the defense of the country (and thus exempt from conscription). On November 29, 1916, the Russian Chamber of Agriculture appealed for conscription deferments to all still serving agronomists, agricultural instructors, elders, and other categories of agricultural specialists. See "K voprosu ob otsrochkakh prizyva v voiska i ob otpuskakh iz armii lits, neobkhodimykh v sel'skom khoziaistve," *Novyi Ekonomist*, no. 50 (December 10, 1916): 6–9.

78. Specifically, Circular no. 13 (August 29, 1916) of the Main Committee for Granting Conscription Deferments included managers of small credit banks and executives of the unions of consumer cooperatives in the privileged category. Circular no. 17 (January 20, 1917) extended the exemption from conscription to the unions of flax-growing cooperatives (no more than two deferments per union for managers, and no more than ten deferments per union for flax sorters). See NART, f. 638, Kazanskii gorodskoi komitet po predostavleniiu voennoobiazannym otsrochek, op. 1, d. 1, ll. 92, 192.

79. On December 12, 1916, Kazan provincial zemstvo agronomist, Ivan Shtutser (born in 1882), who began his professional career in 1905, signed a request to grant a conscription deferment to his deputy and coeval, agronomist Ivan Malakhov, while Malakhov signed a similar request on behalf of Shtutser. Both requests were satisfied just two days later. Simultaneously, a request was submitted on behalf of the zemstvo orchards pest control technician, Nikanor Tsvetkov, who was six years older than Shtutser and Malakhov. He did not get a deferment because of his better medical condition, the relative unimportance of his job, or his lower social status. See NART, f. 638, op. 1, d. 1, ll. 29, 31, 38–39.

80. *Otchet o meropriiatiiakh po uluchsheniiu sel'skogo khoziaistva v Okhanskom uezde za vremia s 1-go aprelia 1914 g. po 1-e aprelia 1915 g.*, pp. 1–11.

81. See "Spisok lits, prizvannykh v riady voisk, sem'i kotorykh poluchaiut soderzhanie, ili imeiut pravo na poluchenie ego polnost'iu ili v chasti po dolzhnostiam, koi zanimali prizvannye do postupleniia ikh v voiska (Sostavlen Kazanskim Gubernskim otdeleniem Komiteta Eia Imperatorskogo Vysochestva Velikoi Kniagini Elisavety Feforovny," a printout held in Kazan University Library, pp. 4, 8, 9, 13, 16, 17, 19.

82. Fridolin, *Ispoved' agronoma*, p. 106

83. Each zemstvo had a pay scale of its own. The Birsk district zemstvo (Ufa province), for instance, at the beginning of the war provided for salary increases of up to 40 percent. Zemstvo employees with the lowest salaries and large families received preferential treatment. See *Sbornik dokladov Birskoi uezdnoi zemskoi upravy i postanovlenii zemskikh sobranii 41 ocherednoi sessii 22 sentiabria 1915 goda i chrezvychainogo sobrania XXIV sessii 19 avgusta 1915 goda* (Ufa, 1916), pp. 26–27.

84. The general policy of the government and provincial zemstvos was to pay the prewar salary to all conscripted employees. The salary of a precinct agronomist came from three sources: government funds, the provincial zemstvo, and the district zemstvo. A unilateral decision of a district zemstvo to reduce its share of payments automatically reduced the amount paid by other parties. See A. V. Portugalov, "O zhalovanii zemskikh uchastkovykh agronomov, prizvannykh v voiska," *Zemskoe delo*, no. 22 (November 20, 1914): 1352–54.

85. See Yaney, *The Urge to Mobilize*, pp. 442–47; Matsuzato, "The Fate of Agronomists in Russia: Their Quantitative Dynamics from 1911 to 1916," pp. 185–89, 193–95; Stanziani, *L'Économie en révolution*, p. 175.

86. See Kerans, "Agricultural Evolution," p. 515.

87. Cf. Kimitaka Matsuzato, "The Role of Zemstva in the Creation and Collapse of Tsarism's War Efforts During World War One," *Jahrbücher für Geschichte Osteuropas* 46 (1998): 324.

88. See Kimitaka Matsuzato, " 'Obshchestvennaia ssypka' i voenno-prodovol'st-vennaia sistema Rossii v gody Pervoi mirovoi voiny," *Acta Slavica Iaponica* XV (Sapporo, Japan, 1997): 38–41.

89. When the war broke out, people and institutions in the rear made gestures symbolizing their care for the defenders of the nation. The Bessarabian mid-level college of viticulture donated 300 buckets of wine to military hospitals, while the Moscow Agricultural Institute transformed its flax-growing station into a hospital for 240 men. See Nekludov, "Voina i sel'skokhoziaistvennye shkoly," p. 542. In addition, the Council of the Institute volunteered to donate 3 percent of the Council members' salaries to the wounded, and invited other faculty members to follow suit. See *Otchet o sostoianii Moskovskogo sel'skohoziaistvennogo instituta za 1914 god*, p. 11. One year later, rural educators outlined a complex program of social rehabilitation for war victims by the agricultural specialists: organizing agricultural shelters for orphans (children would be brought up in a healthy countryside environment and would support themselves working on a shelter's farm); teaching wounded soldiers various agricultural trades; and providing female agricultural education. See "Ocherednye zadachi," *Sel'skokhoziaistvennoe obrazovanie*, no. 1 (1916): 14–16.

90. By October 11, 1916, the recently renamed Ministry of Agriculture had authorized 77 agricultural courses for the wounded. Although more than 90 percent of the funds were provided by the state, its share in the organization and staffing of the courses (which also meant control over their curriculum) was minimal. Agricultural societies and cooperatives organized 39 percent of the courses; agricultural schools organized 29 percent; zemstvos—14 percent; state agencies and private sponsors together accounted for 18 percent of the courses. Out of these 77 courses, 16 were dedicated to apiculture; 9—to cooperative accountancy; 8—to stock breeding; 7—to gardening; 7—to agricultural machinery, and so on. See I. M., "Kursy dlia uvechnykh voinov," *Sel'skokhoziaistvennoe obrazovanie*, no. 2

(1917): 54–55. Besides regular and long-term courses, there were also numerous educational lectures given in the hospitals, which were never counted as systematic courses, as was the case with the Kazan Society of Apiculture. See V. I. Loginov, "O chteniiakh v lazaretakh," *Zhurnal Kazanskogo obshchestva pchelovodstva*, no. 1 (1917): 24–26.

91. This dilemma of choosing between considerations of social welfare and socioeconomic efficiency was recognized as a key problem by top cooperative activists during the winter of 1915–16, when they were discussing projects for future cooperative schools and their prospective students. See "Protokol zasedaniia Komissii po soiuznomu stroitel'stvu 6 i 7 ianvaria 1916 goda," *Vestnik kooperatsii*, no. 2 (1916): 94.

92. See *Otchet o sostoianii Moskovskogo sel'skohoziaistvennogo instituta za 1910 god*, p. 33.

93. See *Otchet o sostoianii Moskovskogo sel'skohoziaistvennogo instituta za 1914 god*, p. 93.

94. As agriculture acquired strategic importance and zemstvos rose to power through the All-Russian Union of Zemstvos and through attaining a virtual monopoly on procurement campaigns, it was easier to obtain a conscription deferment working in any agriculture-related zemstvo agency. Even the Department of Agricultural Census of the Kazan provincial zemstvo claimed that it qualified for a conscription exemption as a key defense enterprise. See NART, f. 638, op. 2, d. 1 "Po khodotaistvu otdela sel'skokhoziaostvennoi i pozemel'noi perepisi Kazanskogo gubernskogo zemstva." Professionals in towns faced significantly greater problems. They had to quit profitable jobs and seek protection from powerful Zemstvo and Town Unions. As a result, there appeared lawyers supervising canteens for the poor or assisting junior clerks (NART, f. 638, op. 2, d. 8, ll. 120–121, 234, 410, 441 ob.).

95. See "Vysochaishe utverzhdennoe polozhenie Soveta Ministrov o poriadke predostavleniia voennoobiazannym otsrochek po prizyvam v armiiu v tekushchuiu voinu" (December 6, 1916) in NART, f. 638, op. 1, d. 1, l. 5.

96. The point is substantiated in American historiography by William E. Gleason "The All-Russian Union of Towns and the All-Russian Union of Zemstvos in World War I: 1914–1917" (Ph.D. diss., Indiana University, 1972). Thomas Fallows argued that "it was the government itself which first encouraged zemstvos to become involved in food matters during the war." Thomas Fallows, "Politics and the War Effort in Russia: The Union of Zemstvos and the Organization of the Food Supply, 1914–1916," *Slavic Review* 37, no. 1 (March 1978): 75 and passim. In fact, the government did not have much choice, as there were no alternative structures capable of large-scale socioeconomic mobilization in the country. Kimitaka Matsuzato put forward a powerful and well-supported thesis about the collapse of the tsarist regime in February 1917 as a result of the state's loss of control over socioeconomic mobilization in the country to zemstvos. When local administrative authorities of provincial zemstvos were reinforced by the almost unlimited authority given to zemstvos in collecting and distributing food, the national system of supplies collapsed having been torn apart by the regional interests of the mobilized "local societies" (to use the term of German social historians Lutz Häfner and Guido Hausmann). The hypothesis of Matsuzato corresponds to a conclusion made by Thomas Fallows ("Politics and the War Effort," pp. 89–90), and is supported by recent case studies of local procurement agencies. See Matsuzato, " 'Obshchestvennaia ssypka'

i voenno-prodovol'stvennaia sistema Rossii v gody Pervoi mirovoi voiny," pp. 17–51; idem., "The Role of Zemstva in the Creation and Collapse of Tsarism's War Efforts During World War One," pp. 321–37; Peter Fraunholtz, "State Intervention and Local Control in Russia, 1917–1921: Grain Procurement Politics in Penza Province" (Ph.D. diss., Boston College, 1999), esp. pp. 71–72. Arthur DuGarm also sees the cause of the politcal crisis in the failure of the procurement apparatus (in his case, in Tambov province). See Arthur Delano DuGarm, "Grain and Revolution: Food Supply and Local Government in Tambov, Russia. 1917–1921" (Ph.D. diss., Stanford University, 1998), pp. 23–24.

97. Quoted in Fallows, *Politics and the War Effort*, p. 82.

98. Cf. "Among the possible outcomes of the war, in case of its ending unfavorably for us, would be…the establishment of a…hard economic dependence on Germany…. At this moment, we cannot even approximately define the period of the war's continuation." See M. B-v, "Sel'skoe khoziaistvo i sovremennye obshchestvennye zadachi," *Agronomicheskii zhurnal*, no. 3–4 (1915): 3; "Vnutrennee obozrenie," *Agronomicheskii zhurnal*, no. 7–8 (1915): 113.

99. See "Vnutrennee obozrenie," *Agronomicheskii zhurnal*, no. 5 (1915): 42.

100. In the early spring of 1916, the Pokrovsk district (Vladimir province) Congress of Agriculturists was attended by the guest of honor, a close friend and colleague of Alexander Chaianov in scholarship and cooperative business, Alexander Rybnikov. He argued in favor of a state monopoly on flour distribution until the end of the war. The audience listened to him attentively but no resolution was adopted on Rybnikov's proposal. However, the discussions that followed a couple of other presentations indicated that local cooperative managers and agricultural specialists opposed the idea of a state trade monopoly and fixed prices, or agreed to the latter only under the provision that the prices on all goods, including industrial products, would be fixed. See F. S., "S Pokrovskogo s"ezda sel'skikh khoziaev," *Samarskii zemledelets*, no. 8 (April 15, 1916): 241–44. Apparently, only respect for the status of the Moscow guest prevented local activists from opposing his proposal. By the end of 1916, however, the antimarket mood was already widely expressed by local second element and agricultural specialists. For example, see M. F.[rankfurt], "Kak organizovat' artel'nyi maslodel'nyi zavod," *Samarskii zemledelets*, no. 21 (December 1, 1916): 659–62; Zemskii, "Nash dolg," *Samarskii zemledelets*, no. 24 (December 15, 1916): 671–72.

101. In 1909, the renowned populist economist V. V. (Vorontsov) could only dream about the time when zemstvos would distribute army orders directly between petty producers in the countryside. "Fixing prices not according to the fluctuating moods of the market but according to the strict cost of production,…in agreement with a competent…customer such as the state,…eliminates…any possibility of the speculation and agitation that facilitate the spirit of profiting." See V. V., "Proizvoditel'naia kooperatsiia v industrii," *Vestnik kooperatsii*, no. 1 (1909): 60, 61. By 1916, with the introduction of firm prices and the system of compulsory procurement later that year, this vision of V. V. became reality. Cf. Agr. V., "Kooperativnyi sbyt na armiiu v 1914–1915 g.," *Vestnik kooperativnykh soiiuzov*, no. 2 (February 1916): 75–79.

102. See A. Chaianov, "K voprosu o vosstanovlenii sel'skogo khoziaistva v mestakh razrushennykh voinoi," *Agronomicheskii zhurnal*, no. 1 (1915): 22–26.

103. See Ge-Tan [German Tanashev], "Voina i agrarnyi vopros," *Agronomicheskii zhurnal*, no. 1 (1916): 41–65.

104. K. Matseevich, "Ob"edinennaia kooperatsiia i ee glavnye zadachi," *Agronomicheskii zhurnal*, no. 5 (1915): 4.
105. "Vnutrennee obozrenie," *Agronomicheskii zhurnal*, no. 7–8 (1914): 137. See an almost literal repetition of this call in Ar. S-kii, "Moskovskie s"ezdy i demokratiia," *Novyi kolos*, no. 35 (September 12, 1915): 1.
106. For extensive treatment of the uneasy relationships of Progressivism with socialist and liberal visions of state and society, see Kloppenberg, *Uncertain Victory*.

9 Nation as the People

1. On this epistemological situation, see Ilya Gerasimov, Sergey Glebov, Jan Kusber, Marina Mogilner, Alexander Semyonov, "New Imperial History and the Challenges of Empire," in Ilya Gerasimov et al., eds., *Empire Speaks Out: Languages of Rationalization and Self-Description in the Russian Empire* (Leiden: Brill, 2009), pp. 3–32.
2. Cf. Kimitaka Matsuzato, "General-gubernatorstva v Rossiiskoi imperii: ot etnicheskogo k prostranstvennomy podhodu," in Gerasimov et al., eds., *Novaia imperskaia istoriia postsovetskogo prostranstva*, pp. 427–37 and passim.
3. Dillon, "The Rural Cooperative Movement," p. 404. Quoted from the Ukrainian-language periodical *Dniprovi khvily* 1, no. 16–17 (July 22, 1911): 238.
4. See Glukhov et al., *Mestnyi agronomicheskii personal, 1 ianvaria 1915 g.*, p. 109.
5. *Otchet po Belebeevskoi, Ufimskogo gubernskogo zemstva, nizshei sel'skokhoziaistvennoi shkole*, p. 179.
6. "Agronomicheskii personal zemskikh uchrezhdenii," pp. 621–39; *Trudy 2-go s"ezda predstavitelei kreditnykh tovarishchestv Belebeevskogo uezda, Ufimskoi gubernii, sostoiavshegosia 15–17 iunia 1910 goda v g. Belebee* (Belebei, 1910), p. 9.
7. Alexandrovskii et al., *Mestnyi agronomicheskii personal, 1 ianvaria 1914 g.*, p. 375.
8. See Glukhov et al., *Mestnyi agronomicheskii personal, 1 ianvaria 1915 g.*, p. 419.
9. Ibid., p. 413.
10. For instance, together with Alexander Chaianov, in the class of 1910, 15 students sponsored by the Department of Agriculture graduated from the Moscow Agricultural Institute (12 percent of the class). At least four of them had the same educational background (primary agricultural school) as Taichinov. One example is the Ukrainian Timofei Kvitko, whose prior education was at the Kharkov agricultural school, and who later became the head of an experimental field in Poltava Province. See "Diplomnaia rabota predstavlennaia okanchivaiushimi v techenii 1910 g.," pp. 43–50; Alexandrovskii et al., *Mestnyi agronomicheskii personal, 1 ianvaria 1914 g.*, p. 269.
11. One may suggest that in the predominantly rural Bashkir society, middle-class groups were too few to invest in a career as a rural professional, when so many urban professional occupations were still dramatically understaffed by Tatars and Bashkirs.
12. In the first class of 1899, 16 men graduated from the Belebei agricultural school. Besides Taichinov, there were no more than three Bashkirs (judging from the estate status of the graduates). Next year, only 2 out of 10 graduates could have been Bashkirs (a trader's son and a *meshchanin*), one of them ran his own estate after graduation, and the other served on a gentry estate. In the class of 1906, 19 coevals of Alexander Chaianov graduated from the Belebei school. Eight of them theoretically could be ethnic Bashkirs, but none of them chose a position with the

local zemstvo. In 1907, there were a total of 66 pupils in the school, just two of them (3 percent) were Bashkirs. See *Otchet po Belebeevskoi, Ufimskogo gubernskogo zemstva, nizshei sel'skokhoziaistvennoi shkole*, pp. 9–10, 177–81, 186–89.

13. *Trudy 11-go i 12-go gubernskikh agronomicheskikh soveshchanii pri Ufimskoi gubernskoi zemskoi uprave s 4 po 7 ianvaria i s 1 po 5 marta 1913 g.*, p. 96.

14. *Trudy 2go s"ezda predstavitelei kreditnykh tovarishchestv Belebeevskogo uezda, 15–17 iiunia 1910 goda v g. Belebee*, p. 11.

15. Ibid., p. 48. The periodical's regular annual budget was 1,500 rubles; the introduction of an all-Tatar version would raise the cost of production to 4,000 rubles, which the Ufa provincial zemstvo was not ready to allocate at that time.

16. See Charles Steinwedel, "Invisible Threads of Empire: State, Religion, and Ethnicity in Tsarist Bashkiria, 1773–1917" (Ph.D. diss., Columbia University, 1999), pp. 302–4.

17. Cf. Morachevskii, *Spravochnik po sel'skohoziaistvennoi periodicheskoi pechati*, p. 32.

18. In 1916, the Ufa province instructor of cooperation Bluemental singled out the task of publishing in Tatar and other minority languages as a priority. See E. Iu. Bluemental, "O rasprostranenii kooperativnoi literatury," in *IV Gubernskoe soveshchanie po voprosam kooperatsii. 4–5 sentiabria 1916 g. Protokoly, doklady, postanovleniia* (Ufa: Ufa provincial zemstvo small credit bank, 1916), p. 48.

19. A Birsk district precinct agronomist, A. I. Sedoikin (a 1910 graduate of a mid-level agricultural college), in 1912 predominantly worked in Bashkir villages. Borrowing from the experience of his colleagues working in Russian villages, he tried to introduce experimental crops and technologies on the farms of local clergy hoping to use their authority to propagate new methods of agriculture among the Bashkirs. He was eventually compelled to admit that his experiment had failed, as the local mullahs proved to be extremely inert and conservative farmers. See *Sbornik dokladov Birskoi uezdnoi zemskoi upravy i postanovlenii zemskikh sobranii XXXVIII ocherednoi sessii 1912 goda, XX chrezvychainoi sessii 10 dekabria 1912 goda i XXI chrezvychainoi sessii 8 fevralia 1913 goda* (Birsk, 1913), pp. 782–97. Sedoikin's colleague, precinct agronomist Burmatov spent at least a quarter of his time in Tatar-speaking villages. He actively developed apiculture in his district, and, among 126 peasants whom he assisted in beekeeping, at least 10 percent were Bashkirs (ibid., pp. 798–826). Their fellow precinct agronomist P. M. Popov, who had graduated from a mid-level agricultural college back in 1895, used the same methods of instruction in Russian and Bashkir villages. In meeting with only a Russian audience he began his talk by reading aloud a brochure on the topic of the presentation, while a conversation with Bashkirs was preceded by a reading of the appropriate brochure in Tatar by someone from the audience (ibid., pp. 756–57).

20. The young precinct agronomist from the Belebei district Anatoly Diakov (who graduated from a mid-level agricultural college in 1912) wrote, "Tatar is the international language here. Not a single Russian, having lived here for a more or less long period, is unable to speak Tatar." A. K. Diakonov, "Iz zemskoi agronomicheskoi deiatel'nosti," *Zemskii agronom*, no. 1 (1915): 9.

21. Characteristically, a publication targeting the 1,500-strong Kazan Jewish community was also published in 500 copies (more than one per family). See NART, f. 420, "Kazanskii vremennyi komitet po delam pechati", d. 158 "S perepiskoi s redaktsiiami i tipografiiami o periodicheskikh pechatnykh izdaniiakh," ll. 2, 104 ob.

22. Steinwedel, *Invisible Threads of Empire*, p. 506.

23. Already in July of 1910, a congress of managers of the Belebei district credit associations asked the local zemstvo board to include special lessons on cooperative credit in the curriculum of the Belebei *medrese*'s so-called Russian classes (*Trudy 2-go s"ezda predstavitelei kreditnykh tovarishchestv Belebeevskogo uezda*, p. 3). Later that year, the board of the Birsk district zemstvo, after heated discussions, for the first time decided to allocate 5,000 rubles in 1911 to support 103 *mektebe* and 171 *medrese* in the same way that it supported less numerous Orthodox parish schools (which received 3,000 rubles per annum). (See *Sbornik dokladov Birskoi uezdnoi zemskoi upravy i postanovlenii zemskogo sobraniia XXXVI ocherednoi sessii 1910 goda*, pp. 235, 237–38.) In 1912, the zemstvo board decided to give 4,202 rubles to Orthodox parish schools and 10,000 rubles to Muslim traditional schools. (*Sbornik dokladov Birskoi uezdnoi zemskoi upravy i postanovlenii zemskikh sobranii XXXVIII ocherednoi sessii 1912 goda*, p. 160.) The Menzelinsk district zemstvo followed suit, allocating 14,700 rubles for *mektebe* and *medrese* of the district, but only those "new-method" (*jadid*): obviously, the zemstvo preferred the most modernized segment of its Muslim constituency (Steinwedel, *Invisible Threads of Empire*, p. 464). In 1915, the Birsk district zemstvo also decided to support only the *jadid* schools that were recognized as perfectly fitting the program of Russian primary public schools. (*Sbornik dokladov Birskoi uezdnoi zemskoi upravy i postanovlenii zemskikh sobranii 41 ocherednoi sessii 22 sentiabria 1915 goda i chrezvychainogo sobraniia XXIV sessii 19 avgusta 1915 goda*, pp. 137–39.)

24. The very first agricultural courses in the Zlatoust district were organized in November 1912 on the topic "Dairy farming and husbandry." Although the language of the courses was Russian, 5 out of 19 students were Bashkirs. See *Otchet o deiatel'nosti Ufimskoi gubernskoi zemskoi upravy po sel'skokhoziaistvennoi chasti za 1912 god* (Ufa: Ufa Provincial Zemstvo, 1913), p. 51.

25. In 1916, 28 percent of consumer cooperatives in the Sterlitamak district were organized by Bashkirs. See A. A. Kuks, "O potrebitel'skoi kooperatsii v Sterlitomakskom uezde," in *IV Gubernskoe soveshchanie po voprosam kooperatsii. 4–5 sentiabria 1916 g.*, p. 20.

26. Ibid. This fact was noticed with some dismay by the zemstvo instructor on consumer cooperatives, A. Kuks. At the same time, he was not surprised by the nationality principle in the organization of cooperatives.

27. See F. G. Turchenko and G. F. Turchenko, *Pivdenna Ukraïna: modernizatsiia, svitova viina, revolutsiia (kinets XIX st.–1921 r.): Istorichni narysy* (Kiïv: Geneza, 2003), p. 52.

28. Georgii Kas'ianov, *Ukraïns'ka intelligentsiia na rubezhi xix–xx stolit'* (Kyiv: Libid', 1993), p. 43.

29. Ukrainian newspaper *Rada*, no. 187, 1907, quoted from Kas'ianov, *Ukraïns'ka intelligentsiia*, p. 64.

30. See Dillon, "The Rural Cooperative Movement," pp. 400–403.

31. *Dniprovi khvili* 1, no. 9 (February 16, 1911): 127. Quoted from Dillon, "The Rural Cooperative Movement," p. 401.

32. Eremeeva, "O nedelimosti melkogo zemlevladeniia," pp. 13–14.

33. A negative result is difficult to substantiate documentally, but it seems worth mentioning that neither the Kharkov zemstvo commission for the organization of public lectures in 1909 nor the 1912 agronomist conference at the Kiev zemstvo board discussing agricultural education in the zemstvo schools even mentioned the Ukrainian language problem. *Otchet o deiatel'nosti Khar'kovskoi komissii po ustroistvu narodnykh chtenii za 1909 god* (Kharkov, 1910); *Trudy iv agronomicheskogo*

soveshchaniia pri Kievskoi gubernskoi zemskoi uprave 10–14 dekabria 1912 goda (Kiev, 1913), pp. 52–55 et seq.

34. Alexander Dillon made this "infusion of nationalism into the [cooperative] movement" in Ukraine the main theme of his Ph.D. dissertation, "The Rural Cooperative Movement and Problems of Modernizing in Tsarist and Post-Tsarist Southern Ukraine (New Russia), 1871–1920." For a more skeptical appraisal of the Ukrainization of the pre-1917 cooperative movement in Ukrainian lands, see V. M. Polovets, *Istoriia kooperatsii Livoberezhnoï Ukraïny (1861–1917 rr.)* (Kyiv: Globus, 2001).

35. See Torgashev, "Kooperativnoe dvizhenie v Rossii," *Vestnik kooperatsii*, no. 1 (1909): section 2, 2–3.

36. *Kooperatsiia na Vserossiiskoi vystavke 1913 g. v Kieve* (Kiev: Bureau of the Exhibit's Cooperative Section, 1914), pp. 43, 48, and passim. In the printed account of the congress, a special chapter IV.9 was dedicated to "Ukrainian Cooperation in Galicia (Austria)." This followed chapters IV.4 "Credit Cooperatives in Ukraine," IV.6 "Cooperatives among Jews," and IV.8 "The Moscow People's Bank."

37. Dillon, "The Rural Cooperative Movement," p. 395.

38. D. Mikheev, "Kooperatsiia i kul'tura (Vtoroi vserossiiskii kooperativnyi s"ezd)," *Agronomicheskii zhurnal*, no. 6 (1913): 107.

39. Steven Lan Guthier, "The Roots of Popular Ukrainian Nationalism: A Demographic, Social and Political Study of the Ukrainian Nationality to 1917" (Ph.D. diss., University of Michigan, 1990), pp. 216–19; Dillon, "The Rural Cooperative Movement," pp. 399–400.

40. Dillon, "The Rural Cooperative Movement," pp. 396, 457.

41. Merkulov, "Kooperativnoe dvizhenie v Rossii," 125.

42. Quoted from Dillon, "The Rural Cooperative Movement," p. 474.

43. Already on the eve of the 1913 Kiev cooperative congress, the Odessa-based magazine the *Southern Cooperator* staged a campaign against the expansion of the Moscow People's Bank in Ukraine by means of opening its branches in major cities. See V. G. "Sobranie Moskovskogo narodnogo banka," *Iuzhnyi kooperator*, no. 10 (May 27, 1913): 219. The magazine of the Kiev cooperative credit union, the *Anthill*, explicitly demanded the introduction of a Ukrainian-language cooperative school a few weeks before the February Revolution. Reprinted in "So stranits soiuznoi pechati," *Vestnik kooperativnykh soiuzov*, no. 2 (February 1917): 50.

44. Dillon, "The Rural Cooperative Movement," pp. 466, 576.

45. Ibid., p. 576.

46. Ibid., p. 365.

47. Cf. Guthier, "The Roots of Popular Ukrainian Nationalism," p. 219; Mark Robert Baker, "Peasants, Power, and Revolution in the Village: A Social History of Kharkiv Province, 1914–1921" (Ph.D. diss., Harvard University, 2001), p. 102; Dillon, "The Rural Cooperative Movement," pp. 401, 406; Polovets, *Istoriia kooperatsii*, p. 179.

48. *Kooperatsiia na Vserossiiskoi vystavke 1913 g. v Kieve*, p. 4.

49. Ibid., p. 93, also graphs 12, 13.

50. Ibid., graphs 23–25, 32.

51. Ibid., p. 92.

52. Ibid., graph 16.

53. Ibid., graph 17.

54. Ibid., graph 15.

55. On the ambivalence of the geographic boundaries of "Ukraine" at least until the beginning of the twentieth century, see Natalia Yakovenko, "Zhiznennoe

prostranstvo *versus* identichnost' rus'kogo shliahticha XVII st. (na primere Iana/Ioakima Erlicha)," *Ab Imperio*, no. 4 (2006): 101–36.

56. Dillon, "The Rural Cooperative Movement," p. 457.

57. Ibid., pp. 457–58.

58. Ibid., p. 510. On the split in Ukrainian cooperative movement over the attitude toward the Moscow People's Bank in 1918, see Polovets, *Istoriia kooperatsii*, pp. 180–85.

59. Merkulov, "Kooperativnoe dvizhenie v Rossii," p. 125.

60. For an example of the anti-"Russophile" propaganda in the local cooperative press, see K. Chuzhezemtsev, "Opasnye druz'ia," *Iuzhnyi kooperator*, no. 4 (February 28, 1913): 73–78.

61. That was the case when the Ust-Sysol district zemstvo (Vologda province) announced a search for a veterinary assistant who should necessarily be of the Orthodox faith (*Veterinarnyi fel'dsher*, no. 4 [April 1915], inside cover). An instance of cooperative xenophobia that only indirectly expressed any Russian "national" agenda appeared in a feuilleton published by a Kostroma "cooperative" magazine in January 1917. The author suggested that in the midst of supply shortages, the only things in abundance were the food procurement officials: K. A. Rapp, Baron Osten-Drizen, and N. A. Beretti in Kursk, "Gintsberg—in charge of procuring vegetables for the army . . . Iov—[in charge] of purchasing meat. Iu. B. Shturmer—[in charge] of supplying the population with firewood. . . . Better fewer plenipotentiaries, and more foodstuffs." Nika, "Malen'kii fel'eton 'O tom, chego u nas mnogo,'" *Kostromskoi kooperator*, no. 2 (January 1917): 7, 8. While the author did not comment formally on the foreign-sounding names, obviously his intention was to criticize not so much the bureaucratic procurement apparatus but the "aliens" in its service: Alexander Beretti was a prominent expert on cooperative movement, agronomist and editor of the "cooperative" periodicals *Splotchina* and *Khroniki melkogo kredita*, and was mentioned in the feuilleton only because of his "non-Russian" name.

62. Cf. "To internationalists, for example, to the world Jews, there is nothing inacceptable in the preponderance of Germany, and so on. This all is hard and deadly only to non-Jews, nonproletarians of the International." E. N. Sakharova-Vavilova, "Dnevnikovye zapisi," l. 299.

63. *Kooperativnyi nastol'nyi kalendar' na 1917 g.* (Moscow: Izdanie Moskovskogo soiuza potrebitel'nykh obshchestv, 1916), pp. 13–16.

64. The educated elite in the Baltic provinces and in the Russian partition of Poland was anxious to mobilize peasants in support of their claims for nationhood, as was the case in the Austrian part of Poland: "Emancipation altered the power relations separating lord from peasant, creating the possibility of alliances across social classes. Villagers became the focus of 'Polonization' efforts by the Polish upper classes who set out to attract their support for national liberation movements. The half century following emancipation in Poland saw the 'moblization' of Polish peasants behind a national agenda." See Keely Stauter-Halsted, *The Nation in the Village: The Genesis of Peasant National Identity in Austrian Poland, 1848–1914* (Ithaca, NY: Cornell University Press, 2001), p. 7.

10   Revolutionary Nation

1. "Novogodnie perspektivy," *Novyi Ekonomist*, no. 1 (January 7, 1917): 2.

2. P. Volodin, "Novyi god," *Kostromskoi kooperator*, no. 1 (January 1917): 2.

3. For example, see a similar assessment of the current economic situation by the young agronomist I. Erofeichev, "Potrebitel'skie obshchestva i kreditnye tovarishchestva," *Samarskii zemledelets*, no. 23 (December 1, 1916): 656.

4. Cf. W. Bruce Lincoln, *Passage through Armageddon: The Russians in War and Revolution, 1914–1918* (New York: Simon and Schuster, 1986): 310–12 et seq.

5. On January 15, 1917, the influential neopopulist economist Nikolaii Oganovskii published probably the most sober of his articles in a decade. He acknowledged that the peasants had access to over 90 percent of the arable land in Russia, and hence "besides land, the agricultural culture, the market, and organization of the peasantry by cooperatives now constitute the core of the agrarian question." Oganovskii emphasized the necessity of further intensification of the peasant economy and acknowledged the role of agricultural specialists in modernization of the countryside. See N. Oganovskii, "Sovremennyi agrarnyi vopros," *Novyi kolos*, no. 2 (January 15, 1917): 1.

6. For example, in January 1917, the *Kostroma Cooperative Activist* reprinted from the daily *Russian Word* a strange telegram from Zaraisk (Riazan Province) about the alleged training of the local police in handling machine guns. No serious external or internal enemy could threaten the district capital of Zaraisk, which was located in the heart of Russia and had no strategic importance. Yet, the disturbing images of modern technology placed in the service of the government conspiracy had haunted the Russian public weeks before the rumors spread in Petrograd about policemen with machine guns shooting revolutionary crowds from the roofs. See "Voennaia podgotovka politsii," *Kostromskoi kooperator*, no. 4 (January 1917): 16.

7. For instance, the January issue of the Kazan apiculturists' magazine featured the poem "Obshchestvennost'," signed with the pseudonym "Obshchestvennik." It repeated all of the poetical and ideological clichés characteristic of the radical intelligentsia's belles-lettres that became virtually obsolete in the 1910s:

> We call on [you] to join us and follow us! …
> Our *obshchestvennost'*, unseen,
> Will tie us together
> By a single common destiny.
> *Obshchestvennost'* is the enemy of the gods of profit;
> Its goddess—the fate of laborers,
> And, therefore, its appeals
> Sound louder and louder with victory. […]

The replacement of "obshchestvennost'" by "revolution" would convert this politically moderate verse into a classic revolutionary poem. See Obshchestvennik, "Obshchestvennost'," *Zhurnal Kazanskogo obshchestva pchelovodstva*, no. 1 (January 1917): 15. At the same time, the anticipated introduction of the county (*volostnoe*) zemstvo was hailed with enthusiasm that had a precedent perhaps only in the "banquet campaign" of 1904. For example, see M. S., "Volostnoe zemstvo i kooperativy," *Iuzhnyi kooperator*, no. 1–2 (January 1917): 7–12. A special fortnightly magazine the *County Zemstvo* [Volostnoe zemstvo] was even founded, the first issue was expected to be published in January 1917. See *Samarskii zemledelets*, no. 23 (December 1, 1916): 653. A prominent zemstvo activist and historian, Boris Veselovskii, became the editor of the magazine and its active contributor, while Dmitry Protopopov, member of the Constitutional Democratic Party's Central Committee, was the publisher. See *Volostnoe zemstvo*, nos. 1–22 (1917).

Ideological reasons clearly prevailed over practical considerations among the ardent enthusiasts of the new "truly democratic" zemstvo institution, as has been shown by Kimitaka Matsuzato in his "The Concept of 'Space' in Russian History—Regionalization from the Late Imperial Period to the Present," in Teruyuki Hara and Kimitaka Matsuzato, eds., *Empire and Society: New Approaches to Russian History* (Sapporo, Japan: Slavic Research Center, Hokkaido University), p. 187.

8. P. B. Struve, "Ozdorovlenie vlasti," in Struve, *Patriotica: politika, kultura, religiia*, p. 393.

9. See "Gosudarstvennyi perevorot," *Novyi Ekonomist*, no. 8–9 (February 25–March 4, 1917): 3–6.

10. "Redaktsiia," *Mashina v sel'skom khoziaistve*, no. 4–5 (March 15, 1917): 153–54.

11. See K. Matseevich, "Novye zadachi sel'skokhoziaistvennoi pechati," *Zemledel'chaskaia gazeta*, no. 9–10 (March 18, 1917): 189–91.

12. K. A. Matseevich, "Obshchestvenno-agronomicheskoe znachenie svobodnykh sel'skokhoziaistvennykh organizatsii," *Agronomicheskii zhurnal*, no. 6 (1914): 12.

13. For the most recent narrative of the deportation in English, see Lesley Chamberlain, *The Philosophy Steamer: Lenin and the Exile of the Intelligentsia* (London: Atlantic Books 2006). Valuable documents were published in V. G. Makarov and V. S. Khristoforov, eds., *Vysylka vmesto rasstrela. Deportatsiia intelligentsii v dokumentakh VChK-GPU. 1921–1923* (Moscow: Russkii Put', 2005). See also L. A. Kogan, " 'Vyslat' za granitsu bezzhalostno' (Novoe ob izgnanii duhovnoi elity," *Voprosy filosofii*, no. 9 (1993): 61–84; M. G. Glavatskii, *Filosofskii parohod: Istoriograficheskie etudy. God 1922-i* (Ekaterinburg: Izdatel'stvo Ural'skogo universiteta, 2002); V. G. Makarov and V. S. Khristoforov, "Passazhiry 'filosofskogo parohoda' (sud'by intelligentsii, repressirovannoi letom-osen'iu 1922 g.)," *Voprosy filosofii*, no. 7 (2003): 113–37.

14. "Protokol doprosa N. K. Muravieva ot 25 avgusta 1922 g.," in Makarov and Khristoforov, *Vysylka vmesto rasstrela*, p. 293.

15. Cf. "Forces in the disposal of organized public mind are insignificant in comparison with the might of spontaneous evolution…. The only thing that it [public mind] is capable of and what it ought to do is to place spontaneous evolution under circumstances in which it [the evolution] would come to the social ideals of democracy." Alexander Chaianov, "Real'nye vozmozhnosti zemel'noi reformy," *Vlast' naroda*, no. 87 (1917), cited in *Vestnik sel'skogo khoziaistva*, no. 35–36 (October 1, 1917): 21. Chaianov explicitly (although without acknowledging the source) referred to the notions of "social telesis" or "collective mind" from the sociology of Ward. Cf. Lester F. Ward, "Contributions to Social Philosophy. XII. Collective Telesis," *American Journal of Sociology* 2, no. 6 (May 1897): 801–22.

16. "O s"ezde agronomov," in GARF, f. 102 Departament politsii, D-4, op. 120, d. 181, l. 14.

17. Ibid., l. 15.

18. Gleason, "The All-Russian Union of Towns and the All-Russian Union of Zemstvos, p. 245. A remarkably subtle analysis of the role of the wartime *obshchestvennost'* "megapojects," the Unions of Towns and Zemstvos, can be found in Thomas Fallows, "Politics and the War Effort in Russia." A recent review essay has shown that our understanding of the past does not necessarily progress to a more advanced stage since the 1970s, at least regarding the history of the Unions of Towns and Zemstvos. See Francesco Benvenuti, "Armageddon Not Averted: Russia's War, 1914–21," *Kritika: Explorations in Russian and Eurasian History* 6, no. 3 (Summer 2005): 535–56.

19. As Orlando Figes put it, "The Provisional Government was a government of persuasion … it depended largely on the power of the word to establish its authority. It was a government of national confidence … [it] emphasized the need to govern by consent." Orlando Figes, "The Russian Revolution of 1917 and Its Language in the Village," *Russian Review* 56 (July 1997): 323.

20. See Orlando Figes and Boris Kolonitskii, *Interpreting the Russian Revolution: The Language and Symbols of 1917* (New Haven, CT: Yale University Press, 1999); Boris Kolonitskii, *Simvoly vlasti i bor'ba za vlast': k izucheniiu politicheskoi kul'tury rossiiskoi revolutsii 1917 goda* (St. Petersburg: Dm. Bulanin, 2001).

21. Figes, "The Russian Revolution of 1917 and Its Language in the Village," p. 323.

22. Michael Hickey arrived at similar conclusions in his study of the revolutionary discourse in Smolensk in the spring of 1917: "Kadet legitimacy in the provinces rested on their claim to represent a supraclass public (*obshchestvo*), at the core of which were Russia's urban middle elements: professionals, civil servants…. [L]iberals misapprehended social reality and social symbolism, which helps to explain their political weakness. Liberals in Smolens spoke of building a new Russia through *obshchestvennost'*—the public identification as citizens united in a supraclass center." Michael Hickey, "Discourses of Public Identity and Liberalism in the February Revolution: Smolensk, Spring 1917," *Russian Review* 55 (October 1996): 616.

23. See "S"ezd grazhdan-khleborobov," *Iuzhnyi kooperator*, no. 8 (April 30, 1917): 178.

24. See A. N., "Na krest'ianskom s"ezde (Vpechatleniia uchastnika), *Iuzhnyi kooperator*, no. 8 (April 30, 1917): 179–81; A. F. Lebedev, "O zemle. Doklad krest'ianskomu s"ezdu v izlozhenii uchastnika s"ezda," *Iuzhnyi kooperator*, no. 8 (April 30, 1917): 186.

25. See M. Levenshtam, "Pis'mo iz derevni (Khersonskii uezd)," *Zemledel'chaskaia gazeta*, no. 17 (April 29, 1917): 326.

26. A. Levshin, "Perevesinskii volostnoi ispolnitel'nyi komitet Balashovskogo uezda," *Zemledel'chaskaia gazeta*, no. 17 (April 29, 1917): 325. Levshin joined the profession sometime after 1915.

27. In March 1917 peasants throughout Saratov province, including the Balashov district, removed village elders and replaced county boards with self-proclaimed committees. See Donald J. Raleigh, *Revolution on the Volga* (Ithaca, NY: Cornell University Press, 1986), pp. 90–91.

28. Levshin, "Perevesinskii volostnoi ispolnitel'nyi komitet Balashovskogo uezda," p. 325.

29. Ibid.

30. See K. I., "Agronomy i svobodnaia derevnia (pis'mo iz Khar'kova)," *Zemledel'cheskaia gazeta*, no. 19–20 (May 20, 1917): 389.

31. Polovets, *Istoriia kooperatsii Livoberezhnoï Ukraïny*, p. 128.

32. See V. V. Kabanov, *Krest'ianskoe khoziaistvo v usloviiakh voennogo kommunizma* (Moscow: Nauka, 1988); V. P. Danilov, "Krest'ianskaia revoliutsiia v Rossii, 1902–1922," in *Krestianstvo i vlast'* (Moscow and Tambov, 1996), pp. 4–23.

33. For a systematic attempt to depict peasants as active partners in the creation of the common "revolutionary nation", see Chapter 2 in Aaron B. Retish, *Russia's Peasants in Revolution and Civil War: Citizenship, Identity, and the Creation of the Soviet State, 1914–1922* (Cambridge: Cambridge University Press, 2008), pp. 64–94.

34. See Ilarii Shadrin, "Vsia zemlia- vsemu narodu," *Kazanskaia kooperativnaia zhizn'*, no. 2–3 (June 8, 1917): 15, 16.

35. See "Bor'ba s dezertirstvom," *Kazanskaia kooperativnaia zhizn'*, no. 4 (July 6, 1917): 30.
36. See "Molodye grazhdane," *Kazanskaia kooperativnaia zhizn'*, no. 4 (July 6, 1917): 24–25.
37. See P. P. Migulin, "Prinuditel'nyi trud," *Novyi Ekonomist*, no. 47 (November 19, 1916): 2–5.
38. See P. P. Migulin, "Prodovol'stvennyi vopros v Gosudarstvennoi Dume," *Novyi Ekonomist*, no. 49 (December 3, 1916): 4–8.
39. See "Gosudarstvennyi perevorot," 5.
40. See P. P. Migulin, "Agrarnyi vopros," *Novyi Ekonomist*, no. 19 (May 13, 1917): 1–6.
41. See G. Tanashev, "Agrarnyi vopros i revolutsiia," *Zemledel'chaskaia gazeta*, no. 9–10 (March 18, 1917): 193–94; Ge-Tan, "Revolutsiia i derevnia," *Zemledel'chaskaia gazeta*, no. 11–12 (March 25, 1917): 216–17; idem, "Agrarnye programmy krest'ian," *Zemledel'chaskaia gazeta*, no. 18 (May 6, 1917): 341–43.
42. See A. O. Fabrikant, "Agronomy i velikii perevorot," *Zemledel'chaskaia gazeta*, no. 9–10 (March 18, 1917): 196–97.
43. See K. S. Ashin, "Garantiia posevov," *Zemledel'chaskaia gazeta*, no. 13–14 (April 8, 1917): 245–46.
44. A. Kaufman, "K agrarnomu voprosu," *Zemledel'chaskaia gazeta*, no. 15–16 (April 22, 1917): 242.
45. See an account in *Zemledel'chaskaia gazeta*, no. 19–20 (May 20, 1917): 391.
46. See D. Mikheev, "So stranits pechati," *Vestnik kooperativnykh soiuzov*, no. 3 (1917): 13–18; D. M., "So stranits pechati. Partiinost' i kul'turnaia rabota v derevne," *Vestnik kooperativnykh soiuzov*, no. 3 (1917): 18–23.
47. See "Ot organizatsionnogo komiteta," *Zemledel'chaskaia gazeta*, no. 13–14 (April 8, 1917): 263.
48. For instance, the official notice of the foundation of the League for Agrarian Reform gave three contact addresses in Moscow and St. Petersburg for those interested in the League's activity: the home addresses of B. D. Brutskus, A. V. Chaianov, and K. A. Matseevich. See "Ot organizatsionnogo komiteta," *Zemledel'chaskaia gazeta*, no. 13–14 (April 8, 1917): 263.
49. See L. I. Potapova, "Liga agrarnykh reform," in *Sotsial'no-ekonomicheskaia istoriia derevni kontsa XIX–nachala XX vv.: Sbornik nauchnykh trudov* (Moscow: MGPI im. V. I. Lenina, 1980), p. 138. The account of the League's Executive Committee provided slightly different information: sessions of the First Congress took place only on April 16–17, and were attended by 130 delegates from 19 provinces. See "Otchet o deiatel'nosti Tsentral'nogo rasporiaditel'nogo komiteta Ligi agrarnykh reform za 1917 god," *Vestnik sel'skogo khoziaistva*, no. 21–22 (July 1 [June 18], 1918): 23.
50. Potapova. "Liga agrarnykh reform."
51. Ibid., p. 139. For instance, the Odessa Regional Department was opened on May 1, 1917. See *Iuzhnyi kooperator*, no. 10–11 (1917): 258.
52. With the gradual disintegration of Russian Empire and Russian society, attendance at the League congresses decreased. The second Congress of the League on June 23–25 was attended by 135 experts from 15 provinces. The last, third Congress, convened in the wake of the October Revolution, on November 21–23, 1917. It featured 95 participants from nine provinces. "Otchet o deiatel'nosti Tsentral'nogo rasporiaditel'nogo komiteta," p. 23.
53. See *Trudy komissii po podgotovke zemel'noi reformy*, vyp. 1 (Petrograd: Glavnyi zemel'nyi komitet, 1917), p. 3.

54. Ibid.
55. Ibid., pp. 5–6.
56. See *Trudy komissii po podgotovke zemel'noi reformy*, vyp. 2 (Petrograd: Glavnyi zemel'nyi komitet, 1917), pp. 18–20, 27–30; *Trudy komissii po podgotovke zemel'noi reformy*, vyp. 3 (Petrograd: Glavnyi zemel'nyi komitet, 1917), pp. 26–40, and others.
57. See B. D. Brutskus, "Normy zemel'nogo obespecheniia," in *Trudy komissii po podgotovke zemel'noi reformy*, vyp. 2, pp. 27–30.
58. See *Trudy komissii po podgotovke zemel'noi reformy*, vyp. 2, p. 40.
59. K. Voblyi, "K agrarnomu voprosu v Rossii," *Novyi Ekonomist*, no. 21 (May 27, 1917): 6–9.
60. P. Migulin, "Vopros zemel'nyi i vopros prodovol'stvennyi," *Novyi Ekonomist*, no. 22–23 (June 10, 1917): 7. By August, Migulin had reached the extreme stage of rejection of any state intervention into the economy: he demanded the dissolution of all procurement agencies and restoration of the system of private trade instead. See P. Migulin, "Prodovol'stvennyi krizis," *Novyi Ekonomist*, no. 32 (August 12, 1917): 3.
61. A. Evdokimov, "Ocherki kooperativnoi sel'skokhoziaistvennoi zhizni," *Zemledel'chaskaia gazeta*, no. 23 (June 10, 1917): 451.
62. Quoted from "Liga agrarnykh reform. Vtoroi vserossiiskii s"ezd," *Zemledel'chaskaia gazeta*, no. 25–26 (July 1, 1917): 515.
63. See Ge-Tan, "K voprosu o granitsakh agrarnoi reformy," *Vestnik sel'skogo khoziaistva*, no. 33–34 (September 17, 1917): 6–10.
64. See *Zemledel'chaskaia gazeta*, no. 21 (May 27, 1917): 418.
65. Cf. "Here the organized population in general violates the independence of institutions, created by part of this very same population," while the Ministry of Agriculture did nothing to prevent these violations. See "Moskva 15-go maia 1917 g.," *Vestnik kooperativnykh soiuzov*, no. 4–5 (1917): 3, 4.
66. Evdokimov, "Ocherki kooperativnoi sel'skokhoziaistvennoi zhizni," p. 451.
67. On the pages of his journal and elsewhere, he argued against the economic infantilism of the Socialist Revolutionary Party, against its claims to political leadership, and against any attempts by political parties to speak on behalf of the cooperative movement. See A. E., "Partiinaia taktika i partiinaia bestaktnost'," *Vestnik kooperativnykh soiuzov*, no. 6–7 (1917): 3–10; A. Evdokimov, "Na vzgliad kooperatora," *Vestnik kooperativnykh soiuzov*, no. 6–7 (1917): 10–19.
68. For example, see "Kazan', 6 iulia 1917 goda," *Kazanskaia kooperativnaia zhizn'*, no. 4 (July 6, 1917): 1–4.
69. The decision was made by a majority of 84 votes to 17, which represented worker cooperatives. See L. G. Protasov, *Vserossiiskoe uchrediel'noe sobranie: istoriia rozhdeniia i gibeli* (Moscow: "Rossiiskaia politicheskaia entsyklopediia" (ROSSPEN), 1997), p. 117.
70. S. L. Maslov became minister, A. N. Chelintsev continued to hold the technical position of director of the Department of Agricultural Economy and Statistics, while A. V. Chaianov was in charge of the Department of Agriculture, which had played such an important role during the preceding decade. See GARF, f. 934 Ministerstvo Zemledeliia, op. 1, d. 520 "Prikaz po Ministerstvu Zemledeliia 16 oktiabria 1917 g. N 118"; l. 257.
71. By mid-May, in response to the growing number of complaints from local cooperative associations being harassed by Soviets and various committees, a leading cooperative magazine admitted the parting of the ways of the cooperative

movement and the "organized forces of our *obshchestvennost'* . . . . Here the organized population itself in general violates the independence of institutions created by part of this population." See "Moskva, 15-go maia 1917 g.," *Vestnik kooperativnykh soiuzov*, no. 4–5 (1917): 4. Acknowledging the split with the *obshchestvennost'*, the editorial called for the political mobilization of cooperative activists as representatives of the revolutionary nation (ibid., p. 6).

72. For example, in the Petrograd election district, the list of cooperative candidates to the constituent assembly included only one relatively well-known public figure, Alexander Chaianov. The other six candidates were absolutely obscure. A certain Mikhailov was included in the list even without his first and middle names, and only with the brief annotation, "peasant of the Petergof district, cooperative activist." See "Nashi kandidaty: no. 6," *Golos naroda* (November 12 [25], 1917): 1.

73. In the Petrograd district cooperative list, five of the seven candidates headed procurement committees at district and even provincial levels. See the leaflet "Nashi kandidaty: no. 6," 1. Kimitaka Matsuzato has established a close correlation between the success of an agricultural specialist in popular elections and his or her resignation from procurement duties. See Matsuzato, "The Role of Zemstva in the Creation and Collapse of Tsarism's War Efforts During World War One," p. 337.

74. Quoted in L. G. Protasov, *Vserossiiskoe uchrediel'noe sobranie*, p. 232.

75. See *Iuzhnyi kooperator*, no. 17–18 (September 1917): 484.

76. The leading historian of the Russian cooperative movement, V. V. Kabanov, distinguished six main reasons for the failure of the cooperative movement in the elections: the cooperative political program was too vague; there were no bright political leaders; cooperative leaders were good businessmen but dilettantes in politics; the structure of the cooperative network was unsuitable for an effective political campaign; the cooperative movement superficially united a variety of conflicting social strata; and finally, the cooperative leaders overestimated the political potential of cooperatives. See V. V. Kabanov, "Rossiiskaia kooperatsiia v 1917–1918 godakh," in *Obshchestvennye organizatsii v politicheskoi sisteme Rossii: 1917–1918 gody. Materialy konferentsii* (Moscow: Tverskoi GU, 1992), p. 43.

77. Stanziani, *L'Économie en révolution*, chapter 8, "Les spécialistes au pouvoir: le gouvernement provisoire" (Paris: Albin Michel, 1998); Holquist, *Making War, Forging Revolution*, chapter 2, "Radiant Days of Freedom."

78. See A. V. Chaianov, "Letter to A. A. Rybnikov on 28 October 1917," published in Chaianov, *A. V. Chaianov—chelovek, uchenyi, grazhdanin*, pp. 120–21.

79. See N. Makarov, "Ministry i tovarishchi ministrov po mestam!" *Golos naroda*, (November 12 [25], 1917): 3.

80. P. Rybalov, "Ocherednye voprosy agronomicheskoi pomoshch'i," *Zemledel'cheskaia gazeta*, no. 40–41 (October 21, 1917): 683.

81. N. Buiakovich, "Tretii vserossiiskii s"ezd Ligi Agrarnykh Reform," *Vestnik sel'skogo khoziaistva*, no. 51–52 (December 24, 1917): 12.

82. Ibid.

83. Ibid., p. 15.

84. *Trudy 1-go ocherednogo Vserossiiskogo kooperativnogo s"ezda (18–24 fevralia 1918 goda, Moskva). vyp. 2. Promyshlennost' i torgovlia Rossii i zadachi kooperatsii* (Moscow: Zadruga, 1918), pp. 70–71.

85. Buiakovich, "Tretii vserossiiskii s"ezd Ligi Agrarnykh Reform," p. 15.

11   The Dissolution of the "Imagined Community": Nationalization as Expropriation

1. Redaktsia, "O blizhaishei programme 'Vestnika' na 1918 god," *Vestnik sel'skogo khoziaistva*, no. 1–2 (January 14, 1918): 3.
2. A. Stebut, "Blizhaishaia sud'ba agronomicheskoi pomoshchi naseleniu," *Vestnik sel'skogo khoziaistva*, no. 5–6 (February 18, 1918): 3.
3. The Laishev regional agricultural experimental station in Kazan province was burned by peasants on November 15 (new style), 1917 after they pillaged the neighboring gentry estate, only one week after the October Revolution. See *Laishevskaia raionnaia sel'skokhoziaistvennaia opytnaia stantsiia. Svodnyi otchet o deiatel'nosti Stantsii za vremia 1912–1926 gg.* (Kazan: Izdatel'stvo Narkomzema Tatrespubliki "Ignche," 1928), p. 9.
4. A. Nikolaev, "Kooperatsiia i kul'tura," *Kooperativnaia zhizn'*, no. 1 (April 1918): 19.
5. Cf. "We all are but service people [*sluzhilye liudi*] for the peasantry." D. Mikheev, "Munitsipalizm i kooperatizm," *Vestnik kooperativnykh soiuzov*, no. 4–5 (1917): 7. This was one of the last instances of repetition of this neopopulist mantra in the agropress (published in mid-May 1917).
6. Alexander Stebut mentioned the initiative of former Moscow zemstvo district agronomists to find a special Agricultural Society as a keeper of "agronomist initiatives" of the past. Stebut, "Blizhaishaia sud'ba agronomicheskoi pomoshchi naseleniu," p. 3.
7. As Lewis Siegelbaum has demonstrated, the Soviet regime had survived the civil war only because it managed to employ the expertise of the former third element and the services of protoprofessionals of the fourth element. "Providing badly needed skills and experience, the lower middle strata were important contributors to the state-building process that proceeded far in the course of the civil war." Lewis H. Siegelbaum, *Soviet State and Society Between Revolutions, 1918–1929* (Cambridge: Cambridge University Press, 1992), p. 63. This was already quite clear to contemporary observers. The former tsarist minister of agriculture, Alexander Naumov, was surprised to learn in emigration that his former subordinate, G. F. Chirkin, who was in charge of the peasant resettlement from inner Russia, kept his position under the Bolsheviks. He explained the success of Soviet state-building by the typicality of this case: " 'Chirkins' represent those invisible wheels of the clock mechanism, which are concealed behind the dial, that actually produce the movement of the clock hands. This is what I think happened in Russia: the dial is different but the same 'Chirkins' work behind it.... Is it not the source of... stability of the state mechanism of the former enormous empire, which the Kremlin bosses who are illiterate in governance have managed to preserve." A. N. Naumov, "Iz utselevshikh vospominanii," vol. 9, 2145–46, a manuscript in the Hoover Institution Archive, A. N. Naumov Collection, box no. 24. Gennadii Chirkin was born in 1876 and became a victim of Stalinist purges.
8. A. N. Minin, "Gosudarstvennaia razrukha i kooperatsiia," *Kooperativnaia zhizn'*, no. 1 (1918): 6.
9. The words of Alexander Chaianov at the First All-Russian Cooperative Congress in Moscow in February 1918. See *Trudy 1-go ocherednogo Vserossiiskogo kooperativnogo s"ezda (18–24 fevralia 1918 goda, Moskva), vyp. 2. Promyshlennost' i torgovlia Rossii i zadachi kooperatsii* (Moscow: Zadruga, 1918), p. 53.

10. For example, in late April 1917, Alexander Chaianov was offered the position of director of the Economic Department in the central organization of all Russian cooperatives, the Council of Cooperative Congresses, at a salary of 10,000 rubles per annum. Even taking into account a 200–300 percent rate of inflation since 1913, that was three times the salary of his successful classmate Ivan Barkhatov, who was the chief government specialist in Kazan province, and four times what a district zemstvo agronomist earned in April 1917. See "Postanovlenie pravleniia Soveta s"ezdov ot 28 aprelia 1917 g.," *Izvestiia soveta vserossiiskikh kooperativnykh s"zdov*, no. 2 (August 20, 1917): 8; NART, f. 256, Kazanskoi gubernskoi zemleustroitel'noi komissii, op. 7, d. 43, l. 35; NART, f. 119, Kazanskaia uezdnaia zemskaia uprava, op. 21, d. 81 "Spisok sluzhashchikh Kazanskogo uezdnogo zemstva za 1917 g.," ll. 10, 10 ob.
11. See "Politicheskaia rezolutsiia po tekushchemu momentu Pervogo ocherednogo vserossiiskogo kooperativnogo s"ezda v Moskve 18/5–24/11 fevralia 1918 g.," *Kooperativnaia zhizn'*, no. 1 (1918): 38–39.
12. See "Pervyi ocherednoi vserossiiskii kooperativnyi s"ezd (18–24 fevralia): Materialy k s"ezdu," *Izvestiia soveta vserossiiskikh kooperativnykh s"ezdov*, no. 2 (March 5, 1918): 1–53.
13. N. S-v, "Pervyi ocherednoi vserossiiskii kooperativnyi s"ezd v Moskve 18/5–24/11 fevralia 1918 g.," p. 38.
14. In late 1918, the cooperative activist Alexander Berkenheim (born 1880) led the delegation of the all-Russian cooperative union *Tsentrosoiuz* abroad. With the consent of the Soviet government, this delegation successfully negotiated a number of large-scale trade contracts despite the Entente's economic blockade of Russia. See "Trade with Russia Started by British," *New York Times*, January 21, 1920, p. 4 and also February 1, 1920, p. 5. By mid-1919, Russian cooperatives had agencies in London, Helsinki, Yokohama, Kobe, Constantinople, New York, Stockholm, Oslo, and Shanghai. See "Kontory i agenstva russkikh kooperativnykh organizatsii za granitsei," *Vestnik ob"edinennogo komiteta russkikh kooperativnykh organizatsii v Londone*, no. 3 (October 1, 1919).
15. Back in 1915 it was estimated that cooperatives needed at least 1,200 trained professional "cooperators." See A. Evdokimov, "O tipe kooperativnykh shkol v sviazi s zadachami soiuznogo stroitel'stva," *Vestnik kooperativnykh soiuzov*, no. 11 (1915): 585.
16. The first 116 students began their studies on September 30, 1918. The full course of education took three years and provided specialization in three major areas: cooperative accountancy; cooperative instruction and audit of cooperatives; and cooperative studies. All leading figures of countryside modernization taught at the institute. The first class of students was also the last one, as the institute was closed in 1920. See *Izvestiia soveta vserossiiskikh kooperativnykh s"ezdov*, no. 5 (May, 1918): 1–2; *Kooperativnyi institut Vserossiiskikh kooperativnykh s"ezdov: Prospekt na 1919–20 uchebnyi god* (Moscow, n. d.); S. L. Komlev, "Kon"iunkturnyi institut (sud'ba nauchnoi shkoly N. D. Kondrat'eva): Beseda s A. A. Konusom," in M. G. Iaroshevskii, ed., *Repressirovannaia nauka* (Leningrad: Nauka, 1991), pp. 164–65.
17. Aleksei Gern, who five years earlier in a student presentation argued that the agricultural specialist should be a public activist par excellence, now criticized the Petrograd Institute precisely for its extremely ideological curriculum and a waste of cooperative money. He calculated that almost 160 academic hours were spent on such pseudo-academic topics as "cooperative movement

and socialism" or "social politics, cooperative movement, and the party." Aleksei Gern, "K voprosu o vydelenii agrarno-ekonomicheskogo tsikla na Sankt-Peterburgskikh sel'skokhoziaistvennykh kursakh (ob"iasnitel'naia zapiska)," *Vestnik sel'skogo khoziaistva*, no. 33–34 (October 15 [2], 1918): 8. The program of the Moscow Institute did not include the kind of courses criticized by Gern.

18. See A. Minin, "Kooperatsiia i agronomicheskaia pomoshch' naseleniiu," *Kooperativnaia zhizn'*, no. 2 (May 1918): 46–47.

19. Dillon, "The Rural Cooperative Movement," pp. 466, 576.

20. DAKO, f. R-989, op. 1, spr. 25, ark. 56–59.

21. Alexander Dillon, "The Rural Cooperative Movement," p. 576.

22. The zeal of the censors was often ridiculous but unavoidable. In 1919, the Petrograd Commissariat of Press, Agitation and Propaganda prohibited the Cooperative Publishing House from publishing *The Prince and the Pauper* by Mark Twain, *Prince Serebriannyi* by Aleksei N. Tolstoi, *The Paper Money* by Mikhail Bogolepov, and *The State Bankruptcy* by Veniamin Ziv. See G. Sandomirskii, "Voprosy kooperativnogo izdatel'stva," *Kooperativnaia zhizn'*, no. 7–8–9 (July–August–September 1919): 24. (Veniamin S. Ziv is misprinted as "Zak" in this publication).

23. Cf. "Zhurnal zasedanii Soveta," *Izvestiia soveta vserossiiskikh kooperativnykh s"ezdov*, no. 4 (May 8, 1918): 3–8.

24. Aleksandr G., "Kto vrag truda, tot vrag naroda," *Viatskii kooperator*, no. 1 (February 14/1, 1918): 18.

25. A. Gus-ov, "Iz derevenskikh vpechatlenii," *Viatskii kooperator*, no. 5 (1918): 19.

26. Kollegiia instruktorov Viatskogo kreditnogo soiuza i inspektorov melkogo kredita, "Nado byt' spokoinymi," *Viatskii kooperator*, no. 1 (February 14/1, 1918): 19–20. For a more optimistic assesment of the situation in Viatka countryside, see in Retish, *Russia's Peasants in Revolution and Civil War*.

27. See K. Pyzhenkov, "O blizhaishikh zadachakh kooperatsii," *Derevenskaia kooperatsia* (Chistopol: August 1918): 4.

28. See B. Okushenko, "Sel'skokhoziaistvennaia kooperatsia," *Derevenskaia kooperatsia* (Chistopol: August 1918): 8–10.

29. See A. Zonov, "K pakhariu," *Viatskii kooperator*, no. 5 (1918): 34–35.

30. On May 25, 1918, Nikolai Komarov, an instructor at the cooperative Moscow People's Bank and a former cooperative instructor in the Elabuga district zemstvo, committed suicide after another business trip to Elabuga. Particularly shocking was his firsthand experience with early instances of "red terror." Komarov was born in 1885 and at least in terms of age belonged to the new generation of Russian intelligentsia. Of peasant origin, he was a teacher, a member of the Socialist Revolutionaries Party, but sometime around 1910 shifted gears and dedicated himself to cooperative activism, like Gusakov. N. Boikov, "Nikolai Ivanovich Komarov (nekrolog)," *Kooperativnaia zhizn'*, no. 2 (May 1918): 20–21.

31. See T. K. "K voprosu o goneniiakh na kooperatsiiu," *Kooperativnaia zhizn'*, no. 8–9 (November–December 1918): 49–51.

32. See V. Kul'chinetskii, "Zadacha podlezhashchaia neotlozhnomu razresheniiu," *Mashina v sel'skom khoziaistve*, no. 20–21 (November 1917): 566–569.

33. P. N. Peresypkin, "Agrarnye problemy," *Iugo-vostochnyi khoziain*, no. 1 (January 1, 1918): 5, 6.

34. See "Iuzhno-russkaia artel agronomov," *Iugo-vostochnyi khoziain*, no. 7–8 (April 1, 1918): 15.

35. "Vozzvanie Donskoi trudovoi arteli," *Iugo-vostochnyi khoziain*, no. 15 (August 1, 1918): 13–14.

36. Cf. *Iugo-vostochnyi khoziain*, no. 21 (November 1, 1918).

37. A graduate of Moscow University and Moscow Agricultural Institute, Shorygin served as Bronnitskii district agronomist after the Revolution of 1905, and after a study trip to the United States in 1913 was appointed Moscow district agronomist in 1914. See "Agronomicheskii personal zemskikh uchrezhdenii," pp. 621–39; Kerans, "Agricultural Evolution," p. 440; Glukhov et al., *Mestnyi agronomicheskii personal, 1 ianvaria 1915 g.*, p. 226.

38. See D. Shorygin, "Obzor agronomicheskoi deiatel'nosti za 1918 god," *Vestnik sel'skogo khoziaistva*, no. 1–4 (January 15, 1919): 15–18.

39. See "Iz tekushchei agronomicheskoi deiatel'nosti," *Vestnik sel'skogo khoziaistva*, no. 5–6 (February 15, 1919): 15–16.

40. See "Dekret ob uchete i mobilizatsii spetsialistov sel'skogo khoziaistva," *Izvestiia VTsIK*, no. 21 (January 30, 1919): 6.

41. See S. Sereda, "Natsionalizatsia agronomicheskoi pomoshch'i (Po povodu dekreta ob uchete i mobilizatsii spetsialistov sel'skogo khoziaistva)," *Iskra kommunisticheskogo zemledeliia*, no. 2–3 (Riazan': January 15–February 1, 1919): 1–2.

42. See "Chto sdelano Narkomzemom dlia provedeniia v zhizn' zemel'noi reformy," *Izvestiia VTsIK*, no. 254 (November 13, 1919): 2.

43. See the sample registry card of mobilized agricultural specialists, *Izvestiia VTsIK*, no. 21 (January 30, 1919): 6.

44. The bibliography of studies reconstructing the Bolshevik regime as a ruthless dictatorship, a result of genuine social consent, or a combination of the former two explanations, is immense. Among the classic books, we should certainly mention Leonard Schapiro, *The Communist Party of the Soviet Union* (New York: Random House, 1960); Robert Conquest, *The Soviet Political System* (New York: Praeger, 1968); Sheila Fitzpatrick, *The Russian Revolution* (Oxford: Oxford University Press, 1982); Richard Pipes, *Russia under the Bolshevik Regime* (New York: Knopf, 1993); Martin Malia, *The Soviet Tragedy: A History of Socialism in Russia, 1917–1991* (New York: Free Press, 1994). For a recent and more nuanced contribution to the revisionist literature, see Retish, *Russia's Peasants in Revolution and Civil War*.

45. So read the official ID card issued to mobilized agricultural specialists. See "Prilozhenie 2," *Izvestiia VTsIK*, no. 21 (January 30, 1919): 6.

46. As a warning to agricultural specialists, the famous Petrovsky Park on the grounds of the Moscow Agricultural Institute became the site of public mass executions of "enemies of the people" in September 1918. See S. Kobiakov, "Krasnyi sud," in I. V. Gessen, ed. *Arkhiv russkoi revolutsii*, vol. 7 (Berlin, 1922), pp. 274–75.

47. For example, in April 1919, the precinct agronomist from Kazan province, Valentin Kolosovskii (graduate of a mid-level agricultural college in 1910), extensively developed the idea that the individual farmer would overcome his misery only through the socialist commune and with support from the Soviet government. See V. M. Kolosovskii, "O pereustroistve zemledel'cheskogo khoziaistva," *Kazanskii zemledelets*, no. 15 (April 9, 1919): 673–77.

48. See *Time of Troubles, the Diary of Iurii Vladimirovich Got'e: Moscow, July 8, 1917 to July 23, 1922*, trans., ed., and int., Terence Emmons (Princeton, NJ: Princeton University Press, 1988).

49. The case of Kniazev is of a particular interest for our research because his logic and even imagery are very close to those displayed in the writings of one of the leaders of the agrarian public modernizers, Alexander Chaianov. G. A. Kniazev, "Iz zapisnykh knizhek russkogo intelligenta (1919–1922)," *Russkoe proshloe*, vol. 5 (St. Petersburg: Logos, 1994), pp. 148–242; Gerasimov, *Dusha cheloveka perekhodnogo vremeni*, 142; and others.

50. See Kniazev, "Iz zapisnykh knizhek russkogo intelligenta," 153, 154, 156, 174, et seq.

51. "27/X/1919. The Whites and the Reds … I get sick of this struggle. The common opinion: they might be red, white, black, yellow, violet—if only there would be peace. 14/X/1920. I am tired. Awfully tired. And most awfully—I'm indifferent to many things. Suddenly, all seem to me so unnecessary and insignificant. In particular, this horror of the bloody fight." See Kniazev, "Iz zapisnykh knizhek russkogo intelligenta," pp. 174, 185.

52. For example, see D. V. Fedorov, "Pochemu u nas malo khleba?" *Iugo-vostochnyi khoziain*, no. 1 (January 1, 1918): 4–5.

53. See *Iugo-vostochnyi khoziain*, no. 10 (May 15, 1918): 12–13.

54. See *Iugo-vostochnyi khoziain*, no. 11 (June 1, 1918): 14; no. 14 (July 15, 1918): 16.

55. This incident is mentioned in Baker, "Peasants, Power, and Revolution in the Village," p. 126.

56. See TsDAVOV, f. 27, op. 1, d. 368, ll. 8–11 ob.

57. For example, in 1918 he attempted to prevent nationalization of the cooperative Moscow People's Bank and led the delegation of cooperative leaders during the unsuccessful negotiations with V. I. Lenin. See V. V. Kabanov, "Natsionalizatsiia Moskovskogo Narodnogo banka," *Voprosy Istorii*, no. 4 (1970): 209.

58. See RGAE, f. 731, A. V. Chaianov, op. 1, ed. khr. 73 "Pis'mo kommunisticheskogo universiteta im. Ia. M. Sverdlova v zhilishchno-zemel'nyi otdel gorodskogo raiona ob ostavlenii v prezhnikh zhilishchnykh usloviiakh prepodavatelia A. V. Chaianova," l. 1.

59. See RGAE, f. 731, A. V. Chaianov, op. 1, ed. khr. 74 "Pis'mo Mozemotdela v Moskovskii Uzemotdel o zakreplenii za A. V. Chaianovym doma v der. Barvikhe," l. 1. A. V. Chaianov was the founder of the colony of Moscow intellectuals in Soviet service in Barvikha. When in the early 1920s the government decided to transform Barvikha into a prestigious country resort for top party leaders, Chaianov was compelled to leave Barvikha for a less prestigious place in the village of Nikolina Gora (nowadays, an elite site of gated communities).

60. See RGAE, f. 731, A. V. Chaianov, op. 1, ed. khr. 35 "Raspiska I. Teodorovicha o prieme ot A. V. Chaianova revol'vera sistemy 'Nagan'," l. 2.

61. It is well-known that Chaianov greatly helped Nikolai Kondratiev (a younger representative of the new generation of Russian intelligentsia, who developed into a prominent economist in the 1920s), when during the period of War Communism Kondratiev was arrested several times by the VChK, and did not have a job. Chaianov's loyalty also extended to his old institute classmates. On April 5, 1919, he wrote a letter to the Commissar of Agriculture, S. P. Sereda, asking him to release A. I. Sigirskii, at that time a Smolensk cooperative activist, from mobilization together with other agricultural specialists. See A. V. Chaianov, "Letter to S. P. Sereda," in Chaianov, *A. V. Chaianov—chelovek, uchenyi, grazhdanin*, p. 124. Alexander Sigirskii was Chaianov's classmate in the institute.

62. Lev Kafengauz (1885–1940) contributed to developing the public modernization program, was a deputy minister in the Provisional Government, and taught in

the Cooperative Institute. Iakov Bukshpan (1887–1939?) was also an economist, prominent in zemstvo special committees during the World War I. He was notorious for his open-mindedness in theory and private life: when in 1914 he met his future wife Sara Akhmerova of Kazan, he converted to Islam to participate in the Muslim marriage ceremony *Nikakh*. See B. F. Sultanbekov and S. Iu. Malysheva, *Tragicheskie sud'by* (Kazan: TGZhI, 1996), p. 50.

63. See A. S. Velidov, "Ia. M. Bukshpan, L. B. Kafengauz, 'Rossia posle bolshevitskogo eksperimenta,'" in *Neizvestnaia Rossiia. XX vek*, vol. 1 (Moscow: Istoricheskoe nasledie, 1992), pp. 148–54.

64. See Ia. M. Bukshpan and L. B. Kafengauz, "Programma ekonomicheskogo vozrozhdeniia strany, sostavlennaia 'Natsional'nym tsentrom' v 1919 godu," in *Neizvestnaia Rossiia. XX vek*, vol. 1, pp. 154–55.

65. Ibid., p. 159.

66. See *Zemledel'chaskaia gazeta*, no. 21 (May 27, 1917): 419.

67. See Bukshpan and Kafengauz, "Programma ekonomicheskogo vozrozhdeniia strany," p. 156.

68. Ibid., p. 172.

69. Ibid., p. 164.

70. Fridolin, *Ispoved' agronoma*, p. 108.

71. Ibid., p. 110.

72. This book was written with the clear intention of compromising the very project of public modernization as reactionary and Stolypinist, and to discredit those "old specialists" who occupied key positions in Narkomzem and the system of agricultural education. The arranged character of this surprisingly well-printed book is obvious, both from its content and structure. The author himself felt obliged to account for what (or who) besides the authorities could inspire him to "confess" on behalf of the rural professionals to some alleged misdeeds. And so he began his book with the statement: "Confession is required." On two and a half small pages of his "Prologue," he repeated the word "confession" 10 times, and the word "required" 14 times, but never explained why exactly it was required. See Fridolin, *Ispoved' agronoma*, pp. 3–5. This book by Fridolin can be regarded as one of the earliest instances of the "cultural revolution" campaign against the old intellectual cadres.

73. See "Pis'mo S. S. Maslova E. N. Kuskovoi on February 21, 1923," in GARF, f. 5865 E. D. Kuskova, op. 1, d. 303, l. 14.

74. See RGAE, f. 731, A. V. Chaianov, op. 1, ed. khr. 34 "Protokol rasporiaditel'nogo soveshchaniia kollegii Narkomzema n. 2 o komandirovke A. V. Chaianova v London v rasporiazhenie L. B. Krasina," l. 1.

75. See Gerasimov, *Dusha cheloveka perekhodnogo vremeni*, pp. 69–70.

76. See "Pis'mo I. V. Emel'ianova I. K. Okulichu v Boston ot 6 sentiabria 1920 g."

77. See "Pis'mo 'Russian Agricultural Asssociation of North America' I. V. Emel'ianovu ot 21 oktiabria 1938 g.," in Museum of Russian Culture Archival Collection, I. V. Emel'ianov Collection, box no. 1. For example, Petr Ivanov of San Francisco graduated from Kinel agricultural college on the eve of the World War I, and began working as a precinct agronomist in the Buzuluk district of Samara Province. Another member of the Agricultural Association, Iosif Okulich, received his higher education in Switzerland, and before the war was head of the Administration of State Proprieties (*Upravlenie gosudarstvennykh imushestv*) in Enisei Province. See Glukhov et al., eds., *Mestnyi agronomicheskii personal, sostoiavshii na pravitel'stvennoi i obshchestvennoi sluzhbe 1 ianvaria 1915 g.*,

p. 339; *Spisok lits, sluzhashih po vedomstvy Glavnogo upravlenia zemleustroiistva i zemledeliia na 1912*, col. 318.

78. GARF, f. 5865, E. D. Kuskova, op. 1, d. 548.

79. Sergei Maslov, "Cherez sem' let (K itogam s"ezda "Krest'ianskoi Rossii")," *Vestnik krest'ianskoi Rossii*, no. 1 (June 1928): 7.

80. *"Krest'ianskaia Rossia." Trudovaia krest'ianskaia partiia. Vvedebie. Ideologiia. Programma. Taktika* (Prague: Izdatel'stvo "Krest'ianskaia Rossiia," 1928), p. 38.

81. Both the Pomgol initiative (the All-Russian Committee for Famine Relief in 1921) and the explicitly anti-Soviet Peasant Russia party relied on the practices of public self-organization and activism, but *obshchestvennost'* in both cases was regarded only as a means of getting the needed results (feeding peasants or agitating them against the regime). Cf. E. Kuskova, "Mesiats 'soglashatel'stva'," *Volia Rossii*, no. 3 (1928); "Novye uroki," *Vestnik krest'ianskoi Rossii*, no. 1 (June 1928): 5; Maslov, "Cherez sem' let," p. 8.

82. In summer 1921 a number of public figures prominent during the preceding decade (A. V. Chaianov, A. A. Rybnikov, P. A. Sadyrin, et al.) became initiators and organizers of the All-Russian Committee for Famine Relief (Pomgol). The very foundation of Pomgol was provoked by a dramatic presentation on the famine situation in the Volga region at the Moscow Agricultural Society, made by Artel member and professor A. A. Rybnikov, and cooperative activist M. I. Kukhovarenko from Saratov. The Pomgol initiative ended in the arrest and exile of many of its members.

83. See James W. Heinzen, "'Alien' Personnel in the Soviet State: The People's Commissariat of Agriculture under Proletarian Dictatorship, 1918–1929," *Slavic Review* 56, no. 1 (Spring 1997): 73–100; Markus Wehner, "The Soft Line on Agriculture: The Case of Narkomzem and its Specialists, 1921–27," in Judith Pallot, ed., *Transforming Peasants: Society, State and the Peasantry, 1861–1930. Selected Papers from the Fifth World Congress of Central and East European Studies* (New York: St. Martin's Press, 1998), pp. 210–13.

84. See *Itogi rabot Vserossiiskogo agronomicheskogo s"ezda v g. Moskve (1–11 marta 1922 g.)* (Tula: Tul'skoe gubernskoe zemel'noe upravlenie, 1922), p. 12.

85. See P. Mesiatsev, "Agronomicheskie osnovy zemleustroistva i roli v nem agronomicheskogo personala," in *Trudy III Vserossiiskogo s"ezda agronomov v g. Moskve (1–11 marta 1922 goda)*, vol. 1 (Moscow, 1922), p. 19.

86. See A. G. Doiarenko, "Ocherednye zadachi i perspektivy agronomicheskoi pomoshch'i," in *Trudy III Vserossiiskogo s"ezda agronomov*, pp. 28–29.

87. See I. Stepanov, "Zadachi i metody obshchestvennoi agronomicheskoi raboty," in *Trudy III Vserossiiskogo s"ezda agronomov*, pp. 33–34.

88. See L. L. Nemirovskii, "Regulirovanie raboty uchastkovykh agronomov," in *Trudy III Vserossiiskogo s"ezda agronomov*, pp. 35–36.

89. See A. N. Minin, "Raionnaia agronomicheskaia organizatsiia i puti perehoda k nei," in *Trudy III Vserossiiskogo s"ezda agronomov*, pp. 31–32.

90. See P. K. Gratsianov, "Uluchshenie kolichestvennogo i kachestvennogo sostava agronomicheskogo personala," in *Trudy III Vserossiiskogo s"ezda agronomov*, p. 40.

91. See the list of participants in *Zhurnal Pervogo sobraniia upolnomochennykh Vserossiiskogo soiuza sel'skokhoziaistvennoi kooperatsii, 14–17 oktiabria 1922 g.* (Moscow: Sel'skosoiuz, 1922), pp. 3–7.

92. Ibid., pp. 41–42, 84.

93. See *Itogi rabot Vserossiiskogo agronomicheskogo s"ezda v g. Moskve*, pp. 2–4.

94. P. Stuchka, "Mysli po agrarnomu voprosu," *Kommunisticheskaia revolutsiia*, no. 3 (February 1, 1923): 30.

95. *Trudy pervogo arkhangel'skogo gubernskogo agronomicheskogo soveshchaniia-konferentsii, 15–21 dekabria 1924 goda* (Arkhangelsk: Arkhangel'skoe gubernskoe zemel'noe upravlenie NKZ, 1925), p. 7.

96. V. I. Kopeikin, "Soderzhanie agroraboty blizhaishikh let i perspektivnyi plan razvitiia i oborudovaniia agronomicheskoi organizatsii Samarskoi gub.," in *Trudy 5-go gubernskogo agronomicheskogo soveshchaniia 13–19 sentiabria 1926 g.* (Samara: Samarskoe gubernskoe zemel'noe upravlenie NKZ, 1926), p. 25.

97. See Appendix "Diagramma no. 1. Obshchee chislo agronomicheskogo personala," in Glukhov et al., *Mestnyi agronomicheskii personal, 1 ianvaria 1915 g.*

98. "Shtaty raionnoi agronomicheskoi organizatsii Ukrainy," in TsDAVOVU, fond 27 "Narkomzem USSR," op. 1, d. 368, l. 57.

99. Ibid.

100. "Sostoianie agronomicheskoi organizatsii na Volyni," in TsDAVOVU, fond 27 "Narkomzem USSR," op. 1, d. 368, l. 16.

101. Based on data found in TsDAVOVU, fond 27, op. 1, d. 368 "Materialy gubzemupravlenii Ukrainy ob agroraionakh i lichnye spiski agropersonala. 2 iulia 1923–26 noiabria 1923."

102. The only exception was Illarion Romanenko, who received higher education in 1910 and by 1911 already had a well-paid job in Volokolamsk district (uezd, Moscow province) of the agronomist of the government Land Settlement Commission, a very "politically incorrect" choice from the vantage point of public modernization orthodoxy. As we know, the government agronomist network was totally dismantled in summer 1917, and Romanenko might have moved to his native region. In 1923 he was serving as a "district agronomist" in Mirgorod.

103. See "Svedeniia o mestozhitel'stve, zanimaemoi dolzhnosti, obrazovatel'nom tsenze, prakticheskom stazhe, spetsial'nosti i prodolzhitel'nosti sluzhby v danno meste agropersonala Poltavskoi gubernii," in TsDAVOVU, fond 27, op. 1, d. 368, ll. 2–2 ob.

104. "Vedomost' raspredeleniia agronomicheskikh raionov i agronomicheskogo personala po raionam Ekaterinoslavskoi gubernii (v novykh granitsakh)," TsDAVOVU, fond 27, op. 1, d. 368, ll. 5–7; "Spisok agropersonala, sostoiashchego na sluzhbe v agroraionakh Ekaterinoslavskoi gubernii," Ibid., ll. 76–76 ob.

105. The exception was Andrei Zaitsev, who received higher education in 1914 and became a precinct agronomist in Pskov province, but later moved to Ukraine to become the Petropavlovka district agronomist in 1923. Another migrant, Ivan Kanzi, zemstvo provincial specialist in gardening who had a higher education, moved across Ukraine, from Zhitomir to Nogaisk on the Azov Sea, to become a local district agronomist.

106. More such cases can be found in Podoliia province, where the quality of cadres was generally lower than in other Ukrainian provinces. Karl Damberg, an agricultural elder in Vinnitsa district with primary education, was promoted in 1923 to Vinnitsa "district" agronomist, allegedly now having a mid-level agricultural education. The same happened to Franz Val'chuk: with a primary education, in 1912 he became an instructor at the government Land Settlement Commission in Vilno province. By 1923 he had moved to Ukraine, claimed a mid-level education background, and became a "district" agronomist. Before the war, Nikanor

Kumanskii was an agricultural elder with primary education in Tver province. In 1923 we find him in Ukraine, a stockbreeding instructor, and also with a mid-level (i.e., agricultural college or special school) education. How these three people managed to get special education between 1915 and 1923 is unclear. See "Lichnyi sostav personala Podol'skoi gubernii," TsDAVOVU, fond 27, op. 1, d. 368, ll. 50, 51.

107. Morachevskii, *Spravochnye svedenia o deiatel'nosti zemstv po sel'skomy khoziaistvu,* vol. 11.

108. Glukhov et al., *Mestnyi agronomicheskii personal, 1 ianvaria 1915 g.,* p. 426.

109. Ibid., p. 425.

110. TsDAVOVU, fond 27 "Narkomzem USSR," op. 1, d. 368.

111. See *Spisok lits, sluzhashih po vedomstvy Glavnogo upravlenia zemleustroiistva i zemledeliia na 1912. Ispravlen po 1 noiabria 1912,* col. 277; Alexandrovskii et al., *Mestnyi agronomicheskii personal, 1 ianvaria 1914 g.,* p. 39.

112. See *Trudy pervogo arkhangel'skogo gubernskogo agronomicheskogo soveshchaniia-konferentsii, 15–21 dekabria 1924 goda* (Arkhangelsk: Arkhangel'skoe gubernskoe zemel'noe upravlenie NKZ, 1925), p. 4.

113. See Alexandrovskii et al., *Mestnyi agronomicheskii personal, 1 ianvaria 1914 g.,* p. 240.

114. See *Trudy pervogo arkhangel'skogo gubernskogo agronomicheskogo soveshchaniia-konferentsii,* p. 5.

115. See Alexandrovskii et al., *Mestnyi agronomicheskii personal, 1 ianvaria 1914 g.,* p. 306.

116. See *Trudy pervogo arkhangel'skogo gubernskogo agronomicheskogo soveshchaniia-konferentsii,* p. 4.

117. See Appendix "Diagramma no. 1. Obshchee chislo agronomicheskogo personala," in Glukhov et al., *Mestnyi agronomicheskii personal, 1 ianvaria 1915 g.*

118. See Alexandrovskii et al., *Mestnyi agronomicheskii personal, 1 ianvaria 1914 g.,* p. 76.

119. See *Trudy pervogo arkhangel'skogo gubernskogo agronomicheskogo soveshchaniia-konferentsii,* p. 4.

120. See Alexandrovskii et al., *Mestnyi agronomicheskii personal, 1 ianvaria 1914 g.,* p. 3.

121. See *Trudy pervogo arkhangel'skogo gubernskogo agronomicheskogo soveshchaniia-konferentsii,* p. 4.

122. See Alexandrovskii et al., *Mestnyi agronomicheskii personal, 1 ianvaria 1914 g.,* p. 224; *Trudy pervogo arkhangel'skogo gubernskogo agronomicheskogo soveshchaniia-konferentsii,* p. 4.

123. See *Trudy pervogo arkhangel'skogo gubernskogo agronomicheskogo soveshchaniia-konferentsii,* p. 9.

124. Ibid., pp. 36–37.

125. Ibid., p. 35.

126. On the *vydvizhentsy*—old specialist relationships as an important impetus to the cultural revolution—see Sheila Fitzpatrick, "Cultural Revolution as Class War," in Fitzpatrick, ed., *Cultural Revolution in Russia, 1928–1931* (Bloomington: Indiana University Press, 1978), pp. 3–40; idem., *The Cultural Front: Power and Culture in Revolutionary Russia* (Ithaca, NY: Cornell University Press, 1992), 13; and others.

127. "Sostoianie agronomicheskoi organizatsii na Volyni," l. 16.

128. "Protokol soveshchaniia zaveduiushchikh zemel'nymi upravleniiami Donetskoi gubernii 26 oktiabria 1923 g.," in TsDAVOV, fond 27 "Narkomzem USSR," op. 4, d. 342, l. 204.
129. "Shtaty raionnoi agronomicheskoi organizatsii Ukrainy," l. 57 ob.
130. See "Tipovaia tarifnaia setka," in *Nakaz VTsSPS o provedenii novoi tarifnoi sistemy. Utverzhdeno Prezidiumom VTsSPS 1-go dekabria 1921 g.* (Kazan: Izdanie RIO VTsSPS, 1921), p. 6.
131. One "commodity ruble" in the second "price belt," which included most of Ukraine, converted into 1.3 "chervonets" rubles. The resolution of the Labor and Defense Council (*Sovet truda i oborony*) no. 340, February 1924. Calculated in this relatively stable currency, the consumer index had risen from 1.2 in January 1923 to 1.8 in January 1924 with 1913 cumulative prices taken as 1. *Statistiheskii ezhegodnik 1924 g.*, vyp. 1 (=*Trudy TsSU*, vol. 8, issue 7) (Moscow, 1924), p. 230. This means that in the winter of 1923–24, the chervonets ruble was 1.8 times weaker than the pre-1914 ruble. All calculations are deliberately rounded up and simplified for the sake of clarity in the study, which is not dedicated to budget history. Yet, based on a much more thorough analysis and quite accurate, they give a general impression of the dynamics of living standards.
132. See Kopeikin, "Soderzhanie agroraboty," p. 48.
133. See V. P., "Budzhet spetsialista," *Biulleten' TsSU*, no. 89 (August 27, 1924): 47.
134. Ibid.
135. Ibid., p. 46; V. I. Popov, "Budzhet spetsialista," *Biulleten' TsSU*, no. 97 (February 25, 1925): 68.
136. V. P., "Budzhet spetsialista," p. 48.
137. Ibid.
138. During the spring months, this included (per person, per month): 5.75 pounds of meat, 2.75 pounds of fresh fish, 4.5 eggs, and so on. In summer, the consumption of milk increased three times, and vegetables, ten times, at the expense of cereals and herring. Everyone eat 58 eggs per month in summer. See Popov, "Budzhet spetsialista," p. 72.
139. See V. P., "Budzhet spetsialista," pp. 48, 51.
140. Ibid., p. 49.
141. See Popov, "Budzhet spetsialista," p. 71.
142. In August 1926, the salaries of Samara district agronomists were increased some 27 percent, from 110 to 140 rubles a month. As of October 1, 1926, the precinct agronomists of Nizhegorod province were paid 85 instead of 76 rubles per month. In addition, all agricultural specialists with higher education received a raise of 11 rubles a month after two years of service, and 26 rubles after ten years. Those with mid-level education received the same raise after 5 and 15 years of service, respectively. See "Sodoklad uezdnogo agronoma Samarskogo uezda t. Kolpashchikova," in *Trudy 5-go gubernskogo agronomicheskogo soveshchania 13–19 sentiabria 1926 goda* (Samara: Samarskoe gubersnskoe zemel'noe upravlenie NKZ, 1926), p. 64; O. Pal'sh, "Nam neobhodimo imet' polnotsennogo agronoma," *Sel'skokhoziaistvennaia zhizn'*, no. 49 (1928): 14.
143. Based on data derived from *Gosudarstvennyi apparat SSSR, 1924–1928 g.* (Moscow: Statistichekoe izdatel'stvo, 1926), p. 78.
144. Ibid.
145. Ibid., 44, 45.

146. A. Zubarev, "Kak udovletvoriaetsia potrebnost' v spetsialistakh sel'skogo khoziaistva," *Sel'skokhoziaistvennaia zhizn'*, no. 43 (1928): 12–13.
147. See Kopeikin, "Soderzhanie agroraboty," pp. 14–61.
148. See *Trudy pervogo arkhangel'skogo gubernskogo agronomicheskogo soveshchaniia-konferentsii*, p. 7.
149. Ibid., p. 8.
150. Ibid., p. 9.
151. *Gosudarstvennyi apparat SSSR, 1924–1928 g.*, p. 59.
152. See *Trudy tret'ego sobraniia upolnomochennykh Sel'skosoiuza, 1–6 iiunia 1925 goda* (Moscow: Sel'skosoiiuz, 1925), pp. 196–97.
153. See P. A. Sadyrin, "Sel'skosoiuz i sel'skokhoziaistvennaia kooperatsiia v 1924 godu," in *Trudy tret'ego sobraniia upolnomochennykh Sel'skosoiuza, 1–6 iiunia 1925 goda*, p. 12.
154. In the mid-1920s, 60 percent of the funds in local cooperatives and up to 97 percent of the funds in central cooperative unions were borrowed from the state. See E. V. Serova, *Sel'skokhoziaistvennaia kooperatsiia v SSSR* (Moscow: Agropromizdat, 1991), p. 72. Before 1917, no more than 36 percent of the entire cooperative capital had come as credits of the State Bank. See V. V. Kabanov, "Kooperatsiia kak kanal vzaimodeistviia razlichnykh sotsial'no-ekonomicheskikh ukladov: k voprosu o roli kooperatsii v sotsial'no-ekonomicheskoi strukture kapitalisticheskoi Rossii," in *Voprosy istorii kapitalisticheskoi Rossii: problema mnogoukladnosti* (Sverdlovsk: Izdatel'stvo UralGU, 1972), p. 103. In the 1920s, the system of rural cooperatives was virtually subordinated to the state and its economic interests, thus violating old cooperative principles. Cooperatives purchased up to 40 percent of agricultural products from nonmembers of cooperatives, while peasants (including members of cooperatives) marketed two-thirds of their product to private buyers. See V. Ia. Filimonov, *Gorod i derevnia v 1921–1925 gg.: Po materialam Evropeiskoi Rossii* (Leningrad: Izd-vo LGPI, 1984), p. 53.
155. Ia. Binman, and S. Kheinman, *Kadry gosudarstvennogo i kooperativnogo apparata SSSR* (Moscow: Gosplan SSSR. Gos. Planovo-khoziaistvennoe izdatel'stvo, 1930), pp. 124–25, 136–37, 148–49.
156. Ibid.
157. For a special study of the conflict within the community of rural scholars constituting the top echelon of agricultural specialists, see Susan Gross Solomon, "Rural Scholars and the Cultural Revolution," in Sheila Fitzpatrick, ed., *Cultural Revolution in Russia, 1928–1931* (Bloomington: Indiana University Press, 1978), pp. 129–53; Susan Gross Solomon, *The Soviet Agrarian Debate: A Controversy in Social Science, 1923–1929* (Boulder, CO: Westview Press, 1977).
158. See "Organizatsiia krest'ianstva," *Vestnik kooperativnykh soiuzov*, no. 3 (1917): 9–12.
159. See A. V. Chaianov, *Organizatsiia severnogo krest'ianskogo khoziaistva* (Iaroslavl: Izdanie Iaroslavskogo kreditnogo soiiuza kooperativov, 1918), p. 121; idem., *Puteshestvie moego brata Alekseia Kremneva v stranu krest'ianskoi utopii* (Moscow: Gosudarstvennoe izdatel'stvo, 1920).
160. See Maslov, "Cherez sem' let (K itogam s"ezda 'Krest'ianskoi Rossii')," *Vestnik Krest'ianskoi Rossii*, no. 1 (June 1928): 5–12; *"Krest'ianskaia Rossiia" Trudovaia krest'ianskaia partiia: Vvedenie, Ideologiia, Programma, Taktika* (Prague: Izdatel'stvo "Krest'ianskoi Rossii," 1928).
161. See a recent study M. V. Sokolov, "Iz istorii respublikansko-demokraticheskogo kryla russkoi emigratsii," *Novaia i noveishaia istoriia*, no. 2 (2008): 172–83.

162. See *Krest'ianskaia Rossiia*, no. 5 (Izdanie Dal'nevostochnogo komiteta Trudovoi krest'ianskoi partii, April 1930).

163. See RGAE, f. 731, A. V. Chaianov, op. 1, ed. khr. 77 "Protokoly doprosov A. V. Chaianova v OGPU po delu 'Trudovoi krest'ianskoi partii'," ll. 12, 35, and others.

164. On the prosecution of the alleged leaders of the forged party, see M. L. Galas, "Razgrom agrarno-agronomicheskoi oppozitsii v nachale 1930-kh godov: delo TsK Trudovoi krest'ianskoi partii (po materialam sledstviia)," *Otechestvennaia istoriia*, no. 5 (2002): 89–112.

165. In the Moscow region—68 people, in the Leningrad region—106, in the Northern Caucasus (apparently, including Don and Kuban)—120, in Nizhny Novgorod—24, in Central Russia—132, in the Western regions—174, in the Middle Volga—107, in Western Siberia—35, in the Crimea—26, in Ukraine—143, in the Urals—26, in the Ivanovo region alone—96, and in the Lower Volga—56. See V. Goncharov and V. Nekhotin, eds., *Prosim osvobodit' iz tiuremnogo zaklucheniia* (Moscow: Sovremennyi pisatel', 1998), pp. 173–77.

166. *Zhertvy politicheskogo terrora v SSSR: Compact Disk Database*, 3rd edition. Disk 1: *Baza dannykh o zhertvakh politicheskogo terrora v SSSR* (Moscow: Izdatel'stvo Zven'ia, 2004).

Postscript

1. E. N. Sakharova-Vavilova, "Dnevnikovye zapisi," ll. 300–301.

2. A. Ufimskii, "Sotsial'nye inzhenery," *Vestnik kooperatsii*, no. 21–22 (November 1915): 309.

3. Marc Bloch, *The Historian's Craft* (New York: Knopf, 1953), p. 6.

4. E. N. Sakharova-Vavilova, "Dnevnikovye zapisi," ll. 307.

5. GARF, f. 5865, op. 1, d. 548, l. 3.